International Symposium on Optics and Its Applications

(OPTICS-2011)

Journal of Physics: Conference Series Volume 350

Yerevan and Ashtarak, Armenia
5-9 September 2011

ISBN: 978-1-62276-696-3
ISSN: 1742-6588

Printed from e-media with permission by:

Curran Associates, Inc.
57 Morehouse Lane
Red Hook, NY 12571

Some format issues inherent in the e-media version may also appear in this print version.

Printed by Curran Associates, Inc. (2013)

Institute of Physics
Dirac House, Temple Back
Bristol BS1 6BE UK

Phone: 44 1 17 929 7481
Fax: 44 1 17 920 0979

techtracking@iop.org

Additional copies of this publication are available from:

Curran Associates, Inc.
57 Morehouse Lane
Red Hook, NY 12571 USA
Phone: 845-758-0400
Fax: 845-758-2634
Email: curran@proceedings.com
Web: www.proceedings.com

TABLE OF CONTENTS

Journal of Physics: Conference Series **350** (2012) 011001 doi:10.1088/1742-6596/350/1/011001

PREFACE

The papers selected for this volume were reported at the International Symposium 'Optics and its applications' (OPTICS-2011, Yerevan & Ashtarak, Armenia, 5–9 September, 2011), http://www.ipr.sci.am/optics2011/. The Symposium was organized by the SPIE Armenian Student Chapter and major Armenian R&D organizations, universities and industrial companies working in the field of basic and applied optics: Institute for Physical Research of the National Academy of Sciences of Armenia, Yerevan State University, Russian-Armenian (Slavonic) University, and LT-PYRKAL Closed Joint Stock Company.

OPTICS-2011 was primarily intended to support and promote the involvement of students and young scientists in various fields of modern optics, giving them the possibility to attend invited talks by prominent scientists and to present and discuss their own results. Furthermore, the Symposium allowed foreign participants from 14 countries to become acquainted with the achievements of optical science and technology in Armenia, which became a full member of the International Commission for Optics (ICO) in 2011. To follow this concept, the Symposium sessions were held in various host institutions. The creative and friendly ambience established at OPTICS-2011 promoted further international collaboration in the field and motivated many students to take up research in optics and photonics as a career.

This volume of *Journal of Physics: Conference Series* covers thematic sections of the Symposium (both oral and poster), which represent the main fields of interest in optics for Armenian scientists: quantum optics & information, laser spectroscopy, optical properties of nanostructures, photonics & fiber optics, and optics of liquid crystals. Such wide coverage is consistent with general scope of the Symposium, allowing all the students involved in optics to present, discuss and publish their recent results, and for those who are making their first steps in science to choose the direction of their further studies.

We are confident that the publication of the Symposium proceedings in *JPCS*, a worldwide-known open access journal, will help to disseminate and promote current activities in optics, thus facilitating international cooperation and the integration of Armenian scientists into the worldwide optical community.

We would like to thank the sponsors of the Symposium: National Foundation of Science and Advanced Technologies (NFSAT), The Abdus Salam International Centre for Theoretical Physics (ICTP), LT-PYRKAL, State Committee of Science of Armenia, Russian-Armenian (Slavonic) University, and Devout Generation Foundation. We also express our gratitude to the members of Program Committee for their organization of the manuscript reviewing. Special thanks go to Narine Gevorgyan, Lilit Mantashyan and Paytsar Mantashyan for their invaluable assistance in the compilation of this issue.

The Editors,

Aranya B Bhattacherjee University of Delhi, India
Maria L Calvo Universidad Complutense de Madrid, Spain
Eduard M Kazaryan Russian-Armenian (Slavonic) University, Armenia
Aram V Papoyan Institute for Physical Research of NAS, Armenia
Hayk A Sarkisyan Russian-Armenian (Slavonic) University, Armenia

Journal of Physics: Conference Series **350** (2012) 011001 doi:10.1088/1742-6596/350/1/011001

PHOTO GALLERY

Participants of OPTICS-2011 in front of Yerevan State University.

From left to right: Gagik Buniatyan (LT-PYRKAL CEO), Narine Gevorgyan (Symposium Coordinator), Maria L Calvo (ICO President, OPTICS-2011 Co-Chair), Brian Culshaw (SPIE Past President), Aram Papoyan (IPR Director, OPTICS-2011 Co-Chair).

From left to right: Pargev Avetisyan (Vice-President for Science of RAU), Armen Darbinyan, (President of RAU), Hayk Sarkisyan (Dean of Department of Physics and Technology)

Armen Darbinyan, the President of RAU, awards diploma of Honorable Doctor to Maria L Calvo

Katarina Swanberg, the President of SPIE

Vladimir Lazarev, the President of Bauman Moscow State Technical Univ. SPIE Student Chapter

Maria L Calvo, the President of ICO

Laboratory of Laser Spectroscopy, IPR. From left to right: David Sarkisyan, Avinash Pandey, Nora Sandor, Ute Neugebauer, Alexei Chizhov.

OPTICS-2011 Poster Session, YSU

Participants of OPTICS-2011 in Khor Virap

Participants of OPTICS-2011 in Noravanq

Journal of Physics: Conference Series **350** (2012) 011002 doi:10.1088/1742-6596/350/1/011002

Organizers

DIRECTORS

Maria L Calvo	President of International Commission for Optics, Spain
Aram V Papoyan	Director of Institute for Physical Research of NAS, Armenia

HEADS OF PROJECT

Tigran Dadalyan	YSU, Armenia
Artsrun Martirosyan	IPR, Armenia

COORDINATOR

Narine Gevorgyan	IPR, Armenia / ICTP, Italy

MANAGERS

Paytsar Mantashyan	IPR, Armenia
Karen Vardanyan	IPR, Armenia

INTERNATIONAL ADVISORY COMMITTEE

Marcis Auzinsh	Latvia
Roland Avagyan	Armenia
Tapash Chakraborty	Canada
Yuri Chilingaryan	Armenia
Eduard Kazaryan	Armenia
Albert Kirakosyan	Armenia
Radik Kostanyan	Armenia
Avinash Pandey	India
Marat Soskin	Ukraine

INTERNATIONAL PROGRAM COMMITTEE

David Sarkisyan (Chair)	Armenia
Roman Alaverdyan	Armenia
Dan Apostol	Romania
Levon Aslanyan	Armenia
Aranya Bhattacherjee	India
Gagik Buniatyan	Armenia
Vigen Chaltykyan	Armenia
Roldao Da Rocha	Brazil
Miltcho Danailov	Italy
Vladimir Gerdt	Russia
Samvel Gevorgyan	Armenia
Gayane Grigoryan	Armenia
Rafik Hakobyan	Armenia
Takayuki Miyadera	Japan
Levon Mouradian	Armenia
Atom Muradyan	Armenia
Simon Rochester	USA
Hayk Sarkisyan	Armenia
Aleksandr Vardanyan	Armenia

LOCAL ORGANIZING COMMITTEE

Narek Aghekyan	Armenia
Anahit Gogyan	Armenia
Melanya Grigoryan	Armenia
Armen Hovhannisyan	Armenia
Lilit Hovhannisyan	Armenia
Tatevik Khachatryan	Armenia
Astghik Kuzanyan	Armenia
Satenik Kuzanyan	Armenia
Vladimir Lazarev	Russia
Lilit Mantashyan	Armenia
Hripsime Mkrtchyan	Armenia
Pavel Muzhikyan	Armenia
Wahi Narsisian	Armenia
Sahak Ordukhanyan	Armenia
Anna Reymers	Armenia
Narine Torosyan	Armenia

The Symposium was organized by

YSU & NAS SPIE Armenian Student Chapter
Institute for Physical Research (IPR) of National Academy of Sciences (NAS)
Russian-Armenian (Slavonic) University (RAU)
LT-PYRKAL cjsc
Yerevan State University (YSU)

Official Sponsors of the Symposium

LT-PYRKAL cjsc

Russian-Armenian University

YSU & NAS SPIE Student Chapter

Further sponsors

National Foundation of Science and Advanced Technologies

The Abdus Salam International Centre for Theoretical Physics

State Committee of Science of Armenia

Devout Generation

Electromagnetically induced transparency and photon-photon interactions with Rydberg atoms

D Petrosyan
Institute of Electronic Structure & Laser, FORTH, 71110 Heraklion, Crete, Greece
E-mail: david.petrosyan@iesl.forth.gr

M Fleischhauer
Fachbereich Physik und Forschungszentrum OPTIMAS, Technische Universität Kaiserslautern, D-67663 Kaiserslautern, Germany

Abstract. We discuss electromagnetically induced transparency in a cold ensemble of atoms optically excited to the Rydberg states. Strong dipole-dipole or van der Waals interactions between the atomic Rydberg states translate into large nonlocal nonlinearities for a propagating probe field. In the case of weak quantum fields, this can be used to attain a conditional phase shift of π between pairs of single photons realizing a deterministic photonic phase gate. For stronger fields, long-range interactions between the atoms constrain the medium to behave as a collection of superatoms, each comprising a blockade volume that can accommodate at most one Rydberg excitation. The propagation of a probe field is affected by its two-photon correlations within the blockade distance, which are strongly damped due to low saturation threshold of the superatoms. Our calculations reproduce and interpret the results of recent experiments.

PACS numbers: 42.50.Gy, 42.65.-k, 32.80.Ee, 03.67.Lx

Submitted to: *J. Phys.: Conf. Ser.*

1. Introduction

Strong dipole–dipole (DD) or van der Waals (VdW) interactions between highly excited Rydberg states of atoms [1] constitute the basis for a number of promising schemes for quantum information applications [2] and interesting many-body effects [3, 4, 5, 6, 7]. Some of these schemes employ the dipole blockade mechanism [8] which suppresses multiple Rydberg excitations within a certain interaction (blockade) volume. Employing electromagnetically induced transparency (EIT) [9, 10] in a ladder configuration, one can translate the DD or VdW interactions between optically excited Rydberg states of atoms into sizable interactions between the photons [11, 12, 13]. Several experiments on EIT with Rydberg atoms were recently performed [14, 15, 16, 17].

Due to the strong, long-range interactions of the Rydberg atoms, the theoretical description of EIT in Rydberg media is a highly non-trivial many-body problem [11, 13, 17, 18, 19]. Here we outline our recent work on this subject, which elucidated two regimes of field intensities: interactions of weak, single-photon pulses [11], which can be used to implement universal photonic logic gates [20, 21]; and propagation and attenuation of strong input fields studied in a recent prominent experiment [17], whose theoretical treatment reveals the importance of two-photon correlations [19].

2. Interaction of weak quantum fields via Rydberg atoms

EIT can translate strong DD interactions between the Rydberg states of atoms into sizable interactions between single photons [11], as described below.

2.1. Equations of motion

Consider an ensemble of cold alkali atoms with level configuration as in Fig. 1. All the atoms are initially prepared in the ground state $|g\rangle$. Two distinguishable, weak (quantum) fields $E_{1,2}$ propagate in the opposite directions along the z axis and resonantly interact with the atoms on the transitions $|g\rangle \to |e_{1,2}\rangle$, respectively. The intermediate states $|e_{1,2}\rangle$ are resonantly coupled by two strong (classical) control fields

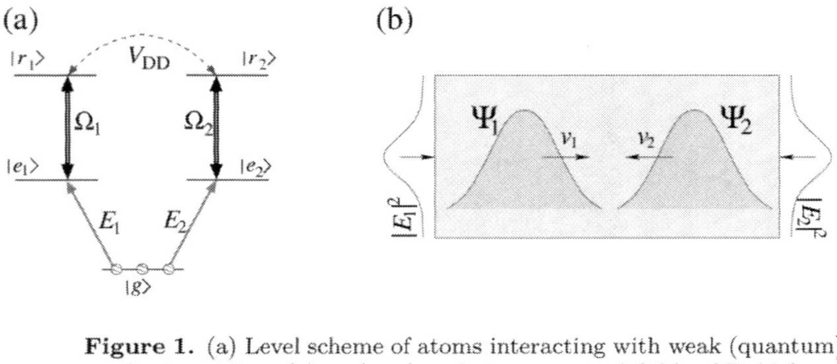

Figure 1. (a) Level scheme of atoms interacting with weak (quantum) fields $E_{1,2}$ on the transitions $|g\rangle \to |e_{1,2}\rangle$ and strong control fields of Rabi frequencies $\Omega_{1,2}$ on the transitions $|e_{1,2}\rangle \to |r_{1,2}\rangle$, respectively. V_{DD} denotes the DD interaction between pairs of atoms in Rydberg states $|r\rangle$. (b) Upon entering the medium, each field is converted into the corresponding polariton $\Psi_{1,2}$ representing a coupled excitation of the field and atomic coherence. These polaritons propagate in the opposite directions with slow group velocities $v_{1,2}$ and interact via the DD interaction.

with Rabi frequencies $\Omega_{1,2}$ to the highly excited Rydberg states $|r_{1,2}\rangle$. In a static electric field $E_{\rm st}\mathbf{e}_z$, the Rydberg states $|r\rangle$ possess large permanent dipole moments $\mathbf{p} = \frac{3}{2}nqea_0\mathbf{e}_z$, where n and $q \equiv n_1 - n_2$ are, respectively, the (effective) principal and parabolic quantum numbers, e is the electron charge, and a_0 is the Bohr radius [1]. A pair of atoms i and j at positions $\mathbf{r}_i$ and $\mathbf{r}_j$ excited to states $|r\rangle$ interact with each other via the DD potential $\hbar\Delta(\mathbf{r}_i - \mathbf{r}_j) = \hbar C_3(1 - 3\cos^2\vartheta)|\mathbf{r}_i - \mathbf{r}_j|^{-3}$, where ϑ is the angle between vectors $\mathbf{e}_z$ and $\mathbf{r}_i - \mathbf{r}_j$, and $C_3 \propto \wp_r^2$ is a constant proportional to the product of atomic dipole moments $\wp_r = \langle r_l|\,\mathbf{p}\,|r_l\rangle \propto n^2$ assumed the same for both states $|r_{1,2}\rangle$. In the frame rotating with the frequencies of the optical fields, the interaction Hamiltonian $H = V_{\rm af} + V_{\rm DD}$ contains the atom-field interaction $V_{\rm af} = -\hbar\sum_{l=1,2}\sum_j^N [g_l^j\hat{\mathcal{E}}_l\hat{\sigma}^j_{e_l g} + \Omega_l\hat{\sigma}^j_{r_l e_l} + \text{H.c.}]$, and the DD interaction $V_{\rm DD} = \hbar\sum_{i<j}^N \hat{\sigma}^i_{rr}\Delta(\mathbf{r}_i - \mathbf{r}_j)\hat{\sigma}^j_{rr}$, where $N = \rho V$ is the total number of atoms of density ρ in the (quantization) volume V, $\hat{\sigma}^j_{\mu\nu} \equiv |\mu\rangle_{jj}\langle\nu|$ is the transition operator of the jth atom, $\hat{\mathcal{E}}_l$ is the slowly-varying operator corresponding to the electric field E_l $(l = 1, 2)$, and g_l^j is the corresponding atom-field coupling constant on the transition $|g\rangle_j \to |e_l\rangle_j$.

We assume that the transverse profile of both weak fields E_l is described by a Gaussian $e^{-r_\perp^2/w^2}$ of width w, where $r_\perp = |\mathbf{r}_\perp|$ is the distance from the field propagation z-axis. Then the (transverse-averaged) coupling constants are given by $g_l = (\wp_{ge_l}/\hbar)\sqrt{\hbar\omega/2\epsilon_0 V}$, $\wp_{ge_l}$ being the dipole matrix element on the transition $|g\rangle \to |e_l\rangle$, $V = \pi w^2 L$, and L the medium length. The Rabi frequencies of strong control fields Ω_l are assumed uniform over the entire volume V. Using Hamiltonian H, we derive the Heisenberg-Langevin equations for the atomic operators $\hat{\sigma}_{ge_l}(\mathbf{r})$, $\hat{\sigma}_{gr_l}(\mathbf{r})$ and propagation equations for the quantum fields $\hat{\mathcal{E}}_l(z)$. Solving for the atomic operators perturbatively in the small parameters $g_l\hat{\mathcal{E}}_l/\Omega_l$ and in the adiabatic approximation [9, 10, 11], and after substituting into the equations for the fields, we obtain the following propagation equations for the dark-state polaritons $\hat{\Psi}_l = \sqrt{c/v_l}\,\hat{\mathcal{E}}_l$ [9],

$$(\partial_t \pm v_l\partial_z)\,\hat{\Psi}_l(z,t) = -i\sin^2\theta_l\hat{S}(z,t)\hat{\Psi}_l(z,t), \tag{1}$$

the sign "+" or "−" corresponding to $l = 1$ or 2, respectively, $v_l = c\cos^2\theta_l$ is the group velocity of the corresponding field in the EIT medium, and the mixing angles θ_l are defined through $\tan^2\theta_l = g_l^2N/|\Omega_l|^2$. Operator $\hat{S}(z,t)$ is responsible for the self- and cross-phase modulation between the fields,

$$\hat{S}(z,t) = \frac{1}{L}\int_0^L dz'\Delta(z-z')\big[\sin^2\theta_1\hat{\mathcal{I}}_1(z',t) + \sin^2\theta_2\hat{\mathcal{I}}_2(z',t)\big], \tag{2}$$

where $\hat{\mathcal{I}}_l \equiv \hat{\Psi}_l^\dagger\hat{\Psi}_l = (c/v_l)\hat{\mathcal{E}}_l^\dagger\hat{\mathcal{E}}_l$ are the polariton intensity (excitation number) operators in the EIT medium, which correspond to the photon number operators outside the medium ($v_l = c$) [9], while the effective one-dimensional DD potential $\Delta(z-z')$ result from $\Delta(\mathbf{r}-\mathbf{r}')$ upon double integration over the transverse coordinate,

$$\begin{aligned}\Delta(z-z') &= \frac{1}{(\pi w^2)^2}\int d^2r_\perp\int d^2r'_\perp e^{-(r_\perp^2+r_\perp'^2)/w^2}\Delta(\mathbf{r}-\mathbf{r}') \\ &= \frac{C_3}{\sqrt{2}w^3}\left[2|\zeta| - \sqrt{\pi}(1+2\zeta^2)e^{\zeta^2}\text{erfc}(|\zeta|)\right], \qquad \zeta \equiv (z-z')/\sqrt{2}w. \end{aligned}\tag{3}$$

As seen in Fig. 2(a), $\Delta(\zeta)$ is sharply peaked around $\zeta = 0$.

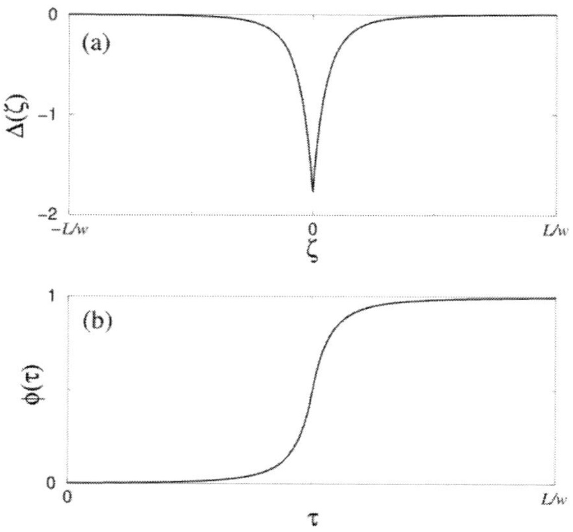

Figure 2. (a) One-dimensional DD potential $\Delta(\zeta)$ of Eq. (3) as a function of dimensionless distance ζ, in units of $C_3/\sqrt{2}w^3$ Hz. (b) The resulting phase-shift $\phi(\tau) \equiv \phi(vt, L - vt, t)$ of Eq. (6) as a function of dimensionless time $\tau = vt/w$, in units of $C_3/(vw^2)$ rad.

It follows from Eq. (1) that the intensity operators $\hat{\mathcal{I}}_l$ are constants of motion: $\hat{\mathcal{I}}_l(z,t) = \hat{\mathcal{I}}_l(z \mp v_l t, 0)$, the upper (lower) sign corresponding to $l = 1$ ($l = 2$). The solution for the field operators then reads

$$\hat{\Psi}_l(z,t) = \exp\left[- i \sin^2\theta_l \int_0^t dt' \hat{S}(z \mp v_l(t-t'), t') \right] \hat{\Psi}_l(z \mp v_l t, 0). \quad (4)$$

The validity of this dissipation-free solution hinges on the following assumptions: (i) The duration T of the pulses exceeds the inverse of the corresponding EIT bandwidth $\delta\omega = |\Omega_l|^2/(\gamma_{e_l}\sqrt{\kappa_l L})$, where γ_{e_l} is the transversal relaxation rate and $\kappa_l \simeq 3\lambda_l^2/(2\pi)\rho$ is the resonant absorption coefficient on the transition $|g\rangle \to |e_l\rangle$. With $v_l = 2|\Omega_l|^2/(\kappa_l \gamma_{e_l})$, this yields the condition $(\kappa_l L)^{-1/2} \ll T v_l/L < 1$ which requires a medium with large optical depth $\kappa_l L \gg 1$ [10]. (ii) The DD induced frequency shifts lie within the EIT bandwidths, $\langle \hat{S}(z) \rangle < \delta\omega$, $\forall\, z \in [0, L]$. (iii) The propagation/interaction time of each pulse $t_{\mathrm{out}} = L/v_l$ is limited by the relaxation rate γ_{r_l} of the Rydberg state coherence via $t_{\mathrm{out}}\gamma_{r_l} \ll 1$.

For simplicity of notation, we set $\theta_{1,2} = \theta$, i.e., $g_1^2 N/|\Omega_1|^2 = g_2^2 N/|\Omega_2|^2$.

2.2. Photonic phase gate

We now employ Eq. (4) to demonstrate the quantum phase gate between two single-photon pulses. We are concerned with the evolution of input state $|\Phi_{\mathrm{in}}\rangle = |1_1\rangle |1_2\rangle$ composed of two single-excitation wavepackets $|1_l\rangle = \left[\frac{1}{L}\int dz f_l(z) \hat{\Psi}_l^\dagger(z)\right]|0\rangle$ whose spatial envelopes inside the medium $f_l(z) = \langle 0| \hat{\Psi}_l(z,0) |1_l\rangle$ are normalized as $\frac{1}{L}\int dz |f_l(z)|^2 = 1$. With the operator solution (4), for the (equal-time) correlation amplitude or the "two-photon wavefunction" $F_{12}(z_1, z_2, t) = \langle 0| \hat{\Psi}_1(z_1,t)\hat{\Psi}_2(z_2,t) |\Phi_{\mathrm{in}}\rangle$ [21] we obtain

$$F_{12}(z_1, z_2, t) = f_1(z_1 - vt) f_2(z_2 + vt) \exp[i\phi(z_1, z_2, t)], \quad (5)$$

$$\phi(z_1, z_2, t) = -\sin^4\theta \int_0^t dt' \Delta\big(z_1 - z_2 - 2v(t - t')\big). \tag{6}$$

Hence, the two polaritons counterpropagate in a shape-preserving manner with group velocities $\pm v$. Since $\hat{\mathcal{I}}_l \hat{\Psi}_l |1_l\rangle = 0$, the self-interaction within each pulse is absent, while the cross-interaction between the pulses results in the phase-shift (6) shown in Fig. 2(b). Assume that at $t = 0$ the first pulse is centered at $z_1 = 0$ and the second pulse at $z_2 = L$, while after the interaction, $t_{\text{out}} = L/v$, the coordinates of the two pulses are $z_1 = L$ and $z_2 = 0$, respectively. The accumulated phase-shift is then $\phi(L, 0, L/v) = -\sin^4\theta/v \int_0^L dz' \Delta_{12}(2z' - L)$. To evaluate the integral, we replace the variable $(2z' - L)/\sqrt{2}w \to \zeta'$ and extend the integration limits to $L/\sqrt{2}w \to \infty$, obtaining a *spatially uniform* $\phi = C_3/(vw^2)$ ($\sin\theta \simeq 1$). The state of the system at t_{out} is then $|\Phi_{\text{out}}\rangle = e^{\phi}|\Phi_{\text{in}}\rangle$. Since for input states $|m_1\rangle|n_2\rangle$ ($m, n = 0, 1$) there is no phase shift when $m + n < 2$, the conditional two-photon phase shift $\phi = \pi$ is equivalent to the CPHASE gate $|\Phi_{\text{out}}\rangle = (-1)^{mn}|m_1\rangle|n_2\rangle$ [20].

2.3. Experimental considerations

To relate the foregoing discussion to a realistic experiment, we assume that the quantum fields are confined in a hollow-core waveguide of length $L \sim 1$ cm with the lowest transverse mode of width $w \simeq 2\mu$m [22]. The waveguide is filled with $N \simeq 5 \times 10^4$ cold Rb atoms at density $\rho \simeq 2 \times 10^{11}$ cm^{-3}. For the two quantum fields tuned to the D1 and D2 transitions $|g\rangle \to |e_{1,2}\rangle$ ($\lambda_1 = 795$ nm, $\lambda_2 = 780$ nm), the corresponding optical depths are $\kappa_1 L \simeq 600$ and $\kappa_2 L \simeq 580$. With $\gamma_{e_1} \simeq 1.8 \times 10^7\,\text{s}^{-1}$, $\gamma_{e_2} \simeq 1.9 \times 10^7\,\text{s}^{-1}$, and taking $\Omega_1 \simeq 7.35 \times 10^6$ rad/s, $\Omega_2 \simeq 7.43 \times 10^6$ rad/s, the group velocities are $v_{1,2} = 100$ m/s. The bandwidth of the pulses $T^{-1} \gtrsim v/L = 10^4\,\text{s}^{-1}$ is smaller than the EIT bandwidth $\delta\omega \simeq 1.2 \times 10^5$ rad/s. To realize the CPHASE gate, we choose the Rydberg states $|r_{1,2}\rangle$ with $\wp_{r_1} = \wp_{r_2} = 315ea_0$ (quantum numbers $n = 15$ and $q = n - 1$), leading to the conditional phase shift $\phi = \pi$. We have verified that the DD frequency shift is within the EIT window $\delta\omega$.

3. Strong field EIT with Rydberg atoms

We now consider propagation of a single (multiphoton) probe field $\hat{\mathcal{E}}_p$ of variable intensity through a ladder EIT medium with the VdW interactions between the atomic Rydberg states, as shown in Fig. 3(a). This system was studied experimentally in Ref. [17], where increasing the probe field amplitude led to reduction of its transmission within the EIT window, which, quite surprisingly, was accompanied by negligible broadening and indiscernible shift of the EIT line. We have developed an efficient theoretical model for EIT with Rydberg atoms [19], whose predictions fully reproduce the experimental observations [17], as detailed below.

3.1. Equations of motion

In the frame rotating with the frequencies of the probe ω_p and control ω_c fields, the system Hamiltonian $H = H_{\text{a}} + V_{\text{af}} + V_{\text{VdW}}$ now contains the unperturbed atomic part, $H_{\text{a}} = -\hbar \sum_j^N [\Delta_p \hat{\sigma}_{ee}^j + (\Delta_p + \delta_c)\hat{\sigma}_{rr}^j]$, where $\Delta_p = \omega_p - \omega_{eg}$ and $\delta_c = \omega_c - \omega_{re}$ are the corresponding detunings; and the atom-field and VdW interactions, $V_{\text{af}} = -\hbar \sum_j^N [\hat{\Omega}_p(\mathbf{r}_j)\hat{\sigma}_{eg}^j + \Omega_c \hat{\sigma}_{re}^j + \text{H.c.}]$ and $V_{\text{VdW}} = \hbar \sum_{i<j}^N \hat{\sigma}_{rr}^i \Delta(\mathbf{r}_i - \mathbf{r}_j)\hat{\sigma}_{rr}^j$, where $\hat{\Omega}_p = g\hat{\mathcal{E}}_p$ is the probe Rabi frequency, while $\hbar\Delta(\mathbf{r}_i - \mathbf{r}_j) = \hbar C_6 |\mathbf{r}_i - \mathbf{r}_j|^{-6}$ is the VdW

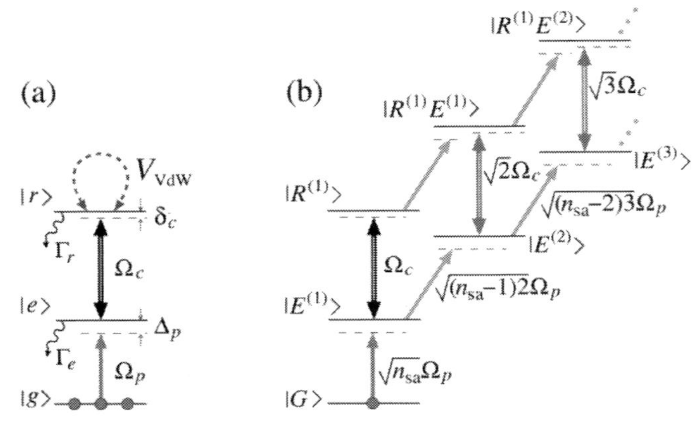

Figure 3. (a) Level scheme of atoms interacting with the probe Ω_p and control Ω_c fields on the corresponding transitions with detunings Δ_p and δ_c, respectively. Γ_e and Γ_r are the (population) decay rates of states $|e\rangle$ and $|r\rangle$, and V_{VdW} denotes the VdW interaction between the atoms in the Rydberg state $|r\rangle$. (b) Truncated level scheme of a superatom, composed of n_{sa} atoms, with the corresponding transition amplitudes due the probe and control fields.

potential between a pair of atoms i and j at positions $\mathbf{r}_i$ and $\mathbf{r}_j$. We consider stationary propagation of the probe field along the z axis, and assume uniform, undepleted control field Ω_c. Using Hamiltonian H, we derive Heisenberg-Langevin equations for the field $\hat{\mathcal{E}}_p(\mathbf{r})$ and atomic $\hat{\sigma}_{\mu\nu}(\mathbf{r})$ operators. For moderate Rabi frequency $\Omega_p < \gamma_e$ and number density of probe photons $\rho_{\mathrm{phot}} \ll \rho$, we can assume linear response of individual atoms to the applied field and arrive at the propagation equation for the probe field amplitude, $\partial_z \hat{\mathcal{E}}_p = i\frac{\kappa}{2}\hat{\alpha}\hat{\mathcal{E}}_p$ with the polarizability

$$\hat{\alpha}(\mathbf{r}) = \frac{i\gamma_e}{\gamma_e - i\Delta_p + |\Omega_c|^2[\gamma_r - i(\Delta_2 - \hat{S}(\mathbf{r}))]^{-1}}, \tag{7}$$

where $\hat{S}(\mathbf{r}) \equiv \int d^3r' \rho(\mathbf{r}')\, \Delta(\mathbf{r}-\mathbf{r}')\hat{\sigma}_{rr}(\mathbf{r}')$ is the total VdW induced shift of level $|r\rangle$ for an atom at position $\mathbf{r}$. Since $\hat{S}(\mathbf{r})$ involves integration over all spatial coordinates $\mathbf{r}' \in V$, Eqs. (7) is highly nonlocal. We therefore need to contrive an efficient method to evaluate the VdW shift $\hat{S}(\mathbf{r})$.

The strong, long-range interactions between the Rydberg atoms suppresses multiple Rydberg excitations within a blockade volume $V_{\mathrm{sa}} = \frac{4\pi}{3}R_{\mathrm{sa}}^3$ where $R_{\mathrm{sa}} \simeq \sqrt[6]{C_6\gamma_e/|\Omega_c|^2}$ is the blockade radius [8]. We therefore call the ensemble of $n_{\mathrm{sa}} = \rho V_{\mathrm{sa}}$ atoms "superatom" (SA) [19]. Each SA can contain only one Rydberg excitation delocalized over V_{sa}. The level scheme of SA is shown in Fig. 1(b): $|G\rangle = |g_1, g_2, \ldots, g_{n_{\mathrm{sa}}}\rangle$ is the ground state, and $|R^{(1)}\rangle = \frac{1}{\sqrt{n_{\mathrm{sa}}}}\sum_j^{n_{\mathrm{sa}}} |g_1, g_2, \ldots, r_j, \ldots, g_{n_{\mathrm{sa}}}\rangle$ is the single collective Rydberg excitation state, while $|E^{(k)}\rangle$ are the properly symmetrized (Dicke) states with k atoms in $|e\rangle$. The corresponding transition amplitudes follow from $\langle E^{(1)}|\, V_{\mathrm{af}}\, |G\rangle = \sqrt{n_{\mathrm{sa}}}\hat{\Omega}_p$, $\langle R^{(1)}|\, V_{\mathrm{af}}\, |E^{(1)}\rangle = \Omega_c$, etc. We then obtain for the SA operators $\hat{\Sigma}_{GR} \equiv |G\rangle\langle R^{(1)}| = \Omega_c\sqrt{n_{\mathrm{sa}}}\hat{\Omega}_p\hat{\Sigma}_{GG}/[(\Delta_p + i\gamma_e)\Delta_2 - |\Omega_c|^2]$ and $\hat{\Sigma}_{RR} = \hat{\Sigma}_{RG}\hat{\Sigma}_{GR}$. To account for possible saturation of transition $|G\rangle \to |R^{(1)}\rangle$, we take $\hat{\Sigma}_{GG} + \hat{\Sigma}_{RR} = 1$, which finally yields

$$\hat{\Sigma}_{RR} = \frac{|\Omega_c|^2 n_{\mathrm{sa}}\hat{\Omega}_p^\dagger\hat{\Omega}_p}{|\Omega_c|^2 n_{\mathrm{sa}}\hat{\Omega}_p^\dagger\hat{\Omega}_p + [|\Omega_c|^2 - \Delta_p\Delta_2]^2 + \Delta_2^2\gamma_e^2}. \tag{8}$$

We can now treat the medium as a collection of $N_{sa} = \rho_{sa} V$ SAs at positions $\mathbf{r}_j$, which implies a spatial coarse-graining with the grain size $2R_{sa}$ [23, 24]. The total VdW shift $\hat{S}(\mathbf{r})$ at position $\mathbf{r}$ can then be expressed as

$$\hat{S}(\mathbf{r}) \approx \sum_j^{N_{sa}} \Delta(\mathbf{r}-\mathbf{r}_j)\hat{\Sigma}_{RR}(\mathbf{r}_j) = \bar{\Delta}\hat{\Sigma}_{RR}(\mathbf{r}) + \hat{s}(\mathbf{r}). \tag{9}$$

The physical meaning of the first term on the rhs of Eq. (9) is that an excited SA at $\mathbf{r}_j \simeq \mathbf{r}$ $[\hat{\Sigma}_{RR}(\mathbf{r}) \to 1]$ induces divergent VdW shift averaged over the SA volume: $\bar{\Delta} \simeq \frac{1}{V_{sa}} \int_{V_{sa}} \Delta(\mathbf{r}')d^3r' \to \infty$. The last term $\hat{s}(\mathbf{r}) \equiv \sum_{j \neq j_r}^{N_{sa}} \Delta(\mathbf{r}-\mathbf{r}_j)\hat{\Sigma}_{RR}(\mathbf{r}_j)$ describes the VdW shift induced by the external SAs outside the volume $V_{sa}^{(\mathbf{r})}$ centered at $\mathbf{r}$. It can be evaluated by replacing the summation by an integration over the entire volume V, excluding the SA at $\mathbf{r}$, which, upon using the mean-field approximation, yields a small shift $\langle \hat{s}(\mathbf{r}) \rangle = \frac{w}{8} \langle \hat{\Sigma}_{RR}(\mathbf{r}) \rangle$.

Using $\hat{\Sigma}_{RR}(\mathbf{r})$ as a projector onto the Rydberg excitation of SA at $\mathbf{r}$ [while $\bar{\Delta} \gg \gamma_e$], we now recast the polarizability of Eq. (7) as

$$\hat{\alpha}(\mathbf{r}) = \hat{\Sigma}_{RR}(\mathbf{r}) \frac{i\gamma_e}{\gamma_e - i\Delta_p} + [1 - \hat{\Sigma}_{RR}(\mathbf{r})] \frac{i\gamma_e}{\gamma_e - i\Delta_p + |\Omega_c|^2[\gamma_r - i(\Delta_2 - \langle \hat{s}(\mathbf{r}) \rangle)]^{-1}}. \tag{10}$$

Here the first fraction is the polarizability α_{TLA} of a two-level atom, which applies when the SA at position $\mathbf{r}$ contains a Rydberg excitation $[\hat{\Sigma}_{RR}(\mathbf{r}) \to 1]$, while the second fraction, barring the small mean-field shift $\langle \hat{s}(\mathbf{r}) \rangle$, is the usual EIT polarizability α_{EIT} [10] acting when no Rydberg excitation is present $[\hat{\Sigma}_{RR}(\mathbf{r}) \to 0]$.

The expectation value of the probe field intensity obeys the equation

$$\partial_z \langle \hat{\mathcal{E}}_p^\dagger(\mathbf{r})\hat{\mathcal{E}}_p(\mathbf{r}) \rangle = -\kappa \langle \hat{\mathcal{E}}_p^\dagger(\mathbf{r}) \mathrm{Im}[\hat{\alpha}(\mathbf{r})] \hat{\mathcal{E}}_p(\mathbf{r}) \rangle. \tag{11}$$

Note that factorizing out $\mathrm{Im}[\langle \hat{\alpha}(\mathbf{r}) \rangle]$ in a mean-field sense would amount to neglecting the essential two-particle quantum correlations [13] originating from nonlinear response of the atoms to the Rydberg excitations. We therefore replace $\hat{\alpha}(\mathbf{r})$ in Eq. (11) by its expectation value *conditioned* upon the presence of photon at $\mathbf{r}$, denoted by $\langle \cdot \rangle_{\mathbf{r}}$,

$$\langle \hat{\alpha}(\mathbf{r}) \rangle_{\mathbf{r}} = \langle \hat{\Sigma}_{RR}(\mathbf{r}) \rangle_{\mathbf{r}} \alpha_{TLA} + [1 - \langle \hat{\Sigma}_{RR}(\mathbf{r}) \rangle_{\mathbf{r}}] \alpha_{EIT}. \tag{12}$$

The conditional Rydberg population $\langle \hat{\Sigma}_{RR} \rangle_{\mathbf{r}}$ of the SA at $\mathbf{r}$ is obtained from Eq. (8) by the replacement $\hat{\Omega}_p^\dagger(\mathbf{r})\hat{\Omega}_p(\mathbf{r}) \to \langle \hat{\Omega}_p^\dagger(\mathbf{r})\hat{\Omega}_p(\mathbf{r}) \rangle \, g_p^{(2)}(\mathbf{r})$, where the probe field intensity correlation function $g_p^{(2)}(\mathbf{r}) = \frac{\langle \hat{\mathcal{E}}_p^\dagger(\mathbf{r})\hat{\mathcal{E}}_p^\dagger(\mathbf{r})\hat{\mathcal{E}}_p(\mathbf{r})\hat{\mathcal{E}}_p(\mathbf{r}) \rangle}{\langle \hat{\mathcal{E}}_p^\dagger(\mathbf{r})\hat{\mathcal{E}}_p(\mathbf{r}) \rangle \langle \hat{\mathcal{E}}_p^\dagger(\mathbf{r})\hat{\mathcal{E}}_p(\mathbf{r}) \rangle}$ quantifies the probability of having simultaneously at least two photons in the blockade volume $V_{sa}^{(\mathbf{r})}$. Note that linear, e.g. bare EIT, response of the medium does not change the correlation function of the propagating field, and only nonlinear, i.e. conditional, absorption $\propto \mathrm{Im}[\langle \hat{\alpha}(\mathbf{r}) \rangle - \alpha_{EIT}]$ modifies $g_p^{(2)}$, which therefore obeys the equation of motion

$$\partial_z g_p^{(2)}(\mathbf{r}) = -\kappa \langle \hat{\Sigma}_{RR}(\mathbf{r}) \rangle \mathrm{Im}[\alpha_{TLA} - \alpha_{EIT}] g_p^{(2)}(\mathbf{r}). \tag{13}$$

We note that our treatment involves only single transverse mode of the probe field, which is effectively defined by the SA cross-section.

Given the input field "intensity" $I_p \equiv \langle \hat{\Omega}_p^\dagger \hat{\Omega}_p \rangle$ and its correlation function $g_p^{(2)}$ [for "classical" coherent field $g_p^{(2)}(0) = 1$], we then use the following stochastic procedure to integrate the coupled Eqs. (11)-(13) for $z \in [0, L]$: We divide the propagation distance L into $L/(2R_{sa})$ intervals corresponding to SAs, and for z within each SA we determine via Monte-Carlo sampling of $\langle \hat{\Sigma}_{RR}(\mathbf{r}) \rangle_{\mathbf{r}}$ whether the SA is excited, $\hat{\Sigma}_{RR}(\mathbf{r}) \to 1$, or not, $\hat{\Sigma}_{RR}(\mathbf{r}) \to 0$. We then average over several independent realizations. The limit of infinitely many such realizations corresponds to continuous polarizability of Eq. (12).

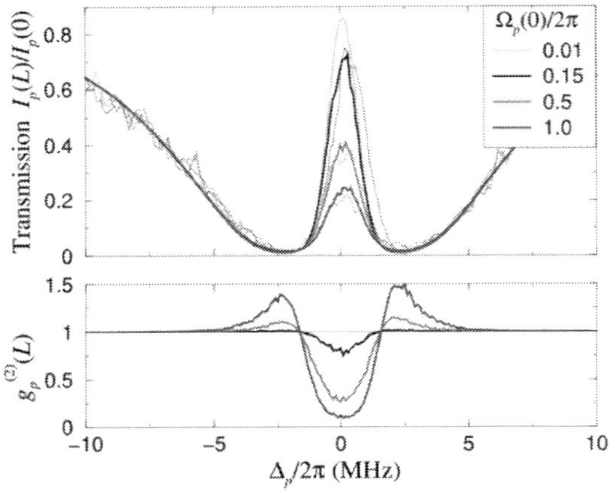

Figure 4. Top: Probe field transmission $I_p(L)/I_p(0)$ versus detuning Δ_p, for input intensities corresponding to $\Omega_p(0)/2\pi = 0.01, 0.15, 0.5, 1.0$ MHz. Thin lines are experimental curves from [17], thicker lines are obtained via stochastic simulations of Eqs. (11)-(13) averaged over 10 independent realizations. Bottom: The corresponding intensity correlation functions $g_p^{(2)}(L)$.

3.2. Numerical simulations and comparison with the experiment

We employ our theory to simulate the experiment of Ref. [17] with an ensemble of cold ^{87}Rb atoms: $|g\rangle \equiv 5S_{1/2}|F = 2, m_F = 2\rangle$, $|e\rangle \equiv 5P_{3/2}|F = 3, m_F = 3\rangle$ with $\Gamma_e = 3.8 \times 10^7\ \mathrm{s}^{-1}$, and $|r\rangle \equiv 60S_{1/2}$ with $\Gamma_r = 5 \times 10^3\ \mathrm{s}^{-1}$ and $C_6/2\pi = 1.4 \times 10^{11}\ \mathrm{s}^{-1}\mu\mathrm{m}^6$ corresponding to repulsive VdW interactions. $\gamma_{e,r}$ also include the one- and two-photon laser linewidths $\delta\omega_{1,2}/2\pi \simeq (5.7, 11) \times 10^4\ \mathrm{s}^{-1}$. The atomic density $\rho = 1.2 \times 10^7\ \mathrm{mm}^{-3}$ and the medium length $L = 1.3\ \mathrm{mm}$ lead to the resonant optical depth of $\kappa L = 4.524$. The control field $\Omega_c/2\pi = 2.25 \times 10^6\ \mathrm{s}^{-1}$‡ is slightly detuned by $\delta_c/2\pi = -10^5\ \mathrm{s}^{-1}$. The corresponding blockade radius is $R_{\mathrm{sa}} \simeq 6.6\ \mu\mathrm{m}$ and each SA contains on average $\bar{n}_{\mathrm{sa}} \simeq 14.7$ atoms.

In Fig. 4 we compare the transmission spectra for different input probe intensities with the corresponding plots of Ref. [17]. The weak field of $\Omega_p/2\pi \lesssim 0.01$ MHz leads to linear EIT response of the medium; the VdW interaction induced nonlinearities become important only at higher intensities. The agreement between our stochastic simulations and the experiment is remarkable. We also show the local intensity correlation $g_p^{(2)}(L)$ at the exit from the medium.

Figure 5 summarizes the results of our simulations involving the continuous polarizability of Eq. (12). Increasing the input probe intensity leads to lesser transmission through the EIT window ($\Delta_2 \sim 0$) and to small mean-field shift and broadening of the EIT line. This is due to the higher probability of two or more photons, exciting Rydberg states $|r\rangle$, to be at the same SA. The induced large VdW level shift $\bar{\Delta}(0)$ results in strong photon absorption, simultaneously reducing the photon coincidence probability within the SA volume V_{sa}. Hence, both $I_p(z)$ and $g_p^{(2)}(z)$ decay, but once $g_p^{(2)}(z) \ll 1$, the attenuation of the probe field intensity $I_p(z)$ slows down. Eventually I_p saturates at a value corresponding to less than one photon per SA, $\rho_{\mathrm{phot}} \lesssim \rho_{\mathrm{sa}}$, with vanishing coincidence probability. With $\rho_{\mathrm{phot}} = \hbar\epsilon_0 c I_p/(2\wp_{ge}^2\omega_p v)$, where the probe group velocity $v = 2|\Omega_c|^2/(\kappa\gamma_e)$ ($\simeq 6000$ m/s),

‡ Our definition of the Rabi frequencies $\Omega_{p,c}$ differ from that in [17] by a factor of $\frac{1}{2}$

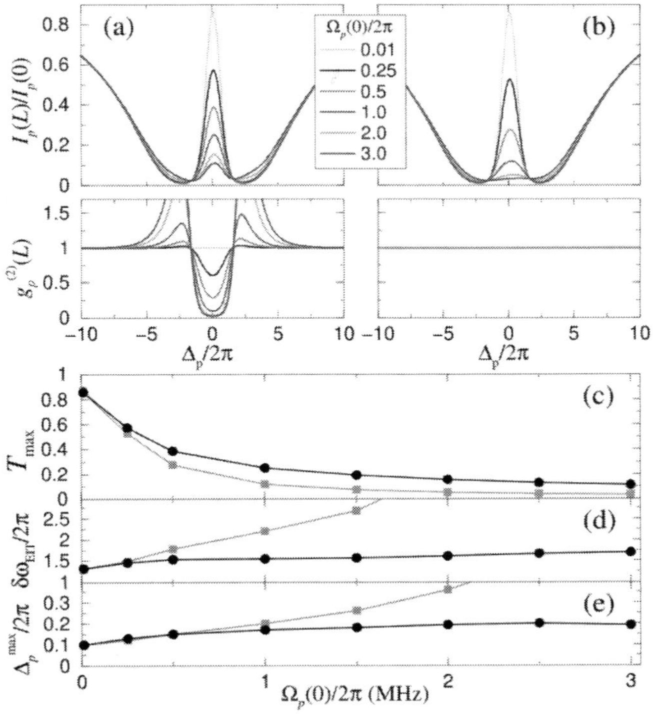

Figure 5. (a) Transmission (top) and intensity correlation (bottom) spectra of the probe field, for various input intensities. (b) The same, but setting $g_p^{(2)}(z) = 1 \, \forall z \in [0, L]$. (c) Peak probe transmission $T_{\rm max}$ around the EIT line center, (d) EIT linewidth $\delta\omega_{\rm EIT}$ (FWHM), and (e) detuning $\Delta_p^{\rm max}$ at the maximum $T_{\rm max}$, versus the input probe Rabi frequency $\Omega_p(0)$. The black lines (circles) correspond to case (a) and the brown lines (squares) to (b). $\Delta_p, \delta\omega_{\rm EIT}, \Omega_p$ are in MHz.

we have that $\rho_{\rm phot} = (\rho/4)\langle\hat{\Omega}_p^\dagger\hat{\Omega}_p\rangle/|\Omega_c|^2$ and the maximal saturation intensity is $\langle\hat{\Omega}_p^\dagger\hat{\Omega}_p\rangle \simeq (4\rho_{\rm sa}/\rho)|\Omega_c|^2$. In the medium the photons are anticorrelated (antibunched) within the temporal window of $\delta t \simeq 2R_{\rm sa}/v$ ($\simeq 1.6$ ns), which does not change when they leave the medium for free space.

Had we not taken into account the probe field intensity correlation, equivalent to setting $g_p^{(2)}(z) = 1 \, \forall z \in [0, L]$, Fig. 5(b), we would have faster, exponential decay of $I_p(z)$, unrestrained by the buildup of avoided volume between the photons, as well as sizable shift and broadening of the EIT line, which contradict the observations of [17].

Outside the EIT window, around the Autler-Townes doublet $\Delta_2 \sim \pm\Omega_c$, the probe is strongly absorbed, $\mathrm{Im}[\langle\alpha\rangle] \simeq 1$, but the correlation function is amplified, since in Eq. (13) $\mathrm{Im}[\alpha_{\rm TLA} - \alpha_{\rm EIT}] < 0$. In other words, linear absorption is larger than the conditional absorption, which results in photon bunching but very low flux.

4. Conclusions

Electromagnetically induced transparency in a medium of strongly interacting Rydberg atoms offers novel, highly non-linear regimes of field propagation. At the low-light level, the induced giant cross-phase modulation enables the realization of universal phase gate between single photon pulses. For stronger input fields, EIT via atomic Rydberg states is suppressed by collective excitations of SAs, which depend on the local probe field intensity and its two-particle correlation within the SA (blockade) volume. The buildup of anticorrelations between the photons upon propagation through the medium leads to the saturation of transmitted field intensity to a value corresponding to one photon per blockade volume. The transmitted field then corresponds to a train of non-overlapping single-photon pulses with the temporal separation δt of a few ns.

Acknowledgments

We are grateful to G. Kurizki and J. Otterbach for many useful discussions and fruitful collaboration. We thank J.D. Pritchard and C.S. Adams for sharing with us the experimental details. This work was supported by the Humboldt Foundation (D.P.) and SFB TR49 (M.F.).

References

[1] T.F. Gallagher, *Rydberg Atoms* (Cambridge University Press, Cambridge, 1994)
[2] M. Saffman, T.G. Walker, and K. Mølmer, Rev. Mod. Phys. **82**, 2313 (2010)
[3] H. Weimer, R. Löw, T. Pfau, and H.P. Büchler, Phys. Rev. Lett. **101**, 250601 (2008)
[4] J. Honer, H. Weimer, T. Pfau, and H.P. Büchler, Phys. Rev. Lett. **105**, 160404 (2010)
[5] H. Weimer and H.P. Büchler, Phys. Rev. Lett. **105**, 230403 (2010)
[6] J. Schachenmayer, I. Lesanovsky, A. Micheli, and A.J. Daley, New J. Phys. **12**, 103044 (2010)
[7] T. Pohl, E. Demler, and M.D. Lukin, Phys. Rev. Lett. **104**, 043002 (2010)
[8] M.D. Lukin *et al.*, Phys. Rev. Lett. **87**, 037901 (2001)
[9] M. Fleischhauer and M.D. Lukin, Phys. Rev. Lett. **84**, 5094 (2000)
M. Fleischhauer and M.D. Lukin, Phys. Rev. A **65**, 022314 (2002).
[10] M. Fleischhauer, A. Imamoglu, and J. P. Marangos, Rev. Mod. Phys. **77**, 633 (2005)
[11] I. Friedler, D. Petrosyan, M. Fleischhauer, and G. Kurizki, Phys. Rev. A **72**, 043803 (2005)
E. Shahmoon, G. Kurizki, M. Fleischhauer and D. Petrosyan, Phys. Rev. A **83**, 033806 (2011)
[12] D. Petrosyan and M. Fleischhauer, Phys. Rev. Lett. **100**, 170501 (2008)
[13] A.V. Gorshkov, J. Otterbach, M. Fleischhauer, T. Pohl, and M.D. Lukin, Phys. Rev. Lett. **107**, 133602 (2011)
[14] A.K. Mohapatra, T.R. Jackson, and C.S. Adams, Phys. Rev. Lett. **98**, 113003 (2007)
[15] A.K. Mohapatra, M. G. Bason, B. Butscher, K.J. Weatherill, and C.S. Adams, Nature Phys. **4**, 890 (2008)
[16] A. Tauschinsky, R.M.T. Thijssen, S. Whitlock, H.B. van Linden van den Heuvell, and R.J.C. Spreeuw, Phys. Rev. A **81**, 063411 (2010)
[17] J.D. Pritchard, D. Maxwell, A. Gauguet, K. J. Weatherill, M.P.A. Jones, and C.S. Adams, Phys. Rev. Lett. **105**, 193603 (2010)
[18] C. Ates, S. Sevinçli, and T. Pohl, Phys. Rev. A **83**, 041802(R) (2011)
S. Sevinçli, N. Henkel, C. Ates, and T. Pohl, Phys. Rev. Lett. **107**, 153001 (2011)
S. Sevinçli *et al.*, J. Phys. B **44**, 184018 (2011)
[19] D. Petrosyan, J. Otterbach and M. Fleischhauer, Phys. Rev. Lett. (in press, 2011) [arXiv:1106.1360]
[20] M. Nielsen and I. Chuang, *Quantum Computation and Quantum Information*, (Cambridge University Press, Cambridge, 2000)
P. Lambropoulos and D. Petrosyan, *Fundamentals of Quantum Optics and Quantum Information* (Springer, Berlin, 2006)
[21] D. Petrosyan, J. Opt. B **7**, S141 (2005)
[22] M. Bajcsy *et al.*, Phys. Rev. Lett. **102**, 203902 (2009)
[23] J. Stanojevic and R. Côté, Phys. Rev. A **80**, 033418 (2009)
[24] F. Robicheaux and J.V. Hernandez, Phys. Rev. A **72**, 063403 (2005)
A. Schwarzkopf, R.E. Sapiro, and G. Raithel, Phys. Rev. Lett. **107**, 103001 (2011)

Creation of coherent superpositions between metastable atomic states in Doppler-broadened media

N. Sandor, J.S. Bakos, Zs. Sörlei and G.P. Djotyan

Research Institute for Particle & Nuclear Physics, Budapest, POBox 49, Hungary

sandor.nora@rmki.kfki.hu

Abstract. We propose and analyze a scheme for creation of coherent superposition of metastable states in a tripod-structured atom using frequency-chirped laser pulses with negligible excitation of the atoms. The underlying physics of the scheme is explained using the formalism of adiabatic states. We show that the proposed scheme may be equally efficient in homogeneously and Doppler-broadened media. By numerically solving master equation for the density matrix operator, we analyze the influence of relaxation processes on the efficiency of the creation of superposition states. We show that the proposed scheme is robust against small-to-medium variations of the laser field's parameters.

1. Introduction

Coherent control of the atomic quantum states is a useful tool in numerous fields of optical science and technology. The preparation of the atoms or molecules in coherent superposition states leads to quantum interference between possible quantum paths which may result in drastic variations in the optical properties (absorption and refraction) of the prepared media. As a consequence, significant changes may be experienced in the propagation characteristics of the laser pulses (Electromagnetically Induced Transparency [1]), the efficiency of nonlinear processes can be enhanced[2] and important applications, such as quantum computing[3] or optical information processing [4]-[5] become possible.

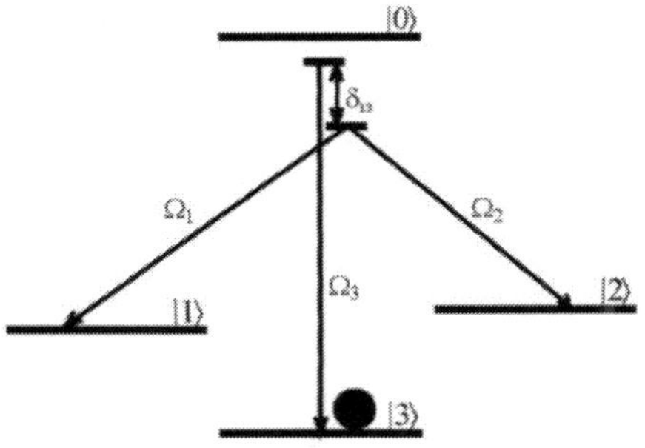

Figure 1: Level structure of tripod-like atom and interacting FC pulses.

In this communication a novel scheme is presented which allows robust creation of large (in certain cases 0.5) coherence between two metastable states of an atom having tripod-like level structure (see Figure 1). It was an important aim to find a scheme that is applicable in presence of spontaneous emission (which means that laser pulses longer than the relaxation time can be used) and in gases of room-temperature. Adiabatic passage schemes based on frequency-modulated (chirped) laser pulses are generally not sensitive to the resonance conditions[4],[6] and allow transition of the populations between metastable states of a multilevel atom without considerable excitation of the atom (in contrary to the other two, widely used adiabatic following methods[7]: stimulated Raman

Published under licence by IOP Publishing Ltd

adiabatic passage(STIRAP)[8]-[9] is sensitive to the two-photon resonance condition between the interacting pulses, while Stark chirped rapid adiabatic passage scheme(SCRAP) is accompanied by temporary excitation of the atom[10]).

The creation of the coherent superposition between two metastable states is reached by coupling the metastable states of the atom to the excited state by three equally chirped laser pulses, two in Raman-resonance, the third one out of it. Initially, all the population is set in the metastable state coupled by the pulse out of Raman-resonance. We show that, as a result of the interaction, the population is transferred into a superposition of the other two ground states with a well-defined relative phase, while the excitation of the atom is suppressed during the whole interaction time, that may minimize the effect of the spontaneous emission. As the scheme is not sensitive to the one-photon resonance condition, it may be applied in Doppler-broadened media.

The remainder of the paper is organized as follows. In section 2, we present the mathematical formalism describing the interaction of three frequency chirped (FC) lasers with the tripod-structured atom based on the master equations for the density matrix operator taking into account longitudinal and transverse relaxation processes. In section 3, the underlying physics of creation of the coherent superposition states is discussed using dressed states analysis and the results are compared with the result of the numerical analysis in case of no relaxation or broadening processes. The effect of Doppler-broadening of the transition lines in an atomic gas of the tripod-structured atoms is analyzed in section 4. In section 5, the influence of longitudinal and transverse relaxation processes is discussed on the efficiency of the coherence creation. The results are summarized in section 6.

2. Mathematical formalism

We consider the interaction of three FC laser pulses with a model atom having tripod-structure of working levels (see Figure 1). Each laser field is acting on corresponding allowed electric-dipole transition in the atom (between a ground state $|n\rangle$, $n\in\{1,2,3\}$ and the excited state $|0\rangle$) according to selection of the carrier frequencies or polarizations of the laser fields. Transitions between the ground states are forbidden in the electric-dipole approximation. The electric field strength of the laser field has the form $E(t)=\sum_{i=1}^{3}E_i(t)\cos\left(\int_{-\infty}^{t}\omega_i(t')dt'\right)$.

A same linear variation in time (linear frequency chirp) of the carrier frequencies of all the interacting laser pulses is assumed in this paper: $\omega_k(t)=\omega_k^{(0)}+\beta t$, where $\omega_k^{(0)}$ is the central carrier frequency of the kth field and β is the speed of the chirp. The central frequency detuning δ_k from the resonance atomic frequency ω_{0k} is $\delta_k=\omega_{0k}-\omega_k^{(0)}$, $k\in\{1,2,3\}$. Since we assume Raman resonance between the fields E_1 and E_2, we have equal single-photon detuning for these waves: $\delta_1=\delta_2=\Delta$, see Figure 1.

As we wish to take the relaxation processes into account, the system can be described by the density matrix $\hat{\rho}$, for which the time evolution is given by the master equation

$$i\hbar\partial_t\hat{\rho}=\left[\hat{H},\hat{\rho}\right]+\hat{R}, \tag{1}$$

where

$$\hat{H}=\begin{pmatrix}0 & \Omega_1 & \Omega_2 & \Omega_3\\ \Omega_1^* & -\Delta & 0 & 0\\ \Omega_2^* & 0 & -\Delta & 0\\ \Omega_3^* & 0 & 0 & -\delta_3\end{pmatrix},\ \hat{R}=i\hbar\begin{pmatrix}-\Gamma\rho_{00} & (-\gamma_{01}+\Gamma/2)\rho_{01} & (-\gamma_{02}+\Gamma/2)\rho_{02} & (-\gamma_{03}+\Gamma/2)\rho_{03}\\ (-\gamma_{01}+\Gamma/2)\rho_{01}^* & \Gamma_1\rho_{00} & -\gamma_{12}\rho_{12} & -\gamma_{13}\rho_{13}\\ (-\gamma_{02}+\Gamma/2)\rho_{02}^* & -\gamma_{12}\rho_{12}^* & \Gamma_2\rho_{00} & -\gamma_{23}\rho_{23}\\ (-\gamma_{03}+\Gamma/2)\rho_{03}^* & -\gamma_{13}\rho_{13}^* & -\gamma_{23}\rho_{23}^* & \Gamma_3\rho_{00}\end{pmatrix} \tag{2}$$

are the Hamiltonian in interaction picture and relaxation operator, respectively. The time-dependent Rabi frequencies are $\Omega_j=\frac{1}{\hbar}E_j(t)d_{0j}\cdot\exp(i\beta t^2)=W_jf(t)\cdot\exp(i\beta t^2)$, with (in general, different) complex amplitudes W_j, ($j=1,2,3$). We assume that the three pulses have the same Gaussian envelope function $f(t)=\exp\left(-(t/\tau_p)^2\right)$. d_{0j} is the dipole moment matrix element for transition between the ground state $|j\rangle$ and the excited state $|0\rangle$. $\Gamma=\Gamma_{01}+\Gamma_{02}+\Gamma_{03}$ is the spontaneous decay rate of the excited state $|0\rangle$ with Γ_{0j} being spontaneous relaxation rate to the ground state $|j\rangle$; γ_{kj} is the dephasing rate for optical coherence between the states $|k\rangle$ and $|j\rangle$,($k,j=0,1,2,3$).

Before presenting and analyzing the results of the numerical simulation of Eq. (1), and to gain an insight into the underlying physics of the processes in the atomic system, let us discuss the behaviour of the system in the case when the relaxation processes may be neglected assuming duration of the pulses much shorter than the relaxations times scales.

For the analysis provided below, it is convenient to introduce new basis functions consisting of the following superpositions of the bare atomic states of the tripod-like atom:

$$|\mathrm{db}_1\rangle = \frac{W_2^*|1\rangle - W_1^*|2\rangle}{\sqrt{|W_1|^2+|W_2|^2}};\ |\mathrm{db}_2\rangle = \frac{W_1|1\rangle + W_2|2\rangle}{\sqrt{|W_1|^2+|W_2|^2}};\ |\mathrm{db}_3\rangle = \frac{W_3|3\rangle}{|W_3|};\ |\mathrm{db}_4\rangle \equiv |0\rangle \tag{3}$$

In the new basis, the Hamiltonian of Eq.(4) can be written as:

$$\hat{H}_{db} = \hbar \begin{pmatrix} 0 & 0 & 0 & 0 \\ 0 & 0 & 0 & f(t)\sqrt{|W_1|^2+|W_2|^2} \\ 0 & 0 & \delta_{13} & f(t)|W_3| \\ 0 & f(t)\sqrt{|W_1|^2+|W_2|^2} & f(t)|W_3| & 2t\beta+\Delta \end{pmatrix}. \tag{4}$$

3. Dressed state analysis

The state vector of the atom in the dark-bright basis can be represented on the basis of the eigenfunctions of the interaction Hamiltonian $\hat{H}_{db}$ as $|\psi(t)\rangle = \sum_k r_k(t)\vec{b}_k(t)\cdot\exp\left(\int_{-\infty}^{t}\lambda_k(t')dt'\right)$, with inital condition $|\psi(-\infty)\rangle = \sum_k r_k(-\infty)\vec{b}_k(-\infty)$ where λ_k and $\vec{b}_k$ are the eigen-values and eigen-vectors of the Hamiltonian $\hat{H}_{db}$.

According to the adiabatic theorem[11], in the adiabatic regime of interaction $r_k(t) \equiv r_k(-\infty)$, $k \in \{1,2,3,4\}$, which means that if the atom is in one (or superposition) of its eigen-states initially (before the interaction process, at $t \to -\infty$), it remains in the same eigen-state (or the superposition of the eigen-states) during the interaction process.

The following eigenvectors can be obtained from the eigen-equation of the Hamiltonian (4):

$$\vec{b}_k = \left(0 \quad \frac{f(t)\sqrt{|W_1|^2+|W_2|^2}}{\sqrt{N}} \quad \frac{\lambda_k(\lambda_k-[\Delta+2\beta t])-f(t)^2(|W_1|^2+|W_2|^2)}{f(t)|W_3|\sqrt{N}} \quad \frac{\lambda_k}{\sqrt{N}}\right)^{(T)}, k\in\{2,3,4\}, \tag{5}$$

and $\vec{b}_1 = (1 \quad 0 \quad 0 \quad 0)^{(T)}$, with the normalization factor

$$N = \left(f(t)^2\left(|W_1|^2+|W_2|^2\right) + \left|\frac{1}{f(t)|W_3|}\left[\lambda_k\left(\lambda_k-[\Delta+2\beta t]\right) - f(t)\left(|W_1|^2+|W_2|^2\right)\right]\right|^2 + \lambda_k^2\right)^{1/2},$$

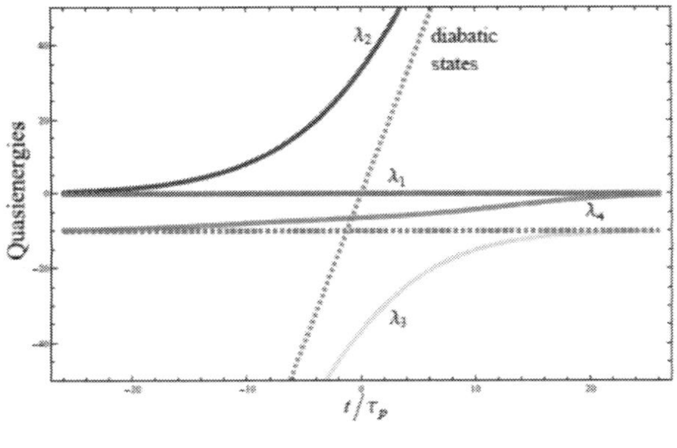

Figure 2: Time dependence of the eigen-values (quasi-energies) of Hamiltonian (4), $|\delta_{13}|\cdot\tau_p = 250$. Parameters are normalized to the pulse duration τ_p.

where λ_k are the eigenvalues of the Hamiltonian, which are plotted in Figure 2. As it can be seen, there is a quasienergy of the system (λ_1) that has zero value at the beginning and during all the interaction process: $\lambda_1 = 0$. This quasienergy corresponds to the dark superposition state $|db_1\rangle$, (see Eq.(3)). In the same time, there is another quasienergy, λ_4 (see Figure 2), which value, in contrast to the other two quasienergies λ_2 and λ_3, does not depend strongly on the Rabi frequencies and is restricted by the value of the Raman detuning δ_{13}:

$$0 \le |\lambda_4| \le |\delta_{13}| \tag{6}$$

The energy of this eigenvalue coinsides with the energy of the bare state $|3\rangle$ in the beginning of the interaction ($t \to \infty$), and the corresponding eigenstate $\vec{b}_4$ coincides with $|3\rangle$. This means that if the atom is prepared is state $|3\rangle$, its time evolution will follow the evolution of $\vec{b}_4$, provided the conditions of adiabaticity are fulfilled. At the end of the interaction $\vec{b}_4$ coincides with the bright superposition of bare states $|1\rangle$ and $|2\rangle$, hence a coherent superposition has been established as a result of the interaction.

It is important to note that if the Rabi frequencies are sufficiently large, the excitation of the atom can be suppressed, namely $\left|b_4^{(4)}/b_4^{(2)}\right| = \left|\lambda_4/\left(|W_1|^2+|W_2|^2\right)^{1/2}\right| \le \left|\delta_{13}/\left(|W_1|^2+|W_2|^2\right)^{1/2}\right|$ (c.f. Eqs. (5) and (6)).

The results of the dressed state analysis are confirmed by the numerical solution of the master equation (1), if we neglect the relaxation processes in the system. We will show in further parts in this communication that this is a good approximation if the characteristic interaction time is shorter than the time of any relaxation (longitudinal and transversal) process of the system. It can be seen in Figure 3, that as a result of the interaction, a maximum coherence of 0.5 has been induced between states $|1\rangle$ and $|2\rangle$ with the complex phase 0, which corresponds to bright state. It is also important to observe that the population of the excited state (ρ_{00}) remains negligible during the whole interaction time.

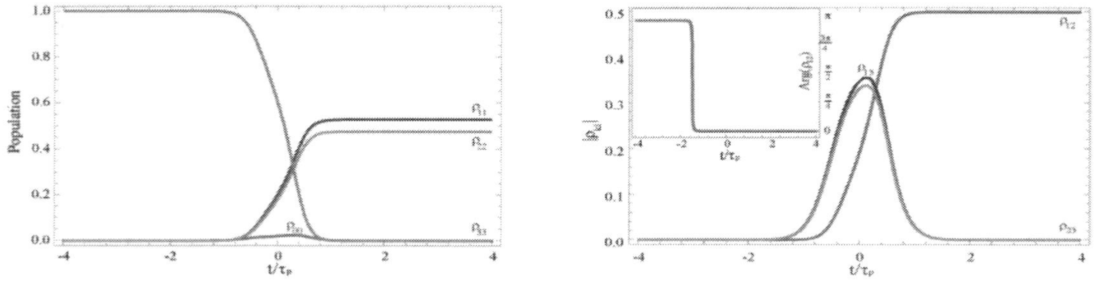

Figure 3: Time evolution of the populations and coherences. The parameters applied are: $W_1\tau_p=500$, $W_2\tau_p=475$, $W_3\tau_p=525$, $\beta\tau_p^2=2500$, $|\delta_{13}|\tau_p=250$.

4. Creation and control of the coherent superposition states in Doppler-broadened media

A Doppler-broadened medium of a gas of tripod-structured atoms is modelled by averaging the created coherence over distribution of the resonance frequencies (velocities) of atoms in the gas at different values of temperature assuming all three FC laser pulses propagating in a same direction.

Considering a gas of ^{87}Rb atoms at temperature T and assuming Maxwell-Boltzmann distribution for the velocities of the atoms, we have for the (normalized) probability distribution $P(\tilde{\Delta})$ for an atom to have single-photon detuning $\tilde{\Delta} = \Delta\tau_p$:

$$P(\tilde{\Delta}) = \left\{mc^2 \Big/ \left[(2\pi)^3 kT\left(f_0\tau_p\right)^2\right]\right\}^{1/2} \exp\left\{-mc^2\left(\tilde{\Delta}\right)^2 \Big/ \left[8\pi^2 kT\left(f_0\tau_p\right)^2\right]\right\}, \tag{7}$$

where k is the Boltzmann constant, $m = 86.909\ u$ is the mass of ^{87}Rb (u being the atomic unit), $f_0 = 384.230\ THz$ is the frequency distance between the excited and the ground states ($F = 1$ and $F' = 0$ hyperfine states in the D_2 line of ^{87}Rb).

The average values of the density matrix elements: $\langle\rho_{kl}\rangle$, $(k,l = 0,1,2,3)$ are calculated numerically. First the master equation (1) is solved numerically for the values of $\tilde{\Delta}$ corresponding to nonzero probability values (see Eq.(7)) in order to obtain the density matrix elements $\rho_{kl}^{final}(\tilde{\Delta})$ at the end of the interaction ($t \to \infty$). One can find that there is a range of values of the Doppler shift $\tilde{\Delta}$, where the final coherences (populations) are independent on $\tilde{\Delta}$. This feature is due to the frequency-chirping of the laser pulses: as long as the Doppler shift is smaller than the frequency range

$[-4\beta\tau_p, 4\beta\tau_p]$ covered by the chirp span during the interaction time (approximately equal to $4\tau_p$), the velocity of motion of the atoms does not have an impact on the resulting population and coherence distribution.

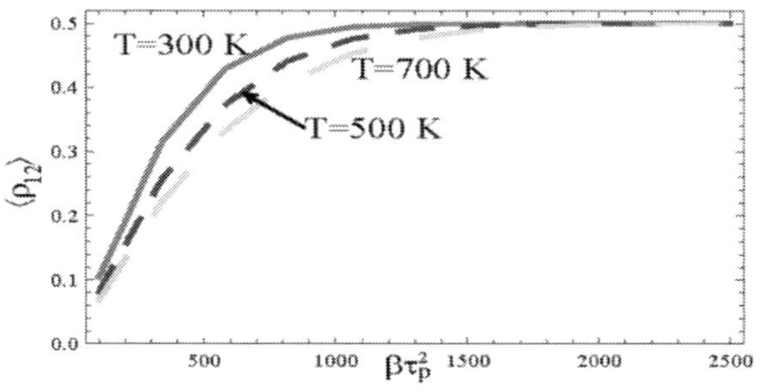

Figure 4: The average value of the final coherence ρ_{12}. Applied parameters: $W_1\tau_p = W_2\tau_p = 250$, $W_3\tau_p = 275$ $\beta\tau_p^2 = 1060$, $\delta_{13}\tau_p = 100$, $\tau_p = 10^{-6} s$

The absolute value of the average induced coherence $|\langle\rho_{12}\rangle|$ established after the interaction with the laser field is presented in Figure 4 as a function of the normalized frequency span $\beta\tau_p^2$ of the laser pulses for different values of the gas temperature. As it can be seen from this Figure, the average value of the induced coherence does not depend on the Doppler-broadening for sufficiently large frequency span of the pulses during the interaction time due, for example to a sufficiently high speed of the frequency chirp.

5. Effect of the relaxation processes

The dependence of the states final populations and of the phase of the created coherence established after the interaction with the laser pulses is shown in Figure 5 as a function of the longitudinal relaxation rate. One could anticipate a negligible influence of the spontaneous relaxation processes on the populations and coherences of the atom when no considerable excitation of the atom takes place. However, the results of the numerical simulations show, that even for the negligible excitation of the atom, the final populations and induced coherences depend on the longitudinal relaxation rate. The reason is linked to the optical coherences ρ_{0k}, (k=1, 2, 3) that are not negligibly small and depend on the longitudinal relaxation rate Γ, (see Eqs. (1) and (2)). As it can be seen from Fig.9, the larger the longitudinal relaxation rate (or longer the laser pulses), the more of the atomic poopulation is transferred (optically pumped) into states $|1\rangle$ and $|2\rangle$ connected by the laser pulses in Raman resonance. Insets in the Figures show behavior of the phase of the induced coherence, which imply that, for longer laser pulses, the atom ends up in the dark superposition in both cases of the positive and negative Raman detuning.

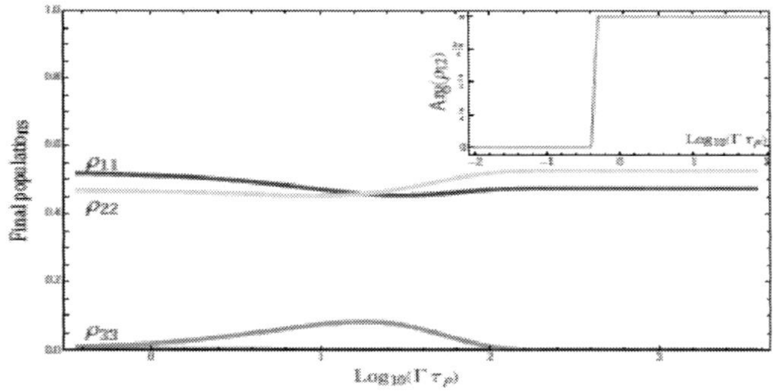

Figure 5: The final populations of the atomic states as a function of the dimensionless longitudinal relaxation rate in the absence of transverse relaxation processes. Insets: The phase of the created coherence ρ_{12}. The parameters used are: $W_1\tau_p$ =500, $W_2\tau_p$ =475, $W_3\tau_p$ =525, $\beta\tau_p^2$=2500, $|\delta_{13}|\tau_p$ =250

Since the quantum interference processes are the basis for the schemes of creation of coherent superposition states, the phase relations between the states probability amplitudes (the values and the phases of the corresponding coherencies) must play an important role in the considered processes. That is why a strong effect of the transverse relaxation (dephasing) processes may be anticipated on creation and control of the coherent superposition states as well as on the population transfer between the atomic states. Our numerical analysis shows that the effect of the transverse relaxation begins to be imperative at $\tau_p \approx 1/(\gamma \cdot 10^{1/2})$. If the interacting pulses are longer, the transverse relaxational processes destroy the adiabatic transfers. At large transverse relaxation rates, all the states are equally populated and there is no coherent superposition created as a result of the interaction.

6. Conclusions

We have presented and investigated a scheme for creation of arbitrary coherent superposition of two ground states in a tripod-like atom without considerable excitation of the atom using FC laser pulses.

We have analyzed the applicability of the scheme in a Doppler-broadened medium of a gas composed of tripod-structured atoms by averaging the induced coherence over the velocity distribution of the atoms in the gas at different temperatures. The results show that the scheme is effective even for relatively large widths of the Doppler – broadened transition lines if the frequency span of the laser pulses due to the chirp exceeds the width of the Doppler-broadening.

We have analyzed the influence of the relaxation processes on the population transfer and creation of the superposition states by numerical simulation of the master equation for the density matrix elements. We have shown that the considered scheme allows minimizing the effect of the spontaneous decay of the excited states by suppression of the population of the excited state. However, even under the condition of negligible excitation of the atom, for longer laser pulses, the longitudinal relaxation may influence the induced coherences and the resulting population distribution among the ground states. For laser pulses longer that the decay time of the excited state, optical pumping of the atom by a pair of the pulses in Raman resonance results in accumulation of the atomic population in dark superposition of the ground states linked by the laser pulses in Raman resonance. It is worth noting at this point that the influence of the spontaneous decay may be minimized by increasing the speed of the population transfer.

As it may be anticipated, the transverse relaxation (dephasing) has a strong destructive effect in the considered scheme. The numerical simulation of the master equation has shown that already at duration of the laser pulses close to the dephasing constant of the medium the transverse relaxation processes destroy the adiabatic transfers. At larger transverse relaxation rates or longer laser pulses, all the states are equally populated as a result of interaction and there is no coherent superposition states created. While the detrimental effect of the transverse relaxation may be avoided by utilizing sufficiently short laser pulses, this effect may be also diminished by increasing the chirp speed of the FC pulses.

The presented adiabatic scheme of coherence creation may find applications in the fields of quantum and nonlinear optics, atomic interferometry and in mapping and long-time storage of optical information in populations and coherences of metastable atomic states.

Acknowledgments: This work was funded by the Research Fund of the Hungarian Academy of Sciences (OTKA) under contracts K 68240, NN 78112; and ELI- 09-1-2010-0010 grant.

References

[1] M. Fleischhauer, A. Imamoglu, J. P. Marangos, Rev. of Mod. Phys., 77, 633 (2005)
[2] Maneesh Jain et al., Phys. Rev. Lett. 77, 4326–4329 (1996)
[3] C.P. Williams and S.H. Clearwater, Explorations in Quantum Quantum Computing (Springer-Verlag, Berlin, 1997)
[4] G.P. Djotyan, J.S. Bakos, Zs. Sörlei, Opt.Express, 4, 113 (1999)
[5] G.P Djotyan, J.S. Bakos, Zs. Sörlei, Phys. Rev. A, 64, 013408 (2001)
[6] G.P. Djotyan, J.S. Bakos, Zs. Sörlei, Phys.Rev.A, **70**, 063406 (2004)
[7] N. V. Vitanov et al, Annu. Rev. Phys. Chem. 52, 763–809 (2001)
[8] F.Vewinger et al., Phys.Rev.Lett., **91**, 213001, (2003)
[9] Z. Kis, N.V. Vitanov, A. Karpati, C. Barthel, K. Bergmann, Phys. Rev A, **72**, 033403 (2005)
[10] L.P. Yatsenko et al., Phys.Rev.A 60, R4237 (1999).
[11] A. Messiah, *Quantum Mechanics*, Vol. II. (North-Holland, 1962).

Journal of Physics: Conference Series **350** (2012) 012003 doi:10.1088/1742-6596/350/1/012003

Chaos in BEC trapped in tilted optical superlattice potential with attractive interaction

Priyanka Verma[1] , Aranya Bhattacherjee[2], ManMohan[1]
[1]Department of Physics and Astrophysics, University of Delhi; Delhi-110007, India
[2]Department of Physics, ARSD College, University of Delhi (South Campus), New Delhi-110021; India

E-mail: bhattach@arsd.du.ac.in

Abstract. The chaos in Bose-Einstein condensate (BEC) in a 1-D tilted optical superlattice potential is studied numerically in this paper. Numerical analysis using the Gross- Pitaevskii (G-P) equation for the system with attractive interaction reveals that as the tilt of the optical superlattice potential is increased the chaos in the BEC increases. This makes the system more unstable and the phase-space orbit increases. The chaos increases as the secondary optical lattice potential increases.

1. Introduction

The phenomenon of chaos in non-linear systems is known to play mainly the role of destruction. But these days focus of researchers is to explore the useful and practical applications [1] of the chaos. There have been intensive studies on the chaos during the past years. There has been a very wide range of different areas in the physics where application of chaos can be explored. The study of chaos in BEC has given a thrust to research in non-linear systems. Theoretically chaos phenomenon has been simply described in the framework of G-P equation [2]. To explore the practical utilization of chaos, restraining chaos in BEC and to develop synchronization in BEC are the subjects of great interest these days [3]. Mostly the signals which drive the dynamical system contain noise and chaos. Therefore a great attention has been devoted in developing the techniques that can synchronize the driven dynamical systems. These techniques can prove to be very useful in secure communications, biomedical systems etc. The phenomenon of chaos have been well explained by the nonlinear Schrodinger equation (NLSE) called G-P equation. Nonlinear resonances, chaos in BEC using the time varying trapping potential and also the dynamics of a weakly open BEC trapped in a double-well potentials revealing chaotic behaviour have also been investigated [4]. Propagation of a bright matter wave soliton and its stabilization in a BEC has also been studied [5]. The chaotic behaviour of the BEC has been widely explored in the last few years for the BEC trapped in tilted optical lattices [6]. In this paper we study the chaos in 1D tilted optical superlattice potential, with attractive interaction between the atoms in a BEC which has not been studied earlier. For the ^{7}Li, the s-wave scattering length in the spin-triplet state is negative which is the cause of attractive interatomic interactions. In earlier literature it has been predicted that BEC does not occur in such kind of systems but now there has been evidence [7] of BEC in such kind of systems. By the adjustment of the wave lengths of the two laser beams, we can observe the effective change in the trajectories in the phase space.

1. Model

The model that we consider here consists of an elongated cigar-shaped BEC containing N number of ^{7}Li atoms each having mass m. The BEC we consider here is quasi-1D system whose order parameter can be described by 1D NLSE or G-P equation given as

$$i\hbar\frac{\partial\psi(x,t)}{\partial t}=-\frac{\hbar^2}{2m}\frac{\partial^2\psi(x,t)}{\partial x^2}+[V_{trap}(x)+g_o\,|\,\psi(x,t)|^2]\psi(x,t) \quad , \tag{1}$$

where $g_o=4\pi\hbar^2a/m$ is the non-linearity in the G-P equation which is basically the interatomic scattering pseudopotential. $V_{trap}(x)$ represents the time-independent tilted optical superlattice potential which is described as

$$V_{trap}(x)=V_{o1}cos(k_1x)+V_{o2}cos(k_2x)+Fx \quad , \tag{2}$$

where V_{o1}, V_{o2} are the amplitudes of the optical superlattice potential and k_1, k_2 are the wave vectors of the laser beams used to create the optical superlattice potential. The tilt of the superlattice potential is described by the term Fx in Eqn. (2) where F is the tilt-coefficient of the superlattice . As the value of F increases the superlattice potential becomes more and more tilted. For the linear part of the Hamiltonian defined as

$$H_o=-\frac{1}{2m}\frac{\partial^2}{\partial x^2}+V_{trap}(x) \quad , \tag{3}$$

the eigenstates are referred to as WS states. To simplify the problem we make use of the time-dependent variational wave function with a trivial phase as

$$\psi(x,t)=\phi(x)e^{(\frac{-i\mu t}{\hbar})} \quad , \tag{4}$$

where μ is the chemical potential and $\phi(x)$ is the real and normalized wave function, where normalization of the wave function is described as $\int\phi^2(x)dx=N$. By substituting above wave function in Eqn. (1), we get following equation

$$-\frac{\hbar^2}{2m}\frac{\partial^2\phi(x)}{\partial x^2}+V_{trap}(x)\phi(x)+g_o\phi^3(x)=\mu\,\phi(x) \quad . \tag{5}$$

Above equation is the stationary state G-P equation in case of zero trivial phase which describes the dynamics of BEC. Above equation can be re-written as

$$-\frac{\partial^2\phi(x)}{\partial x^2}+[\alpha_1cos(k_1x)+\alpha_2cos(k_2x)+\beta x]\phi(x)+\gamma\phi^3(x)=\Gamma\,\phi(x) \tag{6}$$

where the coefficients are $\alpha_1=\frac{2mV_{o1}}{\hbar^2}$, $\alpha_2=\frac{2mV_{o2}}{\hbar^2}$, $\beta=\frac{2mF}{\hbar^2}$, $\Gamma=\frac{2m\mu}{\hbar^2}$. To solve the above non-linear equation we substitute $\frac{d\phi(x)}{dx}=y(x)$ and get following two coupled equations as

$$\frac{d\phi(x)}{dx}=y(x)$$

$$\frac{dy(x)}{dx} = [\alpha_1 cos(k_1 x) + \alpha_2 cos(k_2 x) + \beta x - \Gamma]\phi(x) + \gamma\phi^3(x) \tag{7}$$

Now above two coupled equations can be solved simultaneously in "mathematica" with suitable initial conditions and other parameters.

3. Effect of varying the depth of optical superlattice potential and strength of tilted field on the stability of BEC

In this section we study the effect of varying the amplitude of the laser beams used to create the optical superlattice potential. By varying the amplitude of the secondary laser beam we change the height of the optical superlattice potential, which in turn affects the tunnelling of atoms between the potential wells. As the height of wells increases we should have more stability in the BEC due to decrease in tunnelling of atoms, but here the superlattice potential is tilted due to which the stability and instability of the BEC arises due to an overall effect of the tilt of lattice potential and height of the optical superlattice potential. We fix the value of $k_1 = \frac{2\pi}{850}$ and secondary laser beam is taken such that $k_2 = k_1 + \pi$ with $\Gamma = 0.01$. We are considering the case of $\gamma < 0$ (means attractive interaction between the atoms in a BEC). We are fixing the value of $\gamma = -1.0$. In Fig. (1) we plot the phase space diagrams for a fixed value of $\beta = 0.0001$, $\alpha_1 = 1.0$, with the variation of $\alpha_2 = 0.1, 0.7$. As the value of the coefficient α_2 is increased from 0.1 to 0.7, the motion of the system which was restricted to a small region of the phase space now extends in a large area showing a large instability in the BEC (shown in Fig. 1(b)).

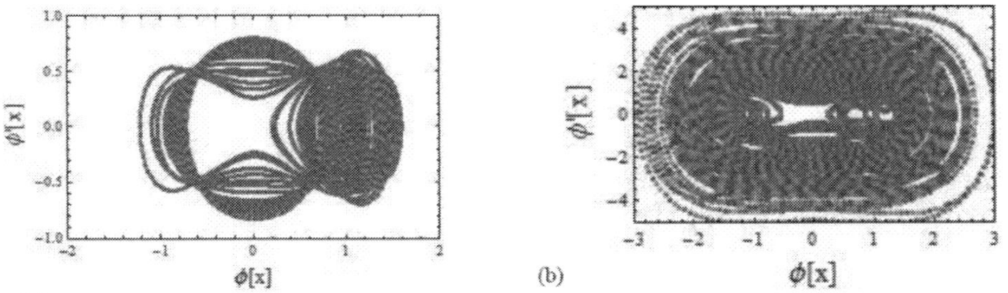

(a)

Figure 1. The phase orbit in the equivalent phase space of the ϕ versus $d\phi / dx$ with $\beta = 0.0001$, $\phi(0) = 0.5$, $\phi'(0.5) = 0.0$ (a) α_2=0.1 (b) α_2=0.7

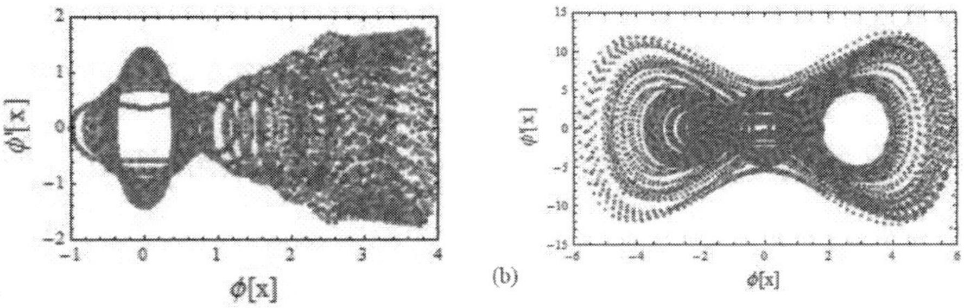

(a) (b)

Figure 2. The phase orbit in the equivalent phase space of the ϕ versus $d\phi / dx$ with $\beta = 0.1$, $\phi(0) = 0.5$, $\phi'(0.5) = 0.0$ (a) α_2=0.1 (b) α_2=0.7

Similarly in the Fig. (2) we show the same plots of phase space diagrams as in Fig. (1) but for $\beta = 0.1$. Here the phase space orbit extends almost upto 6 times when the secondary lattice depth is increased. In all the cases we have taken the initial conditions as $\phi(0) = 0.5, \phi'(0.5) = 0.0$. Corresponding to the above phase space plots, we show in the Fig. (3) and Fig. (4) the spatial evolution of the solution of the coupled Eqn. (7). We observe that as the depth of optical superlattice increases (by increasing the coefficient α_2) with a fixed value of the coefficient β and $\alpha_1 = 1.0$ the amplitude of the wave function increases. This increase is less with the lower value of external field. As the strength of the external field is increased the effect of varying the superlattice depth is more pronounced and the wave function increases abruptly for $\alpha_2 = 0.7$ with $\beta = 0.1$. We observe that the motion is chaotic because on changing the initial conditions slightly, trajectories in the phase space are highly changed. Now we study the effect of changing the initial conditions on the stability of BEC. In the Fig. (5) we show that as the coefficient of tilt is increased from 0.0001 to 0.1 with the initial conditions as $\phi(0) = -2.0$, $\phi'(0.5) = 0.0$ the phase space orbit which appears to be a little periodic in

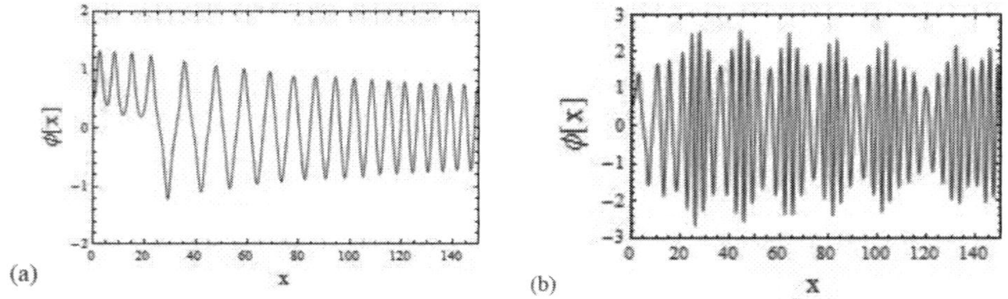

Figure 3. The spatial evolution of the solution $\phi(x)$ with $\beta = 0.0001$, $\phi(0) = 0.5$, $\phi'(0.5) = 0.0$ (a) α_2=0.1 (b) α_2=0.7

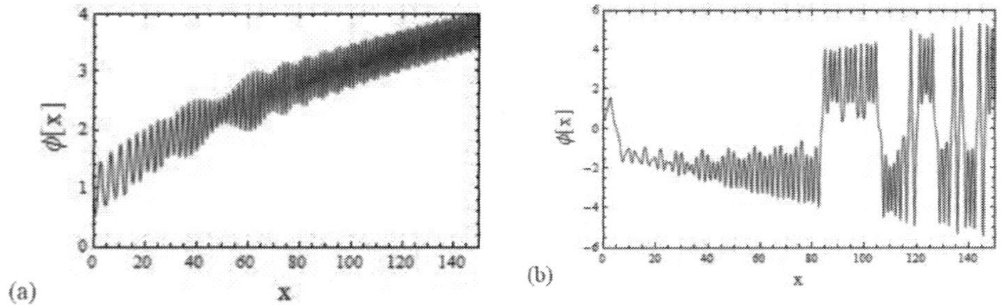

Figure 4. The spatial evolution of the solution $\phi(x)$ with $\beta = 0.1$, $\phi(0) = 0.5$, $\phi'(0.5) = 0.0$ (a) α_2=0.1 (b) α_2=0.7

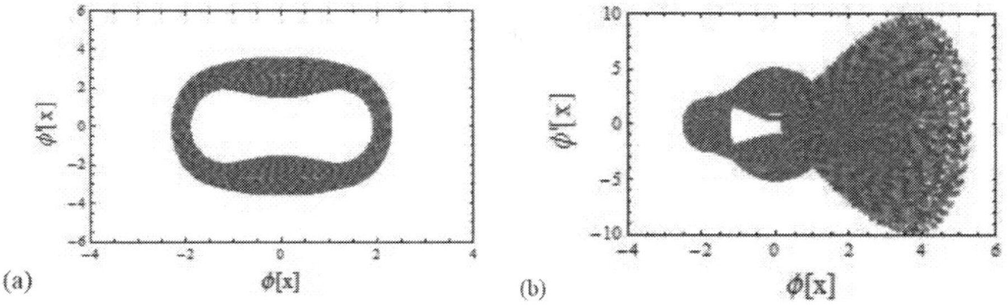

Figure 5. The phase orbit in the equivalent phase space of the ϕ versus $d\phi / dx$ with $\alpha_2 = 0.1$, $\phi(0) = -2.0$, $\phi'(0.5) = 0.0$, (a) $\beta =$ 0.0001 (b) $\beta =$ 0.1

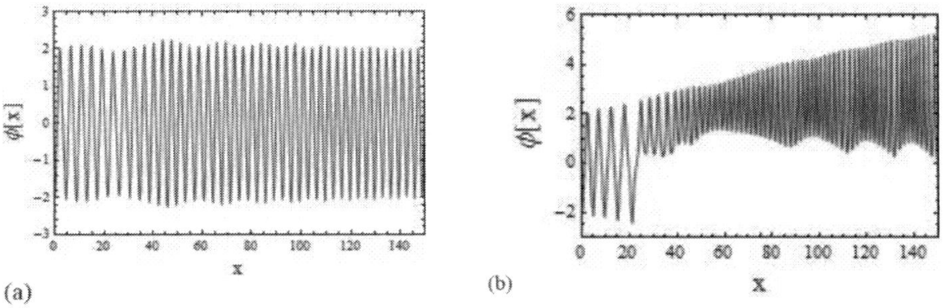

Figure 6. The spatial evolution of the solution $\phi(x)$ with $\alpha_2 = 0.1$, $\phi(0) = -2.0$, $\phi'(0.5) = 0.0$, (a) $\beta =$ 0.0001 (b) $\beta =$ 0.1

Fig. 5(a) with its wave function almost stable (shown in Fig. 6(a)) becomes extremely chaotic on increasing the value of β as shown in Fig. 5(b). The large increase in wave function corresponding to Fig. 5(b) is shown in Fig. 6(b). The effect of changing the initial conditions is clearly seen in Fig. 5(a), 5(b) though the other parameters are same as in Fig. 1(a) and 2(a) respectively. It is observed that the system's trajectory remains almost deterministic for any initial condition, but the instability can also be suppressed to an extent by varying the initial conditions of the system. Also as the value of β is increased the trajectory of the system occupies the more region in phase-space plots which shows that the chaos and instability is increasing in the BEC, but still the trajectory in the phase space remains confined to a particular area showing that the system behaves as chaotic attractor. For higher values of β and α_2 the BEC becomes highly unstable.

4. Conclusions

In this paper we have investigated the chaos in a BEC system trapped in 1D tilted optical superlattice potential. We have shown that even the slightest tilt of the lattice potential leads to great chaos in the system and makes the BEC highly unstable with the abrupt increase in the wave function of the system. We can control this instability of the BEC by adjusting the tilt of the superlattice potential and the ratio of wave lengths of the two laser beams used to create the superlattice. The increase in the amplitude of the superlattice potential further enhances the chaos in the system. The initial conditions of the system also have a deep impact on the chaos in BEC. There are some initial conditions of the system which can make the system's orbit periodic instead of chaotic as in Fig. 5(a). Therefore we can control the instability and chaos in the BEC by adjusting the coefficients β, α_2 and ratio of wave vectors of the laser beams used.

Acknowledgement
One of the authors Priyanka Verma thanks the University Grants Commission, New Delhi for the Junior Research Fellowship.

References

[1] Kim J H and Stringer J 1992 *Applied Chaos (John Wiley and Sons, New York)*
[2] Thommen Q, Garreau J C and Zehnle V 2003 *Phys. Rev. Lett.* **91** 210405
[3] Zhan M, Hu G and Wang X G 2000 *Chin. Phys. Lett.* **17** 332 ; Yang J, Hu G and Xiao J 1998 *Phys. Rev. Lett.* **80** 496 ; Ott E, Grebogi C and Yorke J A 1990 *Phys. Rev. Lett.* **64** 1196
[4] Abdullaev F Kh and Kraenkel R A 2000 *Phys. Rev. A* **62** 023613 ; Coullet P and Vandenberghe N 2001 *Phys. Rev. E* **64** 025202(R)
[5] Abdullaev F Kh and Galimzyanov R 2003 *J. Phys. B* **36** 1099 ; Mayteevarunyoo T, Malomed B A and Krairiksh M 2007 *Phys. Rev. A* **76** 053612
[6] Jian-Shu F and Xiang-Ping L 2011 *Chin. Phys. B* **20** 040310
[7] Bradley C C, Sackett C A, Tollett J J and Hulet R G 1995 *Phys.Rev.Lett.***75**

Alternative techniques of population transfer in multilevel systems

V. Chaltykyan[1], E. Gazazyan[1], G. Grigoryan[1], A. Hovhannisyan[2], O. Tikhova[2]

[1]Institute for Physical Research, 0203 Ashtarak-2, Armenia

[2]Russian-Armenian (Slavonic) University, Yerevan, Armenia

emilgazazyan@gmail.com

Abstract. We consider a five-level atomic system (M-system) driven by four near-resonant laser pulses and study the possibilities of population transfer via different combinations of Stimulated Raman Adiabatic Passage (STIRAP) and bright-STIRAP (b-STIRAP) processes. We examine these combinations analytically and numerically from the point of view of efficiency of population transfer. It is shown, in particular, that the combination of the processes above enhances the efficiency. Analytical expressions are obtained for the quasienergies of the system in two regimes differing by conditions imposed on resonance detunings. Interaction adiabaticity conditions are analyzed. We also study the necessary successions of turning on and off of pulses which ensure efficient transfer of atomic populations. The obtained analytical results agree well with direct numerical calculations.

1. Introduction

A widespread technique for population transfer in Λ-systems employs stimulated Raman process at adiabatic passage of laser pulses (STIRAP) [1]. This method is based on counterintuitive sequence of pulses when the Stokes pulse coupling the final state and the intermediate, excited, state precedes the pump pulse which couples the excited state with the initial one. In multilevel systems the population transfer is realized with use of successive STIRAP processes or STIRAP-chains [2]. .

Recently an alternative efficient method of population transfer where the intuitive sequence of pulses is used with a large one-photon detuning has been predicted and demonstrated experimentally. Unlike the STIRAP for which the dynamics adiabatically projects onto the dark state, i.e., the excited state component is absent, the intuitive sequence follows a bright state. This process has been called bright STIRAP (b-STIRAP) [3]. An essential distinction of b-STIRAP from STIRAP is the rate of population transfer in media which is superluminal [4].

The purpose of the present study was to find other ways of transfer of level populations in multilevel systems through combination of STIRAP and b-STIRAP (STIRAP-b-STIRAP, b-STIRAP-b-STIRAP, b-STIRAP-STIRAP), rather than purely STIRAP-chains. We restrict our study to five-level system which can be considered as two three-level systems with a common level, so it is supposed that STIRAP and b-STIRAP provide more possibilities being combined in different successions. The advantages of one or another combination will be analyzed depending on the configuration of five atomic levels.

2. Atomic system and basic equations

Consider a five-level atomic system with M-type configuration of levels with numbering of levels and Rabi frequencies of incident pulses shown in Fig.1 and the Hamiltonian

$$H = \sum \sigma_{ii}\delta_{i-1} + (\sum \sigma_{ii+1}\Omega_i + h.c.) \tag{1}$$

where the detunings are defined as $\delta_1=\omega_{21}-\omega_1$, $\delta_2=\omega_{31}+\omega_2-\omega_1$, $\delta_3=\omega_{43}-\omega_3+\delta_2$, $\delta_4=\omega_{51}+\omega_4-\omega_3+\omega_2-\omega_1$ with ω_{ik} being the frequency separations between the respective levels, ω_i the frequencies of laser pulses, and Rabi frequencies are defined as usual.

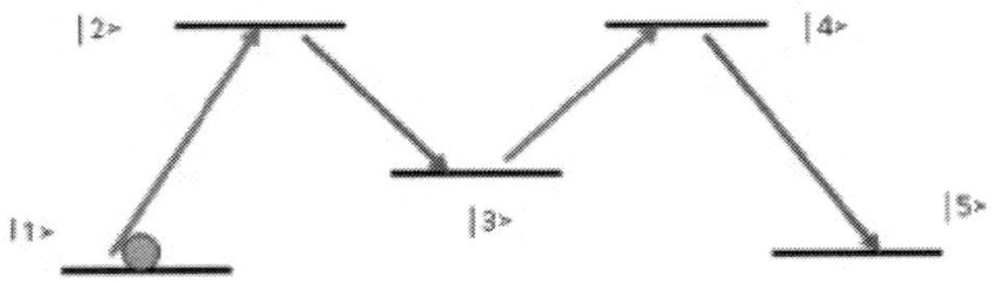

Fig.1. (Color online) Energy level diagram of M-system.

The characteristic equation det($H-\lambda E$)=0 has the following explicit form:

$$\lambda(\lambda-\delta_1)(\lambda-\delta_2)(\lambda-\delta_3)(\lambda-\delta_4)-\Omega_4^2\lambda(\lambda-\delta_1)(\lambda-\delta_2)-\Omega_3^2\lambda(\lambda-\delta_1)(\lambda-\delta_4)+\lambda\Omega_4^2\Omega_2^2$$
$$-\Omega_2^2\lambda(\lambda-\delta_3)(\lambda-\delta_4)-\Omega_1^2(\lambda-\delta_2)(\lambda-\delta_3)(\lambda-\delta_4)+\Omega_4^2\Omega_1^2(\lambda-\delta_2)+\Omega_3^2\Omega_1^2(\lambda-\delta_4)=0 , \tag{2}$$

The conventional STIRAP-STIRAP chain is realized for the eigenstate corresponding to the eigenvalue $\lambda=0$. As follows from Eq.(2), it has a zero root, if the conditions of exact two- and four-photon resonances are met:

$$\delta_2=\delta_4=0. \tag{3}$$

By eliminating this root we obtain a fourth order equation whose roots can be written in not very cumbersome form if some additional conditions are fulfilled. So, under condition $\delta_1=\delta_3$ the roots of characteristic equation are

$$\lambda_0 = 0$$
$$\lambda_{1,2} = \frac{1}{2}\left(\delta_1 \mp \sqrt{\delta_1^2 + 4x_1}\right) \tag{4}$$
$$\lambda_{3,4} = \frac{1}{2}\left(\delta_1 \mp \sqrt{\delta_1^2 + 4x_2}\right)$$

where $x_{1,2}=(1/2)(\Omega^2\pm(\Omega^4-4V^4)^{1/2})$, while the quantities Ω (generalized Rabi frequency) and V are defined as follows:$\Omega^2=\Omega_1^2+\Omega_2^2+\Omega_3^2+\Omega_4^2$, $V^4=\Omega_1^2\Omega_3^2+\Omega_1^2\Omega_4^2+\Omega_2^2\Omega_4^2$.

Adiabaticity of interaction will take place under the conditions:

$$V^4T/\delta_1 \gg 1,\ \Omega^2T/\delta_1 \gg 1,\ \delta_1 T \gg 1,$$

where T is the time of interaction between the system and fields of pulses.

Conditions $\delta_1=-\delta_3$ and $\Omega_1^2+\Omega_2^2=\Omega_3^2+\Omega_4^2$ also lead to a simple appearance of the roots of Eq.(2)

$$\lambda_0 = 0$$
$$\lambda_{1,2,34} = \pm\left(\frac{1}{2}((\delta_1^2+\Omega^2)\mp\left((\delta_1^2+\Omega^2)^2\mp 4V^2\right)^{1/2}\right)^{1/2} \tag{5}$$

Knowledge of eigenvalues (4) and (5) enables us to construct corresponding eigenvectors. However, we will only consider several special cases where an efficient transfer of population from level 1 to level 5 is possible.

3. Combination of b-STIRAP and STIRAP

If $\Omega_1=\Omega_4$, expressions (4) are simplified essentially:

$$\lambda_{1,2}=\frac{1}{2}\left(\delta_1\mp\sqrt{\delta_1^2+4\Omega_1^2}\right)$$
$$\lambda_{3,4}=\frac{1}{2}\left(\delta_1\mp\sqrt{\delta_1^2+4\Omega_s^2}\right) \qquad (6)$$

where the generalized Rabi frequencyΩ_s is defined by the relation $\Omega_s^2=\Omega_1^2+\Omega_2^2+\Omega_3^2$. Let us assume the one-photon detuning to be positive ($\delta_1>0$) and consider the root λ_1 of Eq.(2). An interesting peculiarity of this root is that it is independent of fields Ω_2 and Ω_3. The state vector corresponding to this eigenvalue has the form

$$|\Psi\rangle=\cos\Phi\cos\Theta|1\rangle-\sin\Phi\cos\Theta|2\rangle+\sin\Phi\sin\Theta|4\rangle+\cos\Phi\sin\Theta|5\rangle \qquad (7)$$

where the angles Φ and Θ are defined by relations $\mathrm{tg}\Phi=\lambda_1/\Omega_1$ and $\mathrm{tg}\Theta=\Omega_2/\Omega_3$. The state (7) does not contain the state $|3\rangle$ and the dipole moments of the transitions $1\rightarrow2$ and $4\rightarrow5$ are equal in modulus, but opposite in sign. Figure 2 shows the succession of turning on the pulses and results of direct numerical calculation demonstrating an efficient population transfer. It is seen that the field Ω_1 is turned on prior to Ω_2 (b-STIRAP) and the field Ω_3 prior to Ω_4 (STIRAP).

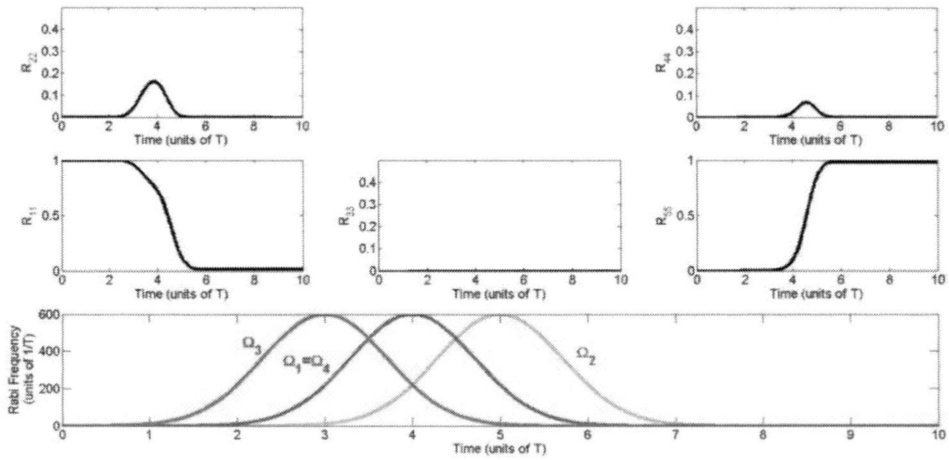

Fig.2. (Color online) Sequence of pulses and dynamics of level populations in case $\Omega_1=\Omega_4$ and $\delta_1>0$.

The same chain may be realized via eigenstate corresponding to the root λ_4. In this case the state vector involves all the bare atomic states:

$$|\Psi\rangle=\cos\Phi_1\cos\Theta_1\sin\Theta|1\rangle-\sin\Phi_1\sin\Theta|2\rangle-\sin\Phi_1\cos\Theta_1|3\rangle$$
$$+\cos\Phi_1\sin\Theta_1|4\rangle+\cos\Phi_1\cos\Theta_1\cos\Theta|5\rangle. \qquad (8)$$

where Φ_1 and Θ_1 are defined similarly to (7): $\mathrm{tg}\Phi=\lambda_2/\Omega_s$ and $\mathrm{tg}\Theta_1=\Omega_s/\Omega_1$ For efficient population transfer through the state (8) pulses of different durations should be used, as shown in Fig.3.

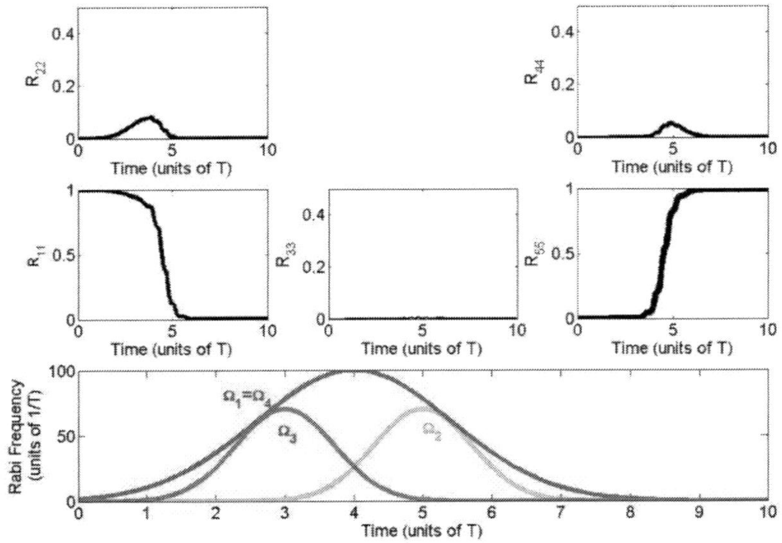

Fig.3. (Color online) Sequence of pulses and dynamics of level populations in case $\Omega_1=\Omega_4$ for pulses of different durations.

4. Combination of STIRAP and b-STIRAP

This chain can be realized when the fields coupling transitions $1\to2$ and $4\to5$ are different, while the transitions $2\to3$ and $3\to4$ are driven by degenerate pumping, i.e., $\Omega_2=\Omega_3$. The succession of pulse switching and the evolution of level populations are depicted in Fig.4. It is seen that this case also provides efficient transfer of population from state 1 to state 5.

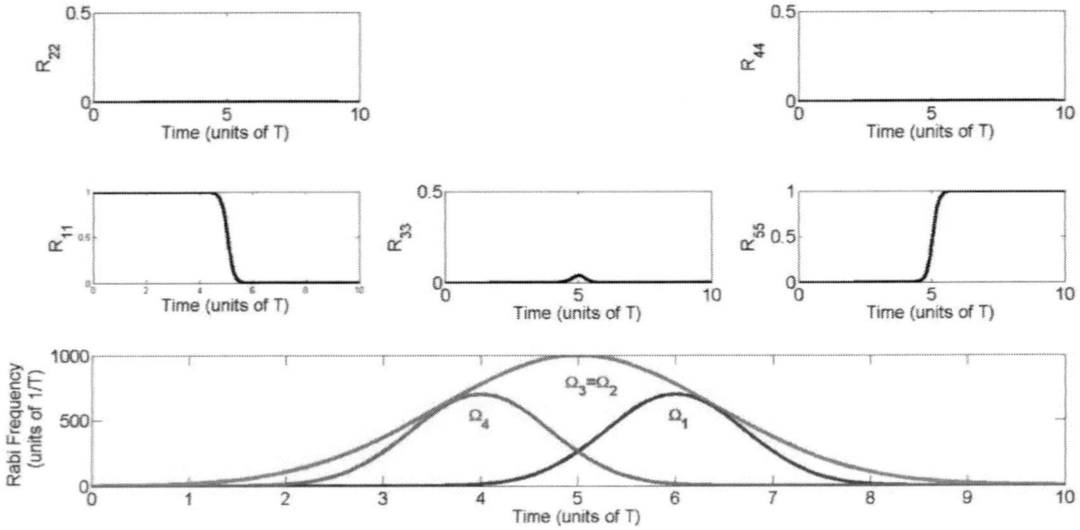

Fig.4. (Color online) Sequence of pulses and dynamics of level populations in case $\Omega_2=\Omega_3$ for pulses of different durations.

Although the advantages of the above-considered schemes as compared with the conventional STIRAP-STIRAP chain are not seen for the studied single-atom problem, in case of atomic medium they may turn to be more efficient from the point of view of the rate of population transfer and from the point of view of increasing the medium length at which the full transfer is possible. Indeed, the main restriction on the medium length in STIRAP-chain transfer is the condition of pumping (field Ω_1) depletion. In the schemes considered above different fields play the role of pumping, so a photon absorbed from the field in one transition will in another transition be emitted back into the same field. However, detailed investigation of the population transfer problem in media is out of scope of the present work and will be performed elsewhere.

5. Combination b-STIRAP-b-STIRAP

Such a sequence, like STIRAP-STIRAP, may be realized by two pulses with Ω_1=Ω_3 and Ω_2=Ω_4. The detailed analysis shows, however, that the population can in this case be transferred efficiently to only the level 3, as depicted in Fig.5.

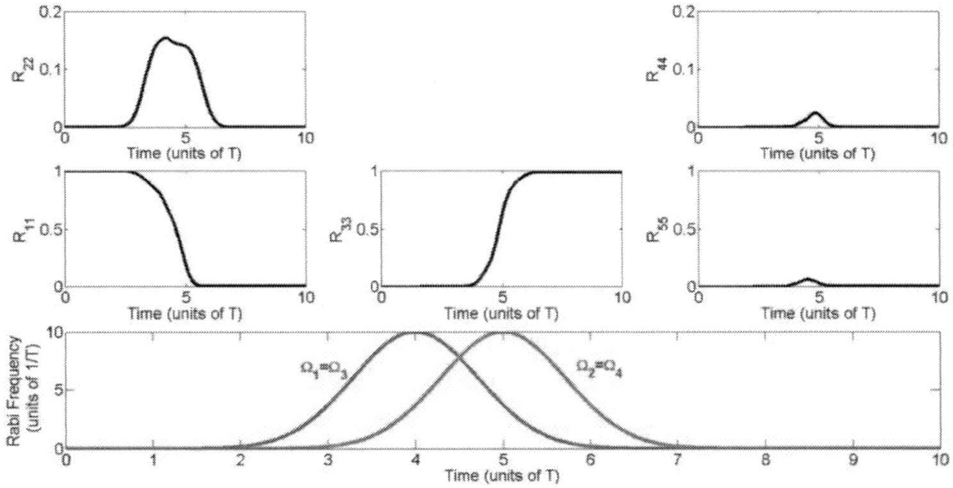

Fig.5. (Color online) b-STIRAP-b-STIRAP chain with two pulses.

Conclusion

The coherent population transfer is at the heart of many quantum processes with applications for quantum information processing, the control of chemical reaction, nonlinear and atomic optics et al.

We have studied a five-level M-type atomic system in the sense of possibilities of population transfer between the levels. We have demonstrated that adiabatic transfer of populations in this system may be realized not only via dark states in traditional STIRAP-chain, but also via other adiabatic states. We have proposed three novel alternative schemes of coherent population transfer between lower levels of an M-type atomic system We found analytically and numerically the possible combinations of parameters and sequences of corresponding laser pulses providing complete transfer of population to a specific level by optimum combinations of STIRAP and b-STIRAP processes. Although the upper levels in considered schemes are during interaction populated provisionally, their populations can be made negligibly low by providing large single-photon detunings. Advantages and shortcomings of proposed schemes as compared with conventional STIRAP-chain are associated with those of the b-STIRAP technique (see [4,5]) and will be analyzed in a subsequent paper in connection with their realization in macroscopic volumes and use for designing logic gates.

Acknowledgment
Work was in part supported by the Ministry of Education and Science of Armenia (MESA), grant no. 11-1c124 of State Science Committee of MESA, IRMAS International Associated Laboratory, and the Volkswagen Stiftung I/84 953.

References

[1] Bergman, H.Theuer, B.Shore, Rev. Mod. Phys. 70, 1003-1025 (2004).
[2] N.V.Vitanov, B.W.Shore,K.Bergman, Adv. Atom.,Mol, Optical Physics, 46, 55-190, (2001).
[3] N.V.Vitanov, B.W.Shore,K.Bergman, Adv. Atom.,Mol, Optical Physics, 46, 55-190, (2001).
[4] G.G. Grigoryan, G. Nikoghosyan, T. Halfmann, Y.T. Pashayan-Leroy, C. Leroy, S.Guerin, Phys.Rev. A 80, 033402 (2009)
[5] M. Scala,B. Militello,A. Messina,N. V. Vitanov, Phys. Rev. A 83, 012101 (2011).

International Symposium on Optics and its Applications (OPTICS2011) IOP Publishing
Journal of Physics: Conference Series **350** (2012) 012005 doi:10.1088/1742-6596/350/1/012005

Method of quasienergies and transparency of multilevel atomic systems

V Chaltykyan[1], G Grigoryan[1], A Hovhannisyan[2], O Tikhova[2]

[1] *Institute for Physical Research, 0203 Ashtarak-2, Armenia*

[2] *Russian-Armenian (Slavonic) University*

E-mail:hovana@mail.ru

Abstract. We consider a medium of multilevel atomic systems interacting with radiation pulses. A relatively simple technique of analytic calculations is proposed, which allows revealing all necessary conditions (with sufficient conditions to be checked separately) imposed on the interaction parameters, for which the mean dipole moment of a multilevel atomic medium vanishes, i.e., the medium becomes transparent via adiabatic interaction. The proposed technique is based on the method of quasienergies and illustrated for three- and five-level atomic systems.

1. Introduction

Adiabatic processes are a specific sub-class of coherent interactions, which rely on adiabatic evolution of the quantum system. As the evolution is slow, the system typically remains in a certain laser-controlled quantum state, which is an eigenstate of the interaction Hamiltonian. This yields a stable situation. An example of such states is the so-called dark-state, formed in three-level systems; it lies in the basis of many widely studied coherent phenomena [1]. Dark-state involves, due to quantum interference, only two of three atomic states; in particular, in a Λ-system it does not involve the upper atomic state leading thus to the medium transparency.

Note that the dark state is not the only state resulting in the medium transparency. For example, for V-systems, a medium is well-known [2] to become transparent when interacting with degenerate pumping; this transparency is not associated with the dark state, but is caused by interference of coherences leading to compensation for the medium dispersion.

In the present work we apply a relatively simple technique of analytic calculations of propagation of light pulses in a medium consisting of multilevel atomic systems; the technique allows us to determine necessary conditions which should be imposed on the interaction parameters, for which the mean dipole moment of the medium vanishes, i.e., the medium becomes transparent via adiabatic interaction. The proposed technique is illustrated for a five-level atomic system.

2. Formalism

We assume that considered pulses have durations much shorter than all times of relaxation and at the same time much longer than the inverse frequency distance between the closest quasienergies, in order to ensure the adiabaticity of interaction.

Adiabatic propagation of pulses in an atomic medium is described by a self-consistent system of truncated equations in running coordinates x and $t-x/c$, which may be represented in the form [3]

Journal of Physics: Conference Series **350** (2012) 012005 doi:10.1088/1742-6596/350/1/012005

$$\frac{\partial E_j}{\partial x} = -i\frac{2\pi N\omega_j\hbar}{c}\frac{\partial\lambda}{\partial E_j^*} \tag{1}$$

Here λ is the eigenvalue of the interaction Hamiltonian, E_j the complex amplitude of the pulse field, and other notations are conventional. If the system has a quasienergy which remains constant during the overall time of interaction, i.e., $\partial\lambda/\partial E_j=0$, then according to (1), the induced dipole moment in such a system is zero and the pulse propagates in such medium without change in shape at the group velocity equal to c. It is known [4] that the quasienergies (eigenvalues) of the system are the roots of characteristic equation

$$\det(H-\lambda I)=0 \tag{2}$$

where I is the unit matrix and H the interaction Hamiltonian which may in the resonance approximation be represented in the form

$$H = \Sigma\sigma_{ii}\delta_{i-1} + (\Sigma\sigma_{ii+1}\Omega_i + h.c.) \tag{3}$$

with σ_{ii} being the projection matrices, Ω_i the Rabi frequencies of corresponding transitions $i\rightarrow i+1$), and δ_{i-1} the $(i-1)$-photon detuning ($\delta_0=0$). By differentiating the characteristic equation with respect to E_j and setting zero the derivatives of quasienergy, we obtain a system of algebraic equations, from which the needed values of interaction parameters may be determined. The problem can, however, be simplified. Really, at turning off the fields the interaction Hamiltonian is diagonalized and the roots of equation (2) go to the following constant values:

$$\lambda_i \rightarrow \delta_i \tag{4}$$

The problem of determination of necessary (but not always sufficient) conditions for medium transparancy is thus reduced to the search of such parameters of adiabatic interaction for which the system quasienergy remains equal to one of resonance detunings always during interaction. By substituting these values successively into equation (2) we obtain the needed conditions for the interaction parameters.

For example, in case of a three-level system the root $\lambda=0$ is realized at exact two-photon resonance $\delta_2=0$ and under the same condition the root $\lambda=\delta_2=0$ is realized. The wave function corresponding to these quasienergies (dark state) is well known [5]:

$$|\psi\rangle = \cos\theta|1\rangle - \sin\theta|3\rangle \tag{5}$$

where $tg\theta = \Omega_1/\Omega_2$. As the population of the level $|2\rangle$ is zero, the dipole moments of $1\rightarrow2$ and $2\rightarrow3$ transitions also vanish.

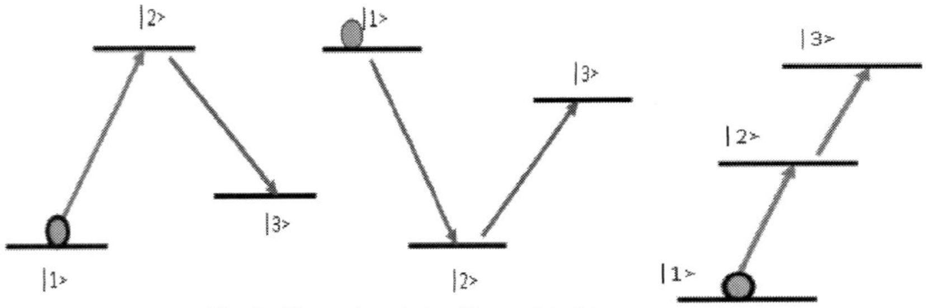

Fig.1. Three-level Λ-, V-, and ladder-systems.

The third root $\lambda=\delta_1$ is realized at degenerate pumping ($\Omega_1=\Omega_2$) and the condition for detunings $\delta_2=2\delta_1$.. As distinct from the previous case, the wave function involves all bare atomic states, however the medium becomes transparent because of destructive interference of dipole moments of transitions $1\to2$ and $2\to3$. Similar scheme of interaction may be realized also in V-system (see Fig.1), if the atom is prepared in state 2. It should be noted that in case of non-degenerate pumping the root $\lambda=\delta_1$ does not provide transparency, since the dipole moments of the transitions $1\to2$ and $2\to3$ do not equal zero individually. So, the condition $\lambda=\delta_1$ is indeed necessary, but not sufficient condition.

3. Five-level atomic system

We use now the proposed technique for a five-level atomic system and the pulses resonant with only the adjacent transitions. Several examples of such diagrams are demonstrated in Fig.2.

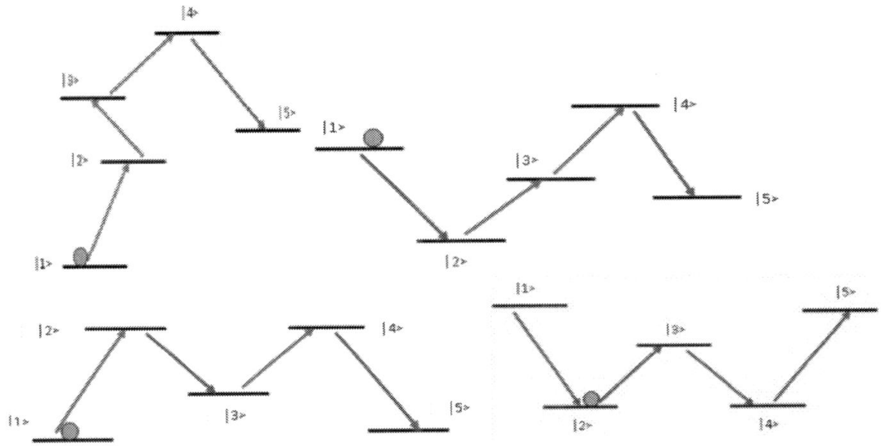

Fig.2. Several five-level configurations and relevant Rabi frequencies.

Under conditions of exact two-photon resonances ($\delta_2=0$, $\delta_4=0$) a dark state is realized in the system (see, e.g., [7]) which corresponds to the root $\lambda=0$:

$$|\psi\rangle = \cos\theta_1 \cos\theta_2 |1\rangle - \sin\theta_1 \cos\theta_2 |3\rangle + \sin\theta_1 \sin\theta_2 |5\rangle \quad (6)$$

where $tg\theta_1 = \Omega_1/\Omega_2; tg\theta_2 = \Omega_3/\Omega_4$

The value $\lambda=\delta_1$ can be realized under conditions $\delta_1=\delta_4=\delta_2/2$ and degenerate pumping at the transition $1\to3$ ($\Omega_1=\Omega_2=\Omega$). The wave function does not contain the level 4 and the dipole moments of transitions $1\to2$ and $2\to3$ are in antiphase leading to their destructive interference:

$$|\psi\rangle = \frac{1}{R}(\sin\Phi\cos\theta_2|1\rangle + \cos\Phi\cos\theta_2|2\rangle - \sin\Phi\cos\theta_2|3\rangle + \sin\Phi\sin\theta_2|5\rangle \quad (7)$$

where $R=(\sin^2\Phi+\cos^2\theta_2)^{1/2}$ and $\mathrm{tg}2\Phi=\Omega/\delta_1$. For realization of this state the atom should be initially prepared in state $|2\rangle$. This state may efficiently be used in atomic level configurations of Fig.2b or of Fig.2c. So, with use of counterintuitive sequence of pulses at transitions $3\to4$ and $4\to5$ with a longer pulse at transition $1\to3$ (see Fig.3), it is possible to transfer the population from level 2 to level 5. In this case a population inversion is produced between the levels 5 and 4, which may be used for amplification of radiation in case of the last diagram of Fig.2.

The value $\lambda=\delta_1$ may be realized also under other conditions, e.g., at $\delta_1=\delta_2=\delta_3$, $\delta_4=2\delta_1$ and degenerate pumpings of transitions $1\to3$ and $3\to5$ ($\Omega_1=\Omega_2$ and $\Omega_3=\Omega_4$). In this case the wave function involves all bare atomic states, but dipole moments of both degenerate transitions are zero:

$$|\psi\rangle=\frac{1}{R}(\sin\Phi_1\sin\Phi_2|1\rangle+\cos\Phi_1\sin\Phi_2|2\rangle-\sin\Phi_1\sin\Phi_2|3\rangle-\sin\Phi_1\cos\Phi_2|4\rangle+\sin\Phi_1\sin\Phi_2|5\rangle \quad (8)$$

where $R=(\sin^2\Phi_1+\sin^2\Phi_2+\sin^2\Phi_1\sin^2\Phi_2)^{1/2}$ and $\mathrm{tg}2\Phi_1=\Omega_1/\delta_1$, $\mathrm{tg}2\Phi_2=\Omega_3/\delta_3$.
The atom should be initially in state $|2\rangle$ and the pulses turn on in counterintuitive sequence (see Fig.3). A similar state may be obtained when $\Omega_1=\Omega_4$ and $\Omega_3=\Omega_2$.

The value $\lambda=0$ also may be realized in different ways. For example, instead of $\delta_2=\delta_4$ we can require vanishing of three-photon detuning, $\delta_3=0$, and $\delta_2=-\delta_4\neq0$. In this case, if pumping of the $3\to5$ transition is degenerate ($\Omega_3=\Omega_4$), we obtain

$$|\psi\rangle=\frac{1}{R}(\sin\Phi_3\cos\theta_1|1\rangle-\sin\Phi_3\sin\theta_1|3\rangle+\cos\Phi_3\sin\theta_1|4\rangle+\sin\Phi_3\sin\theta_1|5\rangle \quad (9)$$

where $R=(\cos^2\Phi_3+\sin^2\theta_1)^{1/2}$ and $\mathrm{tg}\Phi_3=-\Omega_4/\delta_4$. The atom is initially in the state $|4\rangle$, while the state $|2\rangle$ does not enter the superposition.

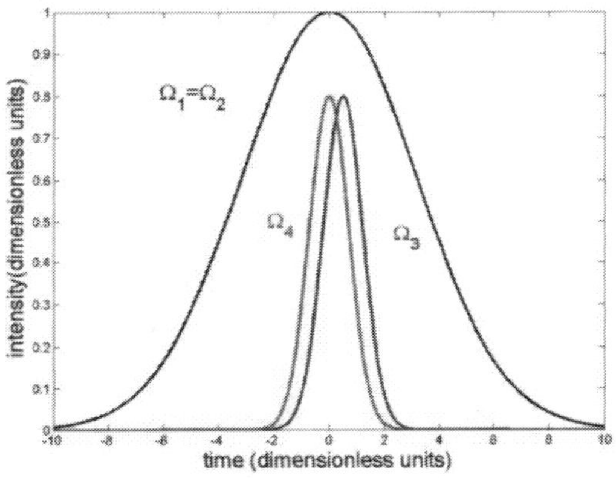

Fig.3. Sequence of pulses for efficient transfer of population from level 2 to 5.

In addition to the cases considered above the value $\lambda=\delta_2$ can be obtained under conditions $2\delta_2=\delta_1+\delta_3=\delta_4$, $\Omega_1=\Omega_4$, $\Omega_3=\Omega_2$ and the value $\lambda=\delta_3$ under conditions $\delta_1=\delta_2$, $3\delta_3=\delta_2+\delta_4$, $\Omega_1=\Omega_2$.

4. Conclusion

We have developed and analyzed, for the example of a five-level system, a relatively simple method for finding necessary conditions under which adiabatic short light pulses can travel in the medium without distortion of the shape and phase. It was shown that the necessary conditions for such bleaching of medium (i.e., vanishing of dipole moments induced in the medium at the frequencies of all interacting fields) is the equality of system quasienergies to one of the values of single- or multi-photon detunings of resonances. In this case the medium bleaching may be caused by both interference of quantum states and interference of dipole moments of different transitions if pumping is degenerate. The existence of such regimes means that at the propagation lengths where the interaction adiabaticity does not break, stable superposition states are produced all over the medium and these states may easily be controlled by adjusting the parameters of pulses. Finding of the sufficient conditions requires determination of roots of algebraic equations which may be done only numerically for the order of equation higher than three. However, the sufficiency of found necessary conditions may easily be checked directly by calculation of the induced dipole moments.

Acknowledgment
Work was in part supported by the Ministry of Education and Science of Armenia (MESA), grant no. 11-1c124 of State Science Committee of MESA, IRMAS International Associated Laboratory, and the Volkswagen Stiftung I/84 953.

References

[1] Fleischhauer M., Imamoglu A., Marangos J.P., Rev. Mod. Phys. 77, 633-673 (2005).
[2] Achmanov S., Kovrigin M., Maksimov S., Ogluzdin V., Pis'ma JETFv.15,186 (1972)
[3] Melikyan A.O., Saakyan S.G., Zh.Eksp.Teor.Fiz.76,1530,(1979) (Sov. Phys.-JETF 49, 776 (1979))
[4] Shore B.W.,"Coherent manipulation of atoms using laser light", Acta Phys. 58, 243-486 (2008).

International Symposium on Optics and its Applications (OPTICS2011) IOP Publishing
Journal of Physics: Conference Series **350** (2012) 012006 doi:10.1088/1742-6596/350/1/012006

Quantum superposition principle and generation of short optical pulses

Gevorg Muradyan, Atom Zh. Muradyan
Department of Physics, Yerevan State University, 1 Alex Manookian 375025, Yerevan Armenia

Abstract. We discuss the propagation of laser radiation through a medium of quantum prepared Λ-type atoms in order to enhance the insight into the physics of quantum superposition based pulse train generator suggested in Phys. Rev. A **80**, 035801 (2009). We obtain analytical results which give a qualitatively corerct description for the outcoming optical pulses and show that for the case of alkali vapor medium pulse train generation may be implemented under ordinary experimental conditions.

E-mail: gmurad@ysu.am

1. Introduction

Atomic coherence is a widely accepted resource in atomic and laser physics for modifying the response of coherently prepared atomic systems and thereby canceling or enhancing the light absorption/reflection near atomic resonance [1-4]. Quantum prepared medium in fact is essentially a new state of matter, named in [5] phaseonium. The basic idea behind almost all of such models is to modify the emissive and absorptive profiles and make them non-reciprocal. Among most important applications of quantum prepared medium, both in theory and experimental science,are listed such phenomena as adiabatic population transfer [6-9], amplification and lasing without inversion [10-13], sub-recoil laser cooling [14-17], light storage and stopping [18-21], atomic spectroscopy [22-26], etc.

On the other hand, many problems in atomic physics, such as precision optical frequency metrology, excitation spectroscopy and quantum state engineering [27-30], require a train of high-energy and high-repetition optical pulses. The most convenient method to accomplish this requirement today is the acousto-optic modulator [31], but still new principles and schemes are suggested in literature. A promising approach to this end is the usage of quantum coherency created between atomic levels in both bare and dressed approximations. For example, a considerable attention is drawn to the parametric beating of a weak probe field in the scheme of stimulated Raman scattering [32,33].

In this paper we suggest a new step in this direction. We consider propagation of an incident wave through a medium of Λ-type three-level atoms with a preliminary superposition between the low lying doublet states, a simplest construction for experimental realizations. We carry a detailed analysis of how the superposition nature of atomic states splits the incident monochromatic wave into a train of short (and high power) pulses in a three-level medium. Such

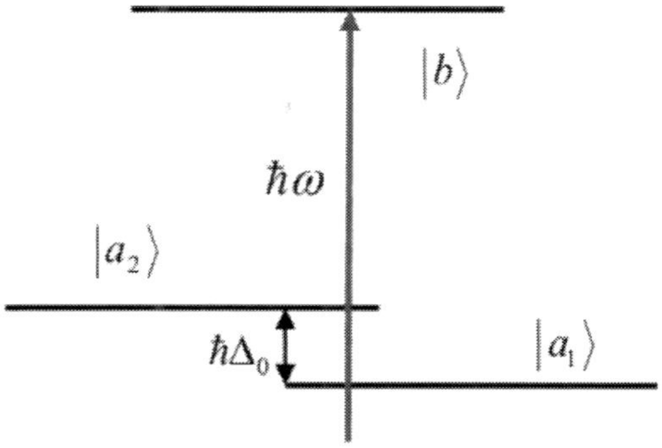

Figure 1. Energy-level diagram for the resonant single-photon process. States $|a1\rangle$ and $|a2\rangle$ are of like parity whereas the upper state $|b\rangle$ is of opposite parity. Energy distance $\hbar\Delta_0$ between the doublet states is much smaller than the photon energy $\hbar\omega$.

a qualitatively new and dramatic effect of superposition principle has already been identified in our previous papers [34,35]. However, it proceeded from a two-level medium and two (probe and pump) optical fields, thus somewhat disguising the inherent ability of superposition principle in this problem. Simultaneously, the technical realization of originally proposed scheme seems stringent as much as it is difficult to onset the preliminary quantum superposition between the energy levels in the optical range of Bohr frequencies. Presented in this paper scheme clarifies the main idea of [34,35] and extends its formalism into much headmost ideological and experimental conditions using Λ-type interaction scheme and only *one* laser travelling wave, as the initial superposition state between the energy levels should be prepared in microwave instead of vizible frequency range, securing much easier implementation and prerservation.

2. Adiabatic states of a Λ-type atom

The basic setup of interaction used throughout this article is presented in Fig. 1. The lower atomic states 1 and 2 are separated in energy by $E_{a_2} - E_{a_1} = \hbar\Delta_0$, and are coupled with the upper state by a plane optical field of frequency ω:

$$\overrightarrow{E} = \overrightarrow{E}_0(z,t)\exp(i\omega t - ikz) + \overrightarrow{E}_0^*(z,t)\exp(-i\omega t + ikz), \tag{1}$$

where $k = \omega/c$ is the vacuum wave number.

The amplitude $\overrightarrow{E}_0(z,t)$ is assumed to vary slowly compared to rapidly oscillating exponential functions: $\left|\partial\overrightarrow{E}_0(z,t)/\partial t\right| << \left|\omega\overrightarrow{E}_0(z,t)\right|$ and $\left|\partial\overrightarrow{E}_0(z,t)/\partial z\right| << \left|k\overrightarrow{E}_0(z,t)\right|$. (Later we will take it as a constant when entering the gas medium, which is the most convenient form for implication purposes). We also assume that the necessary conditions to preserve the coherency are ensured. Dephasing from the upper state is neglected by assuming a far off-resonant interaction, while the relaxation or dephasing between ground levels 1 and 2 lasts milliseconds and is sufficient for the objectives of this paper. Thus we discuss the atom-radiation coupling staying in the limits of Schrödinger equation

$$i\hbar\frac{\partial\Psi(\overrightarrow{r},t)}{\partial t} = \left(\widehat{H}_0 - \widehat{\overrightarrow{d}}\overrightarrow{E}\right)\Psi(\overrightarrow{r},t). \tag{2}$$

The interaction is taken in electric dipole approximation, $\widehat{H}_0$ and $\widehat{\overrightarrow{d}}$ are free atom Hamiltonian and dipole moment operator respectively, and the atomic motion isn't included in the discussion.

We look for the solution of Eq.2 in a form

$$\begin{aligned}\Psi(\overrightarrow{r},t) &= a_1(t)\psi_1(\overrightarrow{r})\exp(-iE_{a_1}t/\hbar) + \\ &\quad a_2(t)\psi_2(\overrightarrow{r})\exp(-iE_{a_2}t/\hbar) + b(t)\varphi(\overrightarrow{r})\exp(-iE_bt/\hbar),\end{aligned} \tag{3}$$

expanded over free atom doublet and excited state eigenstates $\psi_{1,2}(\overrightarrow{r})$ and $\varphi(\overrightarrow{r})$ respectively. $E_{a_{1,2}}$ and E_b are corresponding energies. After substituting (2) into (3) we apply the RWA neglecting terms with rapidly oscillating factor $\exp(\pm i(\omega+\omega_{1,2})t)$ with $\omega_{1,2} = E_b - E_{a_{1,2}}/\hbar$. We want to describe the phenomena under conditions where the reactive part of the atom-field coupling is predominant over the dissipative part so we proceed in the range of resonance detunings much larger than the Doppler and homogeneous widths of optical transitions. This simultaneously allows the first order adiabatic elimination procedure, algebraically presenting the excited state amplitude through the lower lying ones:

$$\begin{aligned}b(t) &= -\frac{\overrightarrow{d}_1^*\overrightarrow{E}_0^*\exp(ikz)}{\hbar(\omega-\omega_1)}\exp(-i(\omega-\omega_1)t)a_1(t) - \\ &\quad \frac{\overrightarrow{d}_2^*\overrightarrow{E}_0^*\exp(ikz)}{\hbar(\omega-\omega_2)}\exp(-i(\omega-\omega_2)t)a_2(t).\end{aligned} \tag{4}$$

Here $\overrightarrow{d}_1$ and $\overrightarrow{d}_2$ are $a1-b$ and $a2-b$ optical transition matrix elements correspondingly.

For simplicity of later calculations we redefine the level 2 probability amplitude

$$A_2(t) = a_2(t)\exp(-i\Delta_0 t), \quad A_1(t) = a_1(t), \tag{5}$$

and rescale the problem parameters to dimensionless ones:

$$\tau \equiv \Delta_0 t, \ \frac{\omega-\omega_1}{\Delta_0} \equiv \Delta_1, \ \frac{\omega-\omega_2}{\Delta_0} \equiv \Delta_2, \tag{6}$$

$$\xi_1(z,\tau) = \frac{\overrightarrow{d}_1\overrightarrow{E}_0(z,\tau)}{\hbar\Delta_0}, \quad \xi_2(z,\tau) = \frac{\overrightarrow{d}_2\overrightarrow{E}_0(z,\tau)}{\hbar\Delta_0}. \tag{7}$$

The equations for doublet amplitudes now will take a relatively simple form

$$\begin{aligned}i\frac{dA_1(\tau)}{d\tau} - \frac{\xi_1^*(z,\tau)\xi_1(z,\tau)}{\Delta_1}A_1(\tau) &= \frac{\xi_1(z,\tau)\xi_2^*(z,\tau)}{\Delta_2}A_2(\tau), \\ i\frac{dA_2(\tau)}{d\tau} - A_2(\tau) - \frac{\xi_2^*(z,\tau)\xi_2(z,\tau)}{\Delta_2}A_2(\tau) &= \frac{\xi_2(z,\tau)\xi_1^*(z,\tau)}{\Delta_1}A_1(\tau).\end{aligned} \tag{8}$$

They take up stationary solutions

$$A_1(\tau) = A_1(0)\exp(-i\lambda_{1,2}\tau), \quad A_2(\tau) = A_2(0)\exp(-i\lambda_{1,2}\tau) \tag{9}$$

with characteristic values

$$\lambda_1 = \frac{1}{2}\left(p + \sqrt{p^2-4q}\right), \quad \lambda_2 = \frac{1}{2}\left(p - \sqrt{p^2-4q}\right), \tag{10}$$

where

$$p = 1 + \frac{\xi_1^*\xi_1}{\Delta_1} + \frac{\xi_2^*\xi_2}{\Delta_2}, \quad q = \frac{\xi_1^*\xi_1}{\Delta_1}. \tag{11}$$

These solutions should be normalized so that $|A_1(0)|^2 + |A_2(0)|^2 = 1$. Thus we obtained adiabatic solutions, mapping the superposition of three bare states on a dressed state. The solutions corresponding to the characteristic values λ_1 and λ_2 have the form

$$\Psi(t,\lambda_{1,2}) = \begin{pmatrix} A_1(0,\lambda_{1,2})\psi_1(\overrightarrow{r})\exp(-iE_{a_1}t/\hbar)+ \\ A_2(0,\lambda_{1,2})\psi_2(\overrightarrow{r})\exp(-iE_{a_2}t/\hbar + i\tau) \\ +b(0,\lambda_{1,2})\varphi(\overrightarrow{r})\exp(-iE_b t/\hbar) \end{pmatrix} \exp(-i\lambda_{1,2}t), \tag{12}$$

where

$$A_1(0,\lambda_{1,2}) = \frac{\xi_2^*\xi_1}{(\Delta_2)(\lambda_{1,2}-q)\sqrt{1+\xi_1^*\xi_1\xi_2^*\xi_2/(\Delta_2)^2(\lambda_{1,2}-q)^2}}, \tag{13}$$
$$A_2(0,\lambda_{1,2}) = \frac{1}{\sqrt{1+\xi_1^*\xi_1\xi_2^*\xi_2/(\Delta_2)^2(\lambda_{1,2}-q)^2}}.$$

Both excited state amplitudes $b(0,\lambda_1)$ and $b(0,\lambda_2)$ are determined from Eq.(4).

3. Light propagation in quantum prepared medium. Simple approximate analytic solution

Now we proceed to the discussion of the field propagation in the medium using wave equation

$$\left(\frac{\partial^2}{\partial z^2} - \frac{1}{c^2}\frac{\partial^2}{\partial t^2}\right)\overrightarrow{E} = \frac{4\pi\rho}{c^2}\frac{\partial^2}{\partial t^2}\left\langle\Psi(t)\right|\widehat{\overrightarrow{d}}\left|\Psi(t)\right\rangle, \tag{14}$$

where the quantity $\rho\partial\left\langle\Psi(t)\right|\widehat{\overrightarrow{d}}\left|\Psi(t)\right\rangle/\partial t$ stands for the density of the bias electric current in dielectric gas medium, ρ is the number of atoms per unit volume and the brackets mean quantum mechanical plus thermal state averaging. The form above in fact assumes that the gas density is low enough and the temperature is high enough for quantum mechanical collective effects to be irrelevant. We make the ansatz

$$\begin{aligned}\left\langle\Psi(t)\right|\widehat{\overrightarrow{d}}\left|\Psi(t)\right\rangle = & \frac{1}{1+\exp\left(-\frac{E_{a_2}-E_{a_1}}{k_B T}\right)}\times \\ & \left\langle\alpha\Psi(t,\lambda_1)+\beta\Psi(t,\lambda_2)\right|\widehat{\overrightarrow{d}}\left|\alpha\Psi(t,\lambda_1)+\beta\Psi(t,\lambda_2)\right\rangle + \\ & \frac{1}{1+\exp\left(\frac{E_{a_2}-E_{a_1}}{k_B T}\right)}\times \\ & \left\langle\overline{\alpha}\Psi(t,\lambda_1)+\overline{\beta}\Psi(t,\lambda_2)\right|\widehat{\overrightarrow{d}}\left|\overline{\alpha}\Psi(t,\lambda_1)+\overline{\beta}\Psi(t,\lambda_2)\right\rangle,\end{aligned} \tag{15}$$

which assumes independent formation of superposition states from the thermally populated levels 1 and 2. The coefficients (α,β) and $(\overline{\alpha},\overline{\beta})$ are the probability amplitudes corresponding to each channel.

Standard transformations in frame of slowly varying amplitude approximation bring us to a reduced form of wave equation (14), which for a better insight of the field dynamics we write in a symbolic form:

$$\left(\frac{\partial}{\partial z} - \frac{1}{c}\frac{\partial}{\partial t}\right)\eta(z,\tau) = i\eta(z,\tau)\times \begin{pmatrix} F\left(|\eta|^2\right) + G_C\left(|\eta|^2\right)\cos((\lambda_2-\lambda_1)\tau)+ \\ G_S\left(|\eta|^2\right)\sin((\lambda_2-\lambda_1)\tau)\end{pmatrix}. \tag{16}$$

The wave amplitude $\eta(z,\tau)$ is defined in a dimensionless, symmetric with respect to both $1-3$ and $2-3$ transitions form

$$\eta(z,\tau)=\frac{2d_1d_2E_0(z,\tau)}{(d_1+d_2)\hbar\Delta_0}. \tag{17}$$

The right-hand side expression in the brackets conditions the response of medium to the coupling field and the nature of field evolution. Note that there is nothing surprising in the fact that the response of medium in superposition state is time dependent, as is seen from Eq.(16), which leads to phase modulation and relevant phenomena. The "hidden" and crucial point here is the existence of an imaginary part in the response function for this non-dissipative medium under consideration. Periodic in time, imaginary part acknowledges the quantum superposition nature of atomic state prior to the atom-field interaction and results in sequential amplification and suppressing of the field intensity during the propagation.This is the reason (or origin) of the birth of series of short or short optical pulses, which we term as quantum superposition pulse train (QSPT) generator.

We will not discuss in this paper the questions connected with phase modulation and proceed to a solution of the problem in terms of the field intensity $|\eta(z,\tau)|^2$. The wave equation for it follows from Eq.(16):

$$\begin{aligned}\frac{\partial}{\partial\widetilde{\zeta}}\left|\eta\left(\widetilde{\zeta},\widetilde{\tau}\right)\right|^2 &= i\frac{2\pi\rho\omega d_1d_2}{\hbar\Delta_0^2}\left(\frac{1}{\Delta_2}-\frac{1}{\Delta_1}\right)\times\\ &\left(\begin{array}{c}\left(\frac{\alpha^*\beta}{1+\exp\left(-\frac{\hbar\Delta_0}{k_BT}\right)}+\frac{\overline{\alpha}^*\overline{\beta}}{1+\exp\left(\frac{\hbar\Delta_0}{k_BT}\right)}\right)\times\\ \exp\left(-i\left(\lambda_2-\lambda_1\right)\left(\widetilde{\zeta}+\widetilde{\tau}\right)\right)-\\ \left(\frac{\alpha\beta^*}{1+\exp\left(-\frac{\hbar\Delta_0}{k_BT}\right)}+\frac{\overline{\alpha}\overline{\beta}^*}{1+\exp\left(\frac{\hbar\Delta_0}{k_BT}\right)}\right)\times\\ \exp\left(i\left(\lambda_2-\lambda_1\right)\left(\widetilde{\zeta}+\widetilde{\tau}\right)\right)\end{array}\right)\times\\ &\left(A_2(0,\lambda_1)A_1(0,\lambda_2)-A_1(0,\lambda_1)A_2(0,\lambda_2)\right)\left|\eta\left(\widetilde{\zeta},\widetilde{\tau}\right)\right|^2,\end{aligned} \tag{18}$$

with $\widetilde{\zeta}=\Delta_0 z/c$, $\widetilde{\tau}=\Delta_0(t-z/c)$. Eq.(18) is the main result of this paper and the basis of later discussion. We should note that in general it is a nonlinear equation and nonlinearity enters through the intensity-dependence of characteristic values λ_1 and λ_2. As we can see they appear in the equation in two forms: $A_2(0,\lambda_1)A_1(0,\lambda_2)-A_1(0,\lambda_1)A_2(0,\lambda_2)$ and $\lambda_2-\lambda_1$. Numerical calculations, see Fig.2, show that the first one of them, denoted $[A_2,A_1]$ with high accuracy may be treated as not depending on wave intensity. The difference $\lambda_2-\lambda_1$ may also be viewed as intensity independent, but only when $\eta(z,\tau)<<1$ and equals to unity $(\lambda_2-\lambda_1=1)$. For this intensity range the problem becomes to a great extend exactly soluble with general solution:

$$\begin{aligned}|\eta(z,t)|^2 &= |\eta(0,t)|^2\times\\ &\exp\left(\begin{array}{c}\frac{2\pi\rho\omega d_1d_2}{\hbar\Delta_0^2}\left(\frac{1}{\Delta_2}-\frac{1}{\Delta_1}\right)\times\\ \left(A_2(0,\lambda_1)A_1(0,\lambda_2)-A_1(0,\lambda_1)A_2(0,\lambda_2)\right)\times\\ \left(\frac{\alpha^*\beta}{1+\exp\left(-\frac{\hbar\Delta_0}{k_BT}\right)}+\frac{\overline{\alpha}^*\overline{\beta}}{1+\exp\left(\frac{\hbar\Delta_0}{k_BT}\right)}\right)\times\\ \frac{1-\exp(-i(\lambda_2-\lambda_1)(\Delta_0z/c))}{\lambda_2-\lambda_1}\exp\left(-i\left(\lambda_2-\lambda_1\right)\Delta_0t\right)+c.c.\end{array}\right),\end{aligned} \tag{19}$$

where the characteristic values λ_1 and λ_2 are determined for the given initial wave-front.

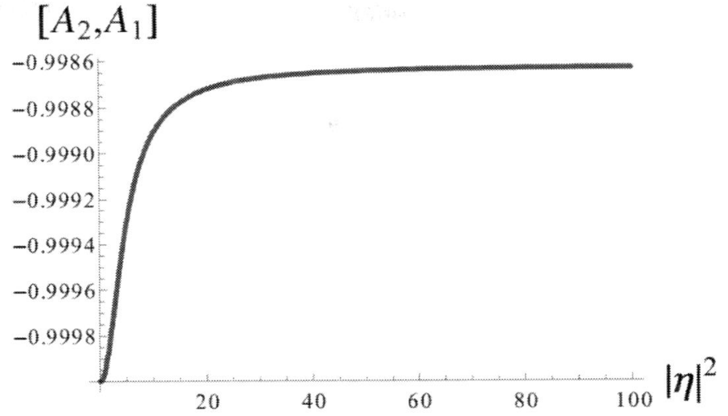

Figure 2. (Color online) Functional form $[A_2, A_1]$ (see the text) as function of scaled field intensity η . Atomic transition frequencies $\omega_1 = 287351\Delta_0$ and $\omega_2 = 287350\Delta_0$ imply sodium atom's spectrum (see paragraph III of the main text), and input laser frequency $\omega = 287360\Delta_0$ is detuned from these an order of magnitude larger than the doublet splitting $\Delta_0 = \omega_1 - \omega_2$.

This formula shows all the immense advantage of the scheme under consideration: the output intensity is modulated at the frequency $\lambda_2 - \lambda_1$, which is the distance between the ground doublet levels. This is a remarkable and a prominent performance index of the presented scheme, since a very wide range of high stability repetition rates can be attained for the pulse trains by simply choosing different atomic species with doublet (or more) ground level structures.

The second prominent advantage of the presented scheme is the presence of a prefactor $1/\Delta_0^2$ instead of $\frac{\Omega/\Delta}{(1+(\Omega/\Delta)^2)^2}\frac{1}{\Delta^2}$ in the original version [34], where Δ is detuning of the intense pump field relative to the two-level optical transition, and Ω is the generalized Rabi frequency. To neglect the spontaneous emission, one usually has to work in the range where detunings are $\Delta > 10^{11}Hz$, which is at least one order of magnitude larger than Δ_0 for alkaline metal gases. At the same time in old version we are limited with $\Omega/\Delta = 0.1$ upper value in the additional coefficient. So the present scheme possesses much more freedom for such parameter values as gas density, and perhaps can be implemented even on weak optical transitions.

Another preferable side of Λ-type interaction scheme is the acquisition of superposition between the closely posed ground states, as it is not a difficult task and may be implemented by means of magnetic fields as well. In case of a preparative radiation field, one has $\alpha^* = \alpha$, $\beta^* = -\beta$, $\overline{\alpha}^* = -\overline{\alpha}$ and $\overline{\beta}^* = -\overline{\beta}$. The expression (19) becomes somewhat simplified and the relative intensity $|\eta(z,t)|^2 / |\eta(0,t)|^2$ (RI) takes the ultimate form

$$\text{Relative intensity} = \exp\left(\begin{array}{c} \frac{4\pi i \rho\omega d_1 d_2}{\hbar\Delta_0^2}\left(\frac{1}{\Delta_2} - \frac{1}{\Delta_1}\right) \times \\ \left(\frac{\alpha\beta}{1+\exp\left(-\frac{\hbar\Delta_0}{k_B T}\right)} - \frac{\overline{\alpha}\overline{\beta}}{1+\exp\left(\frac{\hbar\Delta_0}{k_B T}\right)}\right) \times \\ \frac{A_2(0,\lambda_1)A_1(0,\lambda_2) - A_1(0,\lambda_1)A_2(0,\lambda_2)}{\lambda_2 - \lambda_1} \times \\ \left(\sin((\lambda_2 - \lambda_1)(\tau - \Delta_0 z/c)) - \sin((\lambda_2 - \lambda_1)\tau)\right) \end{array} \right) \quad (20)$$

Expression in the exponent periodically oscillates as a function of time and coordinate around the zero value. Therefore, in principle two types of field envelope transformations are possible: propagation with a periodic amplification and propagation with a periodic suppression. The

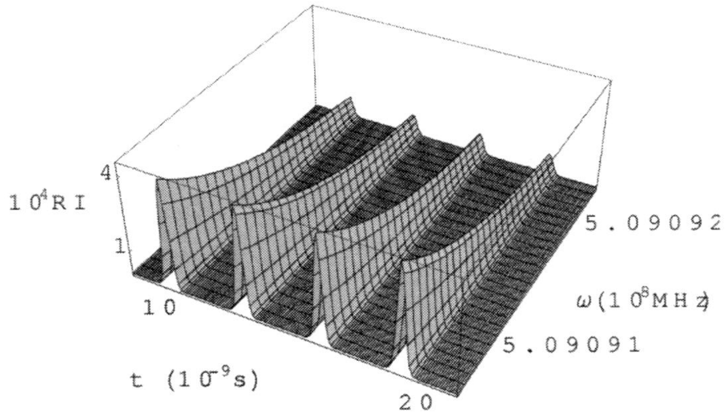

Figure 3. (Color online) Splitting of monochromatic incident wave intensity into a series of short pulses in a quantum prepared three-level atomic medium. Here $z\Delta_0/c = \pi$, $\rho = 4 \times 10^{12} cm^{-3}$, $T = 10^{-6}K$, $\alpha = \overline{\beta} = 0.5$, $\beta = \overline{\alpha} = i\sqrt{0.75}$. All the other parameters are the same as in Fig.2.

answer to the question which process will be realized for a given superposition state is determined through the values of $\omega - \omega_1$ and $\omega - \omega_2$ differences.

4. Numerical analysis of pulse train formation

For quantitative analysis we discuss the case of $3S_{1/2}$– $3P_{1/2}$ transition of ^{23}Na. The hyperfine levels $F = 1$ and $F = 2$ of ground level $3S_{1/2}$ constitute the doublet structure discussed. As the hyperfine splitting in the excited state $3P_{1/2}$ is orders of magnitude smaller than the 1771.6 MHz splitting of $3P_{1/2}$, we discuss the former as a collection of degenerate sublevels of same energy. We are also allowed to neglect all thermal transitions between ground levels $F = 1$ and $F = 2$ during the interaction.

The incident wave is taken circularly polarized and this polarization is preserved during propagation through the medium. Transition matrix elements now should be calculated between the magnetic sublevels, and this should include selection rules of the total moment F and its projection M on the quantization axis aligned with the propagation direction. This, as usually, is realized using common formulas, separating a projection-dependent prefactor from the reduced matrix element $\langle F'|\widehat{d}|F\rangle$ (see, for example [36]) and introducing the oscillator strength $f(F \to F')$ instead of reduced matrix element $\langle F'|\widehat{d}|F\rangle$.

The expression for the light field amplitude is of Eq.(19) form. The depth of modulation increases with atom density and thereby decreases the duration of each individual pulse in the sequence. Pulse duration may be significantly controlled using resonance detuning too. Fig.3 graphically illustrates the situation, plotting the envelope of the field intensity as a function of time for various frequency detunings. Though the parameters are chosen for illustrative purposes, they are realistic for ordinary laboratory conditions. As might have been expected, pulse repetition rate is irrelevant to resonance detuning and mimics the doublet splitting frequency. As to pulse duration,it is around hundred picoseconds for the chosen frequency region and concomitant with power amplification. One will have the same picture for spatial evolution too.

5. Conclusions and outlook

We showed that the gas of quantum prepared Λ-type atoms is very efficient in splitting of monochromatic wave into a sequence of short pulses, enlarging the spectrum of pulse generation mechanisms suggested or implemented so far. We emphasize that the repetition rate of short pulses does not depend neither on the light intensity nor its detuning from the atomic resonance. The sole quantity determining the repetition rate is the frequency separation between the ground doublet levels between which the superposition has been preliminarily established. Duration of each pulse, vice versa, is rigorously dependent on coupling parameters and effectively controllable. These regularities accompanied by mild conditions for the experimental realization presents the QSPT generator as a viable alternative to more conventional methods.

Finally we would like to note, that since the QSPT generator is inevitably grounded on the quantum superposition principle, the studied phenomenon of abrupt shape-reformation of the classical electromagnetic field envelope may be regarded as an example of quantum macroscopic effect.

This work was supported by the Armenian State Committee of Science.

6. References

[1] Arkhipkin V G and Heller Y I 1983 *Phys. Lett. A* **98** 12
[2] Scully M O 1991 *Phys. Rev. Lett.* **67** 1855
[3] Scully M O and Fleischhauer M 1992 *Phys. Rev. Lett.* **69** 1360
[4] Zibrov A S et al 1996 *Phys Rev. Lett.* **76**, 3935
[5] Fleischhauer M et al 1992 *Phys. Rev. A* **46**, 1468
[6] Gaubatz U et al 1998 *Chem. Phys. Lett.* **149**, 463
[7] Unanyan R G et al 1998 *Phys. Rev. A* **57**, 462
[8] Bergmann K et al 1998 *Rev. Mod. Phys.* **70**, 1003
[9] Fleischhauer M et al 2005 *Rev. Mod. Phys.* **77**, 633
[10] Kocharovskaya O and Khanin Ya I 1988 *JETP Letters*, **48**, 630
[11] Harris S E 1989 *Phys. Rev. Lett.* **62**, 1033
[12] Scully M O and Fleischhauer M 1994 *Science* **263**, 337
[13] Wu H et al 2008 *Phys. Rev. A* **78**, 041802(R)
[14] Morigi G et al 2000 *Phys. Rev. Lett.* **85**, 4458
[15] Schmidt-Kaler F et al 2001 *Appl. Phys. B* **73**, 807
[16] Evers J and Keitel C H 2004 *Europhys. Lett.* **68**, 370
[17] Morigi G and Arimondo E 2007, *Phys. Rev. A* **75**, 051404
[18] Fleischhaer M and Lukin M D 2000 *Phys. Rev. Lett.* **84**, 5094
[19] Phillips D F et al 2001 *Phys. Rev. Lett.* **86**, 783
[20] Honda K et al 2008 *Phys. Rev. Lett.* **100**, 093601
[21] Moretti D et al 2009 *Phys. Rev. A* **79**, 023825
[22] Javanainen J 1992 *Europhys. Lett.* **17**, 407
[23] Boozer A D et al 2007 *Phys. Rev. Lett.* **98**, 193601
[24] Weber B et al 2009 *Phys. Rev. Lett.* **102**, 030501
[25] Norris D G et al 2010 *Phys. Rev. Lett.* **105**, 123602
[26] Kampschulte T et al 2010 *Phys. Rev. Lett.* **105**, 153603
[27] Agostini P and DiMauro L F 2004 *Rep. Prog. Phys.* **67**, 813
[28] Hall J L 2006 *Rev. Mod. Phys.* **78**, 1279
[29] Stowe M C et al 2008 *Adv. At. Mol. Opt.* Phys. **55**, 1
[30] Zhdanovich S et al 2009 *Phys. Rev. A* **80**, 063405
[31] Milleret F P et al 2010 *Acousto-Optic Modulator* (VDM Publishing House Ltd: Saarbrücken)
[32] Harris S E and Sokolov A V 1998 *Phys. Rev. Lett.* **81**, 2894
[33] Sokolov A V et al 2005 *J. Mod. Opt.* **52**, 285
[34] Muradyan G and Muradyan A Zh 2009 *Phys. Rev. A* **80**, 035801
[35] Muradyan G and Hovhannisyan M 2011 *Appl. Phys. B* **103**, 649
[36] Landau L D and Lifshitz E M 2003 *Quantum Mechanics. Non-Relativistic Theory* (Elsevier Science Ltd: Oxford)

Novel All-Optical Logic Gates Based on Photonic Crystal Structure

Mortaza Noshad[1,2], Amin Abbasi[2], Reza Ranjbar[2],* Reza Kheradmand[2]
[1] Faculty of Electrical and Computer Engineering, University of Tabriz, Tabriz, Iran
[2] Research Institute for Applied Physics and Astronomy, University of Tabriz, Tabriz, Iran
*Corresponding Email: R_kheradmand@tabrizu.ac.ir

Abstract. We have designed AND, NOT, and NOR logic gates based on photonic crystal structure employing cross-waveguide geometry with nonlinear rods using finite difference time domain (FDTD) method. The logical function is based on the frequency resonance shift of the microcavity caused by Kerr nonlinearity. The proposed devices benefit a simple and small structure, and clear operating principle.

1. Introduction

Photonic crystal (PhC) is a promising candidate as a platform on which to construct devices with dimensions of a few wavelengths of light for future photonic integrated circuits. The photonic crystal concept was proposed in 1987 [1,2], and the first 3D experimental photonic crystal with full band gap was manufactured in 1991. The existence of band gap in PhC structures led to many prominent applications in integrated optics. Photonic crystals are nowadays used for different applications such as filters [3], modulators [4–6], switches [7], beam splitters [8-10], and super prisms [11–13] for multiplexing and demultiplexing for example.

All-optical switching is one of the most important targets for photonics. However, this goal has been considered difficult to achieve because of the inefficiency of optical nonlinearity in materials. All-optical switching utilizes the nonlinear refractive-index change as a function of the electrical field intensity.

Nonlinear photonic crystal microcavities offer unique fundamental ways of enhancing a variety of nonlinear optical processes for optical switching. When ultrasmall and high-quality factor cavities are used as switches, the field is enhanced by Q/V, where V is the mode volume. By taking into account the reduction of frequency shift required for switching by a factor of Q, cavities will generally exhibit a switching power reduction scaled as V/Q^2 [14].

Ultra-compact all-optical logic gates are advanced kinds of optical switches which are used as key elements for real-time optical processing and information communications. As a consequence of recent advantages in nanophotonic fabrication, the amount of compactness and low loss of photonic crystal (PC) structures make them one of the best candidates for constructing ultra-fast optical integrated circuits. So far, several schemes have been investigated to realize various all-optical logic functions such as AND, NOT, NOR, XOR, NAND gates [15-20].

In this paper, we have proposed novel all optical AND, NOT, and NOR gate based on cross-waveguide geometry using nonlinear 2D photonic crystal lattice. The prominent features of these gates in comparison to the former designed gates are their fast switching action (about 10 ps), and structure compactness (it has the dimensions in the order of several wavelengths of light). In addition, since the same structure has the compatibility to be used as AND, NOT, and NOR gates it offers good candidates for all-optical integrated circuits in contrast to previously designed gates [15-20]. Numerical assessment for the bistable switching action in this geometry has been discussed earlier in [21].

2. Design and Simulation

In this work a 2D photonic crystal lattice with nonlinear rods is used. The structure is based on a 15 × 15 square lattice with the lattice constant a = 575 nm. The material used for fabrication of nonlinear rods is AlGaAs which provides the refractive index of 3.5 with instantaneous Kerr coefficient of $n_2 = 1.5\times10^{-17}$ W/m^2. In addition, the background material is taken to be air with refractive index of 1. The radius of the rods is set to 0.2a. As shown in figure 1, two cross waveguides are created by elimination of the rows of rods. These are connected to each other using a asymmetric cavity. Since the radiuses of the elliptical cavity have been taken different, so, we will have two separate modes in the vertical and horizontal directions. The cavity is considered to have the radiuses r_a = 0.27a and r_b = 0.32a. This geometry provides the resonance wavelengths of 1550 nm in X-direction, and 1649 in Y-direction. Only TM modes are considered in the paper, it is well known that Maxwell's equations can be described in scalar forms for TM modes as

$$\frac{\partial E_z}{\partial x} = \varepsilon(r)\frac{\partial H_y}{\partial t} \qquad (1)$$

$$\frac{\partial E_z}{\partial y} = \varepsilon(r)\frac{\partial H_x}{\partial t} \qquad (2)$$

$$\frac{\partial H_y}{\partial x} - \frac{\partial H_x}{\partial y} = -\mu(r)\frac{\partial E_z}{\partial t} \qquad (3)$$

We have also applied numerical FDTD method to simulate the light propagation in the designed structure. To analyze the wave behavior, group velocity, bandgaps and defect modes, we have derived the dispersion diagram by the plane wave expansion method (PWE) shown in Figure 1. The input beam frequency is taken about 1.55 μm which is inside the band gap of the structure for TM mode.

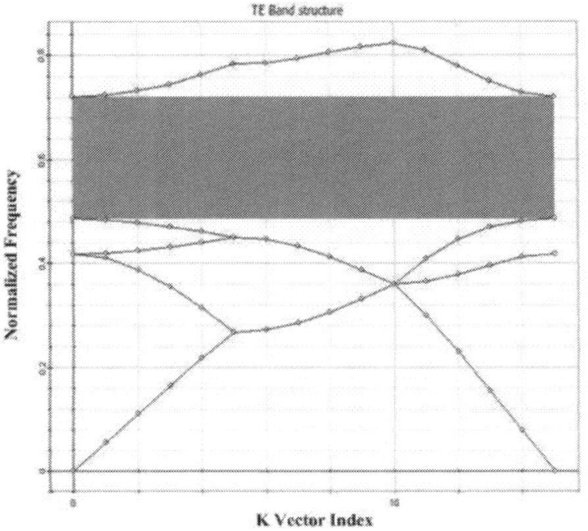

Figure 1. Band structure of the photonic crystal lattice

Figure 2 shows the general structure for proposed logic gates.

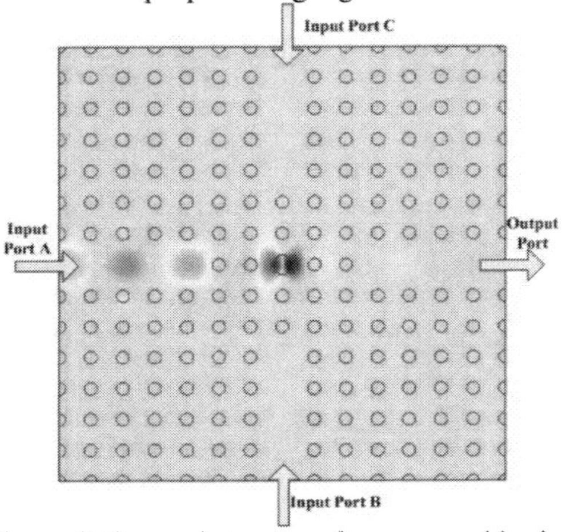

Figure 2. General structure for proposed logic gates

I) AND Gate

Firstly we discuss the AND gate design using this structure. AND gate encompasses of two inputs and an output. As shown in table 1 the output is logically '1' if and only if both of the input values are '1'.

Table 1. Logic table of AND gate

A	B	Output
0	0	0
0	1	0
1	0	0
1	1	1

In this gate we use the input signals A and B (No signal is applied via Input C in this case). We take the frequency of signal A such that normally it does not match with the resonance frequency of cavity. So, the output power in this case is about zero. By applying the input signal B, the resonance frequency will change, and signal A fall at the resonance frequency of cavity. As the result, the output power reaches to about 90% of the input power A. Figure 3 demonstrates the resonance frequency shift of the cavity. ω_c and ω_I represent the cavity resonance and input signal A frequencies respectively. The resonance frequency shift in cavity causes the output power change. For the input B field amplitude of 450 V/m, the resonance wavelength shift is about 5.7 nm. We have taken the amplitude of input signal A so small that it makes no effective influence in resonance shift. Note that existence of both signals A and B are necessary for making nonzero output power. Thus the system performs as an AND gate.

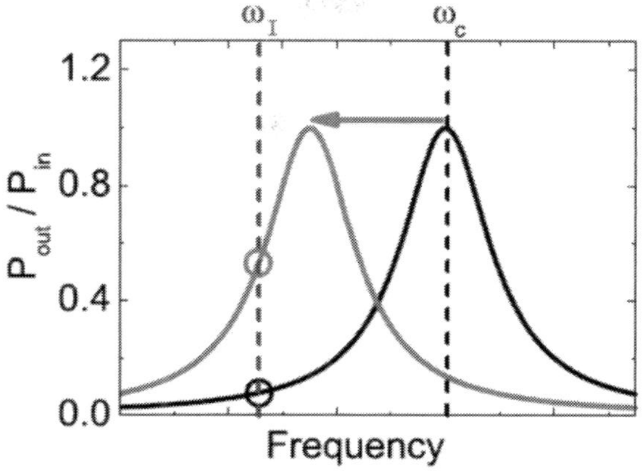

Figure 3. Resonance frequency shift in AND gate

II) NOT Gate

Now we discuss making a NOT gate with this geometry. NOT gate contains an input (as the control signal) and an output. As shown in table 2 the output is logically complement of input.

Table 2. Logic table of NOT gate

B	Output
0	1
1	0

Like as AND gate structure, in this case we apply input A a fixed signal that it's frequency matches with the resonance frequency of the cavity. Note that signal A plays a role of a source for the output. Signal B, the controlling input, is used to manipulate the structure as a NOT gate. So, normally when the input power B is zero, the output power is about 90% of input power A. we attribute the logical '1' to this output state. By applying signal B, resonance frequency of the cavity changes, and the frequency of signal A does not match with the resonance frequency, and output power reaches to zero. Thus the system performs as an NOT gate. Figure 4 represents the performance of NOT gate using the resonance frequency shift.

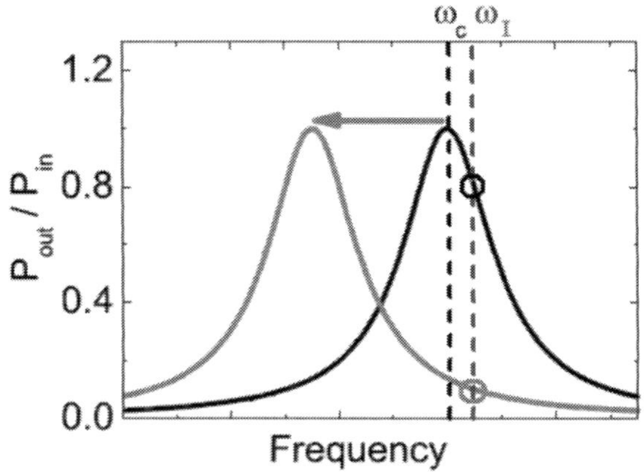

Figure 4. Resonance frequency shift in NOT gate

III) NOR Gate

Finally, we investigate the performance of NOR gate using the cross waveguide structure. NOR gate encompasses of two inputs and an output. As shown in table 3 the output is logically '1' if and only if both of the input values are '0'.

Table 1. Logic table of AND gate

B	C	Output
0	0	1
0	1	0
1	0	0
1	1	0

In this case we apply a fixed signal into the input A, that it's frequency matches with the resonance frequency of the cavity. Note that signal A plays a role of a source for the output. We use the signals B and C as input beams. Both the power and wavelength of the signals A and B are considered equal. When the both input signals B and C are off, the output power is about 90% of input power A (output signal is logically '1'). By applying one or both of the signals B and C resonance frequency of the cavity varies, and it does not match with the frequency of signal A. Therefore the output power reaches to zero. Thus the system performs as an NOR gate.

3. **Conclusion**

In this work we designed AND, NOT, and NOR all optical logic gates based on a cross waveguide geometry using a 2D photonic crystal lattice. The logical function is based on the frequency resonance shift of the microcavity caused by Kerr nonlinearity. Our analyze method for derivation of band structure is based on plane wave expansion, and we used FDTD method to simulate the wave propagation and devices operation. The proposed devices benefit a simple and small structure, and clear operating principle which shows that they can be strong candidates for future photonic integrated circuits.

3. References

[1] Yablonovitch E 1987 "Inhibited spontaneous emission in solid-state physics and electronics" *Phys. Rev. Lett.* **58**, 2059–2062

[2] John S 1987 "Strong localization of photons in certain disordered dielectric superlattices" *Phys. Rev. Lett.* **58**, 2486–2489

[3] Qiu M, Mulot M, Swillo M, Anand S, Jaskorzynak B, Karlsson A, Kamp M and Forchel A 2003 "Photonic crystal optical filter based on contra-directional waveguide coupling" *Appl. Phys. Lett.* **83**, 5121– 5123

[4] Fekete L, Kadlec F, Kuzel P, and Nemec H 2007 "Ultrafast optoterahertz photonic crystal modulator" *Opt. Lett.* **32**, 680–682

[5] Gu L, Jiang W, Chen X, Wang L, and Chen R T 2007, "High speed silicon photonic crystal waveguide modulator for low voltage operation" *Appl. Phys. Lett.* **90**, 071105

[6] Jiang Y, Jiang W, Gu L, Chen T, and Chen R T 2005 "80-micron interaction length silicon photonic crystal waveguide modulator" *Appl. Phys. Lett.* **87**, 221105

[7] Beggs D, White T, Cairns L, O'Faolain L, and Krauss T 2009 "Demonstration of an integrated optical switch in a silicon photonic crystal directional coupler" *Physica E* **41**, 1111–1114

[8] Lin S Y, Chow E, Bur J, Johnson S G, and Joannopoulos S J 2002 "Low-loss, wide-angle Y splitter at 1.6-_m wavelengths built with a two-dimensional photonic crystal," *Opt. Lett.*, vol. **27**, no. 16, pp. 1400–1402

[9] Takeda H and Yoshino K 2003 "Tunable light propagation in Y-shaped waveguides in two-dimensional photonic crystals utilizing liquid crystals as linear defects," *Phys. Rev. B*, vol. **67**, p. 073106
[10] Inoue K, Sugimoto Y, Ikeda N, Tanaka Y, Asakawa K, Sasaki H, and Ishida K 2004 "Ultra-small photonic-crystal-waveguide-based Y-splitters useful in the near-infrared wavelength region," *Jpn. J. Appl. Phys.*, vol. **43**, no. 4A, pp. L 446–L 448
[11] Kosaka H, Kawashima T, Tomita A, Notomi M, Tamamura T, Sato T, and Kawakami S 1998 "Superprism phenomena in photonic crystals" *Phys. Rev. B* **58**, R10096
[12] Amet J, Baida F, Burr G, and Bernal M.-P 2008 "The superprism effect in lithium niobate photonic crystals for ultra-fast, ultra-compact electro-optical switching" *Photonic. Nanostruct.* **6**, 47–59
[13] Diziain S, Amet J, Baida F I, and Bernal M.-P 2008 "Optical far-field and near-field observations of the strong angular dispersion in a lithium niobate photonic crystal superprism designed for double (passive and active) demultiplexer applications" *Appl. Phys. Lett.* 93, 261103
[14] Tanabe T, Notomi M, Mitsugi S, Shinya A, and Kuramochi E 2005 "All-optical switches on a silicon chip realized using photonic crystal nanocavities," *Appl. Phys. Lett.* 87(15), 151112
[15] FragaW B, Menezes J W M, Silva M G, Sobrinho C S and Sombra A S B 2006 All optical logic gates based on an symmetric nonlinear directional coupler *Opt. Commun.* **262** 32
[16] Igarashi K and Kikuchi K 2008 Optical signal processing by phase modulation and subsequent spectral filtering aiming at applications to ultrafast optical communication systems *IEEE J. Sel. Top. Quantum Electron.* **14** 551
[17] Wu Y D, Shih T T and Chen M H 2008 New all-optical logic gates based on the local nonlinear Mach–Zehnder interferometer *Opt. Express* **16** 248
[18] Pereira S, Chak P and Sipe J E 2003 All-optical AND gate by use of a Kerr nonlinear microresonator structure *Opt. Lett.* **28** 444
[19] Li Z, Chen Z and Li B 2005 Optical pulse controlled all-optical logic gates in SiGe/Si multimode interference *Opt. Express* **13** 1033
[20] Fujisawa T and Koshiba M 2006 All-optical logic gates based on nonlinear slot-waveguide couplers *J. Opt. Soc. Am.* B **23** 684
[21] Yanik M F, Fan S, Soljacic M, Joannopoulos J D 2003 "All-optical transistor action with bistable switching in a photonic crystal crosswaveguide geometry", *Opt. Lett.* **28**, 2506-2508

Study of the splitting of electromagnetically induced transparency resonance in strong magnetic field using Rb nano-thin cell

R Mirzoyan, A Sargsyan, A S. Sarkisyan, D Sarkisyan

Institute for Physical Research, NAS of Armenia, Ashtarak, 0203, Armenia

E-mail: rafayelm@gmail.com

Abstract. Recently we have obtained the high-contrast electromagnetically induced transparency resonance on ^{85}Rb and ^{87}Rb D_1 line in nanometric-thin cell of $L=\lambda=794$ nm thickness. In present work we exploit the 40% contrast of EIT resonance to study its splitting in a very wide range of a B-field covering, for the first time, 1 – 1700 G region. Experimental results are fully consistent with the developed theoretical model.

1. Introduction

The continuous interest in coherent processes of electromagnetically induced transparency (EIT) is caused by relative simplicity of forming ultra-narrow optical resonances with a width of less than 1 kHz, several orders of magnitude below the natural linewidth of atomic transitions. The EIT has already found practical applications in a variety of fields such as laser cooling, quantum information, optical magnetometry, laser spectroscopy, etc [1-4]. EIT resonance occurs particularly in a Λ-system with two long-living states and one excited state coupled by two laser fields, and displays a strong reduction in absorption where a maximum is expected in the absence of the coupling laser field. Results of the study of EIT-resonance splitting in a relatively low magnetic field (< 40 G) are presented in Refs. [3,5-9,11].

Recently we have reported the observation of a high contrast EIT resonance on D_1 line of ^{85}Rb and ^{87}Rb in the wavelength-thickness nanometric thin cell (NTC), $L = \lambda = 794$ nm. A remarkable result is that reduction of the cell thickness by 4 orders from conventional centimeter-scale cell to the NTC does not deteriorate formation of a sharp EIT resonance with up to 40 % contrast defined as the ratio of EIT amplitude to the height of the peak absorption shoulders [5]. It is important to note that extremely small thickness of NTC facilitates application of a strong B-field of well-defined strength by use of conventional permanent magnets, which otherwise behave as inhomogeneous magnetic field source.

In the paper we present, for the first time, the results of studies on EIT resonance splitting in an extended range of magnetic field, from 1 G to 1700 G.

2. Experimental

2.1. Design of the nanometric thin cell

The NTC with a thickness smoothly variable in the range between 50 nm and 3 μm is similar to those described in Refs. [10,11]. Rectangular 25 x 30 mm, 2-mm-thick windows were made of monocrystalline sapphire (Al_2O_3) with the *C*-axis perpendicular to the window surface, which significantly reduces the birefringence. Surfaces of the sapphire windows are thoroughly polished (surface roughness ~ 5 nm). A 3 μm-thick platinum strip is placed between the windows in the lower part to form a wedge-shaped gap between the inner surfaces of the NTC. The side extension arm of NTC is filled with a natural mixture of Rb isotopes containing ^{85}Rb (72%) and ^{87}Rb (28%).

2.2. Experimental setup

The experimental arrangement for studying the EIT is shown in Fig. 1. Two independent narrow-band (~ 1 MHz) cw extended-cavity diode lasers (ECDL) are used to form EIT resonance. One of the lasers operates at a fixed frequency ν_C (coupling laser), while the frequency ν_P of the other laser (probe laser) is tuneable. The two laser beams of 1 mm diameter having orthogonal linear polarizations are superimposed by the Glan prism G_1 and are incident on NTC. Faraday isolators (*1*) are used in order to prevent the optical feedback. Part of the probe radiation is branched into the reference NTC *2* with thickness $L = \lambda$ where frequency reference (FR) signal is formed. Velocity selective optical pumping (VSOP) process resulting in appearance of narrow resonances in the NTC transmission spectrum located at atomic transitions was used to form the reference spectrum [11]. The radiations signals are detected by photodiodes *3* followed by operational amplifiers and are displayed on a Tektronix TDS2014B four-channel digital oscilloscope *4*. Part of the coupling-laser radiation at the frequency ν_C is sent to a frequency stabilization system (FS) described in [12]. Prism G_2 blocks the coupling laser so that only the probe radiation is detected.

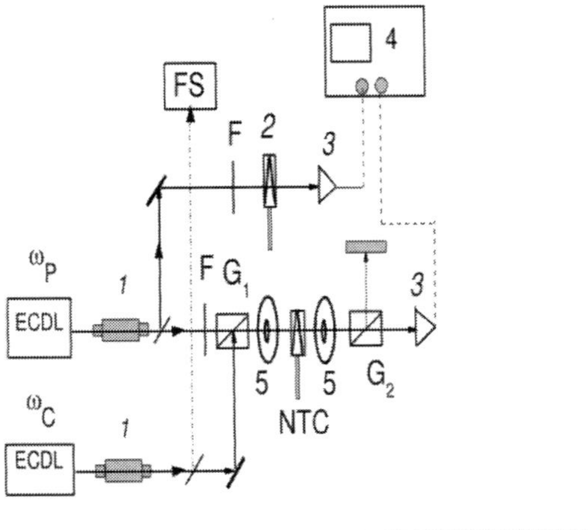

Figure 1. Sketch of the experimental setup. (*1*)- Faraday isolators, $G_{1,2}$- Glan prisms, (*2*)- auxiliary NTC of $L = \lambda$, (*3*)- photodiodes, (4)- oscilloscope; (5)- permanent ring magnets (for details, see text).

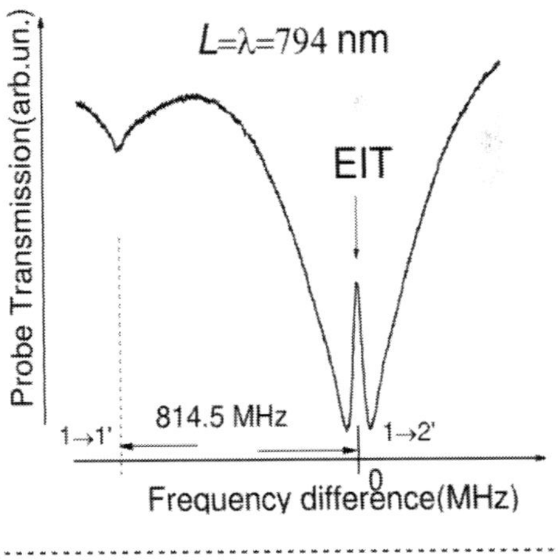

Figure 2. ^{87}Rb D_1 line. EIT resonance of 40% contrast and 30 MHz linewidth. NTC thickness is $L = \lambda = 794$ nm.

System of Helmholtz coils (not shown in Fig. 1) is used to apply a longitudinal magnetic field of relatively small strength 1 - 200 G measured by a calibrated Hall gauge. Among the advantages of NTC is the possibility to apply much stronger magnetic fields using widely available strong permanent ring magnets (PRMs) *5* [13]. In spite of their *B*-field strong non-homogeneity (in our case it can reach 150 G/mm), the variation of *B* inside atomic vapor column is by several orders less than the applied *B* value thanks to small thickness of the NTC (794 nm).

2.3. Experimental results

In Ref. [14] it was shown that for a conventional centimetre-scale cell the excitation of the Rb D_1 line results in greater EIT resonance contrast as compared to that fot D_2 line. We confirm that this statement is correct also in the case of NTC, and for D_1 line we succeeded to obtain ≅ 40% EIT contrast (as is shown in Fig.2), which is ≅ 8 times larger than the EIT contrast observed earlier for D_2 line [11]. The coupling and probe lasers powers are 25 mW and 0.02 mW, correspondingly. The EIT linewidth is 30 MHz FWHM. Decreasing of the coupling laser power results in reduction of both the linewidth and contrast of EIT resonance (for some applications small EIT width is important).

Energy levels of ^{87}Rb D_1 line involved in EIT process are shown in Fig. 3. The coupling laser frequency is resonant with 2 → 2' transition, while the probe laser frequency is scanned across 1 → 1',2' transitions.

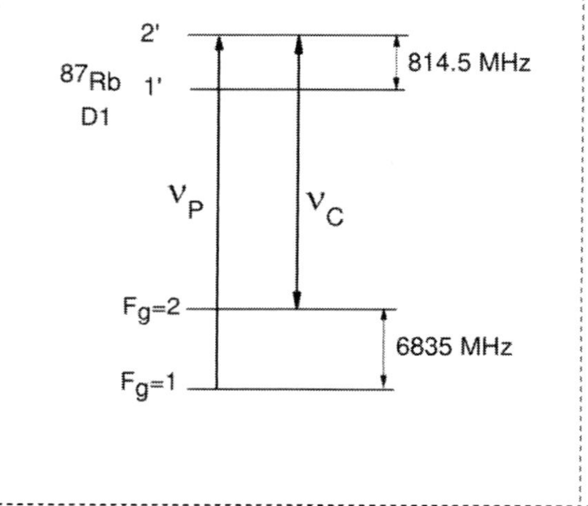

Figure 3. Energy levels of ^{87}Rb D_1 line involved in EIT process. Coupling frequency ν_C is resonant with 2 → 2', probe frequency ν_P is scanned across 1 → 1',2' transitions.

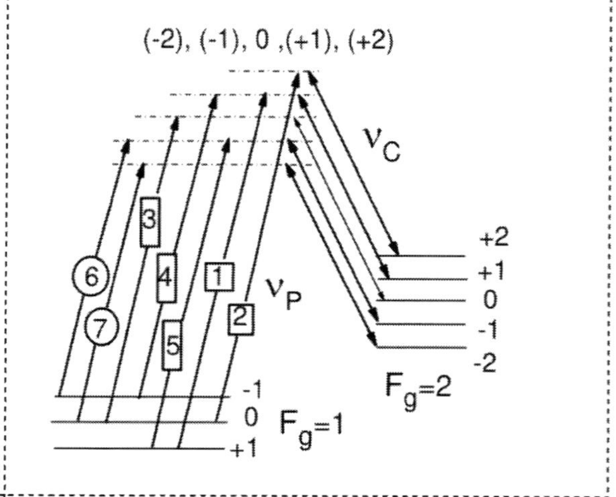

Figure 4. EIT for ^{87}Rb D_1 line. Seven split Λ-systems are formed in magnetic field. For $B < 300$ G EIT components marked by the same symbol (circle, square or rectangle) have the same frequency, i.e. $\nu_6 \cong \nu_7$; $\nu_1 \cong \nu_2$; $\nu_3 \cong \nu_4 \cong \nu_5$.

Since $\nu_P - \nu_C = [E(F_g=2,\, m_F) - E(F_g=1,\, m_F)]/h$, the number of EIT components depends on the Zeeman splitting of the ground levels F_g=1,2 described by quantum number m_F and presented in Fig. 4. Only three EIT components out of seven possible Λ-systems are seen for $B < 300$ G (see Fig. 8): the first component $\nu_6 \cong \nu_7$, which is red-shifted with respect to $B = 0$ EIT resonance; the second component $\nu_1 \cong \nu_2$, which is blue-shifted; and the third component $\nu_3 \cong \nu_4 \cong \nu_5$, which is located in between.

All the possible hyperfine Zeeman transitions are calculated as a function of *B* [3,15]:

$$E(F=2, m_F) = h\nu_{hfs}[-1/8 + 1/2(1 + m_F \times x + x^2)^{1/2}] \quad \text{“}(m_F \neq -2)\text{“}, \tag{1}$$

$$E(F=2,-2) = h\nu_{hfs}(3/8 - x/2), \tag{2}$$

$$E(F=1, m_F) = h\nu_{hfs}[-1/8 - 1/2(1 + m_F \times x + x^2)^{1/2}], \tag{3}$$

where $x = 2\mu_B B/h\nu_{HFS}$, μ_B being the Bohr magneton and $\nu_{HFS} = 6835$ MHz.

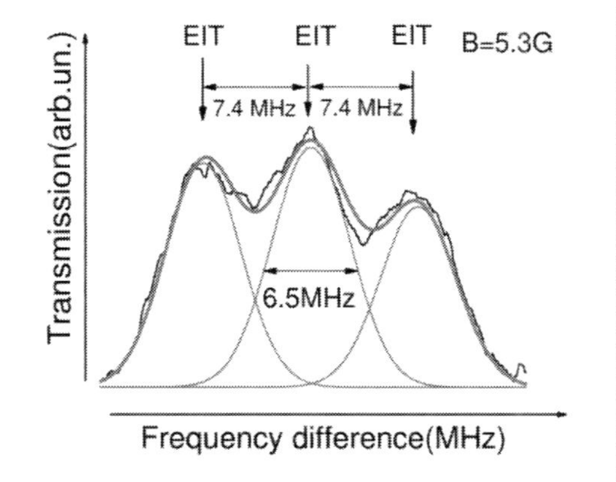

Figure 5. ^{87}Rb D_1 line. EIT splitting into three components, $B = 5.3$ G. NTC thickness $L = \lambda = 794$ nm. Gaussian fitting of EIT components (grey lines) give the linewidth of 6.5 MHz.

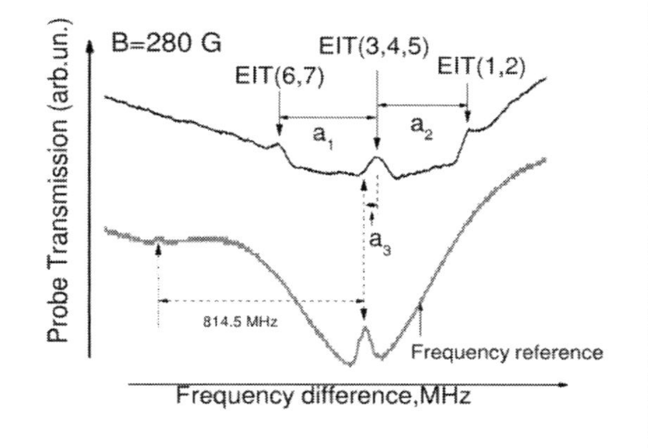

Figure 6. Splitting of EIT resonance into three components (upper curve), $B = 280$ G. The labels denote corresponding transitions shown in Figures 4, 8. The lower grey curve shows frequency reference spectrum.

As is shown in Ref. [11], measuring the splitting of EIT resonances formed on Rb D_2 line in NTC with Rb vapor thickness $L = 2\lambda$ ($\lambda = 780$ nm), one can determine the applied magnetic field. However, due to small contrast of EIT resonance it is impossible to detect the splitting at smaller values of L. The use of D_1 excitation allows to enhance the EIT contrast thus making possible detection of splitting at $L = \lambda$, when longitudinal ($\mathbf{B} \| \mathbf{k}$) magnetic field $B = 5.3$ G is applied to NTC, as presented in Fig. 5. Here $P_C = 2$ mW, $P_P = 0.02$ mW; the side-arm temperature, which determines the Rb atomic vapor density inside the NTC is 120 ^{0}C, while the temperature at the widows is kept 20 degrees higher). Three resolved EIT components are seen as it is expected, with frequency separation 1.4 MHz/G x 5.3 G $\approx$ 7.4 MHz [11]. The Gaussian fitting gives 6.5 MHz linewidth of the components.

So-called "λ -Zeeman technique" (LZT) is introduced in Ref. [13] to investigate atomic transitions in magnetic field. For this technique, Rb atoms are confined in NTC of thickness $L = \lambda$. Narrow (30 - 35 MHz) VSOP resonances are produced in transmission spectrum by a single beam, and they are split into several components in a magnetic field, with frequency positions and transition probabilities depending on B-field. As it is seen from Fig. 5, the EIT resonance splitting provides 5-fold better spectral resolution, than that obtained by LZT [13].

Figure 6 shows EIT components, when 280 G longitudinal magnetic field is applied to NTC with $P_C = 27$ mW, and $P_P = 0.1$ mW. Values of the frequency shifts: a_1, a_2 and a_3 are equal to 399, 356 and 44 MHz, correspondingly. The labels denote corresponding transitions shown in Figures 4, 8. Some broadening of EIT components is caused by non-zero detuning value of the coupling laser frequency from the corresponding intermediate level (see below).

Figure 7 shows three EIT components, when $B = 596$ G longitudinal magnetic field is applied (upper curve). The lower curve shows frequency reference spectrum. It is interesting to note, that in the case of a good adjustment of the coupling and probe beams at the conventional NTC *2* (Fig.2), an EIT resonance appears in the reference spectrum, which is narrower than VSOP resonance presented in the frequency reference spectrum when the coupling is not well adjusted.

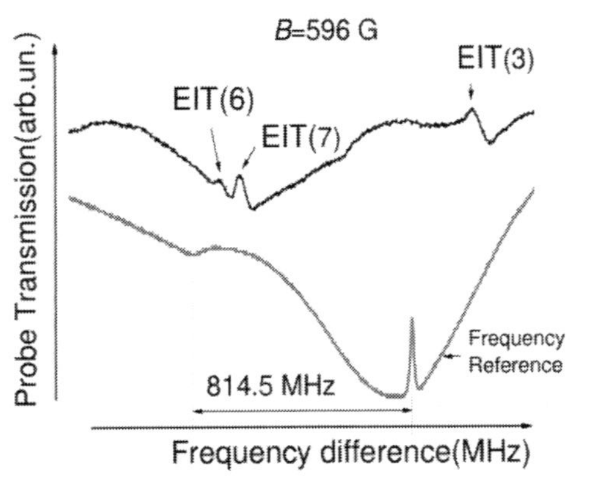

Figure 7. Splitting of EIT resonance into three components. The labels denote corresponding transitions shown in Figures 4, 8. The lower grey curve shows frequency reference spectrum.

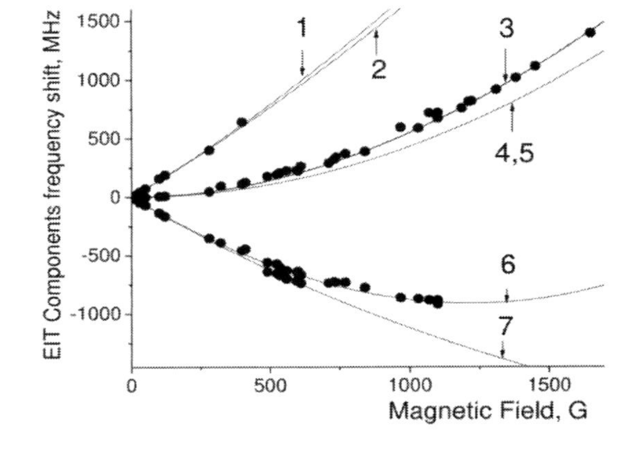

Figure 8. Frequency shift of EIT components as a function of external magnetic field. Ground levels of EIT components labeled 1-7 are shown in Fig.4. Solid lines are plotted using equations (1)-(3); black dots are the experimental results.

Fig. 8 presents frequency shifts of the EIT components as a function of magnetic field (zero shift means the initial frequency position of EIT resonance when $B = 0$). Quantum numbers of the ground levels involved in formation of EIT components labeled 1-7 are shown in Fig.4. Solid lines are plotted using equations (1)-(3); black dots present experimental results. As it is seen, the theory well describes the results of measurements.

By measuring the frequency shift between EIT component and the reference it is possible to determine either homogeneous or non-homogeneous magnetic field B in the range of 1 − 1700 G.

Although in the case of strong B-field six EIT components are expected to exist in the probe radiation spectrum (the components labeled 4 and 5 have the same frequency even at high B-field), however for $B > 700$ G only components 3 and 6 remain (this is mainly caused by the increase of coupling laser frequency detuning from the corresponding intermediate level). The EIT component 3 blue-shifted for high B-field is formed involving two ground states $F_g=1,\ m_F=0$ and $F_g=2,\ m_F=0$, while for formation of the red-shifted component 6 two other ground states $F_g=1,\ m_F=-1$ and $F_g=2,\ m_F=-1$ are responsible.

It is important that the amplitude of EIT component in the case of NTC strongly depends on the detuning Δ of the coupling laser frequency from the appropriate intermediate hyperfine Zeeman sublevel. As it is demonstrated in [5], when the coupling laser frequency is detuned from the corresponding atomic transition by $\Delta > 50$ MHz, the EIT contrast (amplitude) strongly reduces, while the linewidth strongly increases. This causes additional broadening of the EIT components presented in Figures 6, 7, and in some cases the profile can be a dispersive, as it is for EIT 3 shown in Fig. 7. The B-field dependence of coupling frequency detuning Δ from the atomic level $F_e=2\ m_F=+1$ is shown in Fig. 9 for EIT component labeled 3 (for this EIT component $F_g=2\ m_F=0 \rightarrow F_e=2\ m_F=+1$ transition is involved). As it is seen, for $B > 1700$ G the coupling laser detuning Δ exceeds 100 MHz, that is why the component 3 is detectable only when $B < 1700$ G.

We should note that the amplitudes of EIT components strongly depend also on the probability of the atomic transitions forming the Λ-system (see Fig. 3). It is well known that external magnetic field

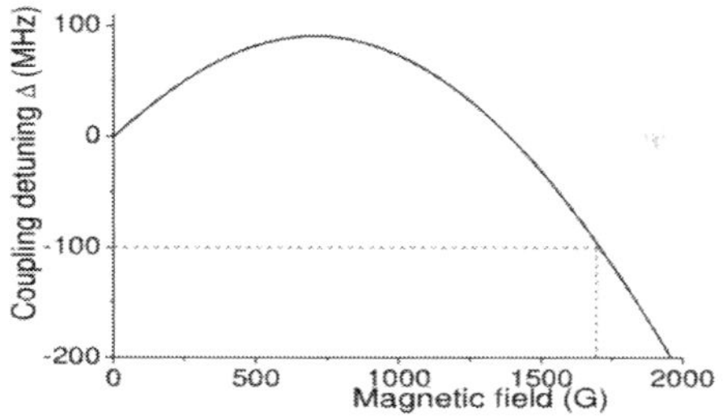

Figure 9. The coupling frequency detuning Δ from the atomic level F_e=2 m_F= +1 vs B-field for the component 3.

causes strong modification of atomic transition probability [13,16]. Consequently, there are two important conditions, which assure existence of EIT component in very strong magnetic field: i) a small value of the coupling laser frequency detuning Δ; ii) high probabilities of atomic transitions involved in the Λ-system formation. Choosing appropriate detuning of the coupling laser frequency in order to achieve $\Delta = 0$, it will be possible to realize selective addressing of the amplitudes increase for separate EIT components at any value of a *B*-field. Well resolved EIT components can be used for optical magnetometry with local nanometric spatial resolution.

Acknowledgement
The authors are grateful for the ANSEF Opt-2428 support. R.M. acknowledges the support from the NFSAT, NAS RA and CRDF (grant no. ECSP-10-10 GRSP).

References
[1] G. Alzetta, A. Gozzini, L. Moi, G. Orriols: Nuovo Cimento Soc. Ital. Fis., B. **36**, 5 (1976).
[2] E. Arimondo : Progress in Optics edited by E. Wolf (Elsevier, Amsterdam) **35**, 257 (1996).
[3] R. Wynands, A. Nagel: Appl. Phys. B: Lasers Opt. **68**, 1 (1999).
[4] M. Fleischhauer, A. Imamoglu, and J. P. Marangos: Rev. Mod. Phys. **77**, 633 (2005).
[5] A.Sargsyan, Y. Pashayan-Leroy, C. Leroy, R.Mirzoyan, A.Papoyan, D. Sarkisyan, Applied Physics B. Lasers and Optics, V.105, pp. 767–774 (2011).
[6] X. Wei, J. Wu, G. Sun, Z. Shao, Z. Kang, Y. Jiang, J-Y.Gao, *Phys. Rev.* **A 72**, 0238061 (2005).
[7] S. M. Iftiquar and V. Natarajan, Phys. Rev. **A79** , 013808 (2009).
[8] G. Katsoprinakis, D. Petrosyan and I. K. Kominis, Phys. Rev.Lett. **97,** 230801 (2006)
[9] V. I. Yudin, A. V. Taichenachev, Y. O. Dudin, *et al,* Phys. Rev. **A 82**, 033807 (2010)
[10] D. Sarkisyan, D. Bloch, A. Papoyan, and M. Ducloy, Opt. Commun. **200**, 201 (2001).
[11] A. Sargsyan, D. Sarkisyan, and A. Papoyan, Phys. Rev.A **73**, 033803 (2006).
[12] A. Sargsyan, A. V. Papoyan, D. Sarkisyan, A. Weis, Eur. Phys. J. Appl. Phys. **48**, 20701 (2009).
[13] A. Sargsyan, G. Hakhumyan, A. Papoyan, *et al,* Appl. Phys. Lett. **93**, 021119 (2008).
[14] M Stähler, R. Wynands, S. Knappe, J. Kitching, L. Hollberg, A. Taichenachev ,V. Yudin, Optics Lett. **27**, 1472 (2002).
[15] K. Motomura and M. Mitsunaga, JOSA, **19**, 2456-2460 (2002).
[16] G. Hakhumyan, C. Leroy, *et al,* Optics Communications, **284** , 4007–4012 (2011).

Journal of Physics: Conference Series **350** (2012) 012009 doi:10.1088/1742-6596/350/1/012009

Collision of atoms with formation of Feshbach resonance in the case of the coupled decay channels with the presence of laser radiation.

E Gazazyan, A Gazazyan and V Chaltykyan
Institute for Physical Research, NAS of Armenia, Ashtarak-2 0203, Armenia.

E-mail: emilgazazyan@gmail.com

Abstract. Resonance scattering with formation of Feshbach resonance in the presence of laser radiation in the case of the decays in open and closed channels is considered. The presence of the resonant laser radiation leads to additional channels of the elastic and inelastic scattering. The cross sections of the elastic and inelastic scattering are obtained which show that they are determined by two interconnected phases of resonance scattering and interference phenomena are observed.

1. Introduction

The Feshbach resonance is displayed when the energy of a bound molecular state in the closed channel is close to the energy of decaying state in the open channel [1-3]. In this case a weak coupling may lead to a strong mixing between the channels. The laser radiation couples the molecular state with the state of the incident atom. In this case a quasibound state of colliding atoms is formed in the open channel. The intermediate bound state has a finite lifetime and can decay in the initial channel or in other channels at interaction with continuum. The energy difference may, in particular, be controlled by means of a magnetic field. An alternative way to control the Feshbach resonance is optical method.

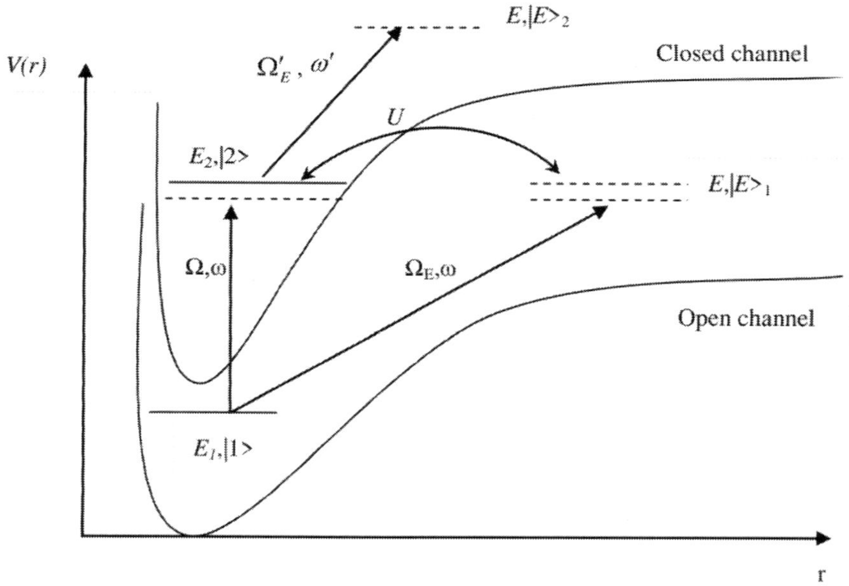

Figure 1. Diagram of formation of Feshbach resonance in the field of laser radiation.

From this point of view the investigation of the Feshbach resonance with formation of a quasibound state (molecule) in the field of laser radiation is of certain interest.
In the present paper we consider collision of atoms with formation of the Feshbach resonance in the field of laser radiation which couples molecular levels. The resonance between the laser field and discrete molecular levels is described by quasi-energy wave functions. On the other hand the one-photon transition from the upper level to other (inelastic) channel (Fig.1) where the molecule decays, under action of laser radiation with the interaction Ω'_E , into two excited atoms which fly away to infinity.

2. The cross sections for the collision of atoms in the laser field.

Laser radiation at the frequency ω_1 couples by the one-photon resonance with the Rabi frequencies Ω the stable state |1> with a quasibound state |2> (Fig.1). Laser radiation at the frequency couples the upper quasistationary state with the second continuum (inelastic channel) with the Rabi frequency Ω'_E. Interaction U_E couples the upper state with the first continuum (initial channel).

$$H = H^{(1)} + H^{(2)} \tag{1a}$$

$$H_1 = E_1 |1><1| + E_2 |2><2| + \Omega(t)|2><1| + \Omega^*(t)|1><2| \tag{1b}$$

$$H_2 = \int E\left(|E>_{11}<E| + |E>_{22}<E|\right)dE + \int \left(\Omega_E(t)|E>_1<1| + \Omega_E^*(t)|1>_1<E| + \right.$$

$$\left. + \Omega'_E(t)|E>_2<2| + \Omega'^*_E(t)|2>_2<E| + U_E|E>_1<2| + U_E^*|2>_1<E|\right)dE \tag{1c}$$

We choose as basis wave functions of the discrete spectrum the quasienergy wave functions $\varphi_1(t)$ and $\varphi_2(t)$ of the Hamiltonian obtained in the one-photon resonance approximation with adiabatic turning-on the periodic perturbations $\Omega(t)$ and $\Omega'(t)$:
Here

$$\bar{\lambda}_k = E_k + \mu_k + \omega, \tag{2}$$

$$|\varphi_k(t)> = e^{-i(\bar{\lambda}_k - \omega)t}\left(\alpha_k|1\rangle + \beta_k e^{-i\omega t}|2\rangle\right), k = 1,2$$

$$\mu_{12} = \frac{1}{2}\left(\nu \mp \sqrt{\nu^2 + 4|\Omega|^2}\right), \tag{3}$$

$$\alpha_{12} = \left(\frac{\mu_{21}}{\mu_{21} - \mu_{12}}\right)^{1/2}, \qquad \beta_{12} = \pm\frac{\Omega}{|\Omega|}\left(\frac{\mu_{12}}{\mu_{12} - \mu_{21}}\right)^{1/2}$$

The solution of the Schrödinger equation has the form

$$\Phi_\lambda^{(j)}(t) = \sum_k e^{i\bar{\lambda}_k t} a_k^{(j)}(\lambda)|\varphi_k(t)> + \int dE\left[\frac{P}{\lambda - E} + z^{(j)}(\lambda)\delta(\lambda - E)\right]$$

$$\times\left(\sum_k \left[W_{1k}(E)|E>_1 + e^{-i\omega' t}W_{2k}(E)|E + \omega'>_2\right]a_k^{(j)}(\lambda)\right): \tag{4}$$

$$W_{\alpha k}(E) = \begin{cases} \alpha_k\Omega_E + \beta_k U_E, \alpha = 1 \\ \beta_k\Omega'_{E+\omega'}, \alpha = 2 \end{cases} \qquad k = 1,2 \tag{5}$$

$$\left\langle \Phi_{\lambda'}^{(j')}(t) \middle| \Phi_\lambda^{(j)}(t)\right\rangle = \delta(\lambda - \lambda')\delta_{j'j} \tag{6}$$

Here $-\left[\frac{2\pi}{z_l^{(j)}(\lambda)}\right], (j=1,2)$ are the eigenvalues of the Hermitian reaction matrix:

$$K(\lambda) = -2\pi W(\lambda)[\lambda I - \Lambda - \Delta(\lambda)]^{-1} W^{+}(\lambda), \tag{7}$$

where

$$\Lambda_{i,k} = \bar{\lambda}_i \delta_{i,k} \tag{8}$$

with eigenvectors $X_\alpha^{(j)}(\lambda)$,

$$Z_L^{(j)}(\lambda) = \frac{1}{2} \frac{1}{\left(\sum_\mu \frac{|\tilde{W}_{1\mu}(\lambda)|^2}{\lambda - \tilde{\lambda}_\mu}\right)\left(\sum_\mu \frac{|\tilde{W}_{2\mu}(\lambda)|^2}{\lambda - \tilde{\lambda}_\mu}\right) - \left|\sum_\mu \frac{\tilde{W}^*_{1\mu}(\lambda)\tilde{W}_{2\mu}(\lambda)}{\lambda - \tilde{\lambda}_\mu}\right|^2}$$

$$\left\{\sum_\mu \frac{|\tilde{W}_{1\mu}(\lambda)|^2 + |\tilde{W}_{2\mu}(\lambda)|^2}{\lambda - \tilde{\lambda}_\mu} \pm \left[\sum_\mu \frac{|\tilde{W}_{1\mu}(\lambda)|^2 - |\tilde{W}_{2\mu}(\lambda)|^2}{\lambda - \tilde{\lambda}_\mu} + 4\left|\sum_\mu \frac{\tilde{W}^*_{1\mu}(\lambda)\tilde{W}_{2\mu}(\lambda)}{\lambda - \tilde{\lambda}_\mu}\right|^2\right]^{1/2}\right\} \tag{9}$$

and

$$\sum_\alpha X_\alpha^{(j)^*}(\lambda) X_\alpha^{(j')}(\lambda) = \frac{\delta_{jj'}}{\pi^2 + z_l^{(j)2}(\lambda)}, \qquad \alpha, j = 1,2. \tag{10}$$

$$\left|X_{1l}^{(j)}(\lambda)\right|^2 = \frac{1}{\pi^2 + z_l^{(j)}(\lambda)^2} \frac{\left(1 - z_l^{(j)}(\lambda)\sum_\mu \frac{|\tilde{W}_{2\mu}(\lambda)|^2}{\lambda - \tilde{\lambda}_\mu}\right)^2}{\left(1 - z_l^{(j)}(\lambda)\sum_\mu \frac{|\tilde{W}_{2\mu}(\lambda)|^2}{\lambda - \tilde{\lambda}_\mu}\right)^2 + z_l^{(j)}(\lambda)^2 \left|\sum_\mu \frac{\tilde{W}^*_{1\mu}(\lambda)\tilde{W}_{2\mu}(\lambda)}{\lambda - \tilde{\lambda}_\mu}\right|^2}$$

$$\left|X_{2l}^{(j)}(\lambda)\right|^2 = \frac{1}{\pi^2 + z_l^{(j)}(\lambda)^2} \frac{\left|\sum_\mu \frac{\tilde{W}^*_{1\mu}(\lambda)\tilde{W}_{2\mu}(\lambda)}{\lambda - \tilde{\lambda}_\mu}\right|^2}{\left(1 - z_l^{(j)}(\lambda)\sum_\mu \frac{|\tilde{W}_{2\mu}(\lambda)|^2}{\lambda - \tilde{\lambda}_\mu}\right)^2 + z_l^{(j)}(\lambda)^2 \left|\sum_\mu \frac{\tilde{W}^*_{1\mu}(\lambda)\tilde{W}_{2\mu}(\lambda)}{\lambda - \tilde{\lambda}_\mu}\right|^2} \tag{11}$$

The sign "~" above the letter denotes a quantity obtained after unitary transformation of diagonalization of the interaction matrix for discrete states.
The cross sections for elastic and inelastic scattering are:

$$\sigma_{el} = \frac{\pi}{k^2}\sum_l (2l+1)|1 - S_l(\lambda)|^2,$$

$$\sigma_{inel} = \frac{\pi}{k^2}\sum_l (2l+1)\left(1 - |S_l(\lambda)|^2\right), \tag{12}$$

where

$$S_l(\lambda)=\frac{X_{1l}^{(1)}(\lambda)X_{2l}^{(2)}(\lambda)S_l^{(1)}(\lambda)-X_{2l}^{(1)}(\lambda)X_{1l}^{(2)}(\lambda)S_l^{(2)}(\lambda)}{X_{1l}^{(1)}(\lambda)X_{2l}^{(2)}(\lambda)-X_{2l}^{(1)}(\lambda)X_{1l}^{(2)}(\lambda)}, \quad (13)$$

and

$$S_l^{(j)}(\lambda)=e^{2i\left(\delta_l+\eta_l^{(j)}(\lambda)\right)} \quad (14)$$

Here δ_l is the phase of the potential (nonresonant) scattering and $\eta_l^{(j)}(\lambda)$ is the phase caused by the resonant scattering with formation of the Feshbach resonance.

$$tg\,\eta_l^{(j)}(\lambda)=-\frac{\pi}{z_l^{(j)}(\lambda)} \qquad (j=1,2) \quad (15)$$

The cross sections for elastic and inelastic scattering are

$$\sigma_{el}=\frac{4\pi}{k^2}\sum_l(2l+1)\left|Sin\delta_l-\left[A_1(\lambda)e^{i\eta_l^{(+)}(\lambda)}Sin\,\eta_l^{(+)}(\lambda)+A_2(\lambda)e^{i\eta_l^{(-)}(\lambda)}Sin\,\eta_l^{(-)}(\lambda)\right]e^{i\delta_l}\right|^2, \quad (16)$$

$$\sigma_{inel}=\frac{4\pi}{k^2}\sum_l(2l+1)\left|A_3(\lambda)\right|^2Sin^2\left(\eta_l^{(-)}(\lambda)-\eta_l^{(+)}(\lambda)\right) \quad (17)$$

Where

$$A_1(\lambda)=\frac{X_{1l}^{(1)}(\lambda)X_{2l}^{(2)}(\lambda)}{X_{1l}^{(1)}(\lambda)X_{2l}^{(2)}(\lambda)-X_{2l}^{(1)}(\lambda)X_{1l}^{(2)}(\lambda)}$$

$$A_2(\lambda)=-\frac{X_{2l}^{(1)}(\lambda)X_{1l}^{(2)}(\lambda)}{X_{1l}^{(1)}(\lambda)X_{2l}^{(2)}(\lambda)-X_{2l}^{(1)}(\lambda)X_{1l}^{(2)}(\lambda)} \quad (18)$$

$$A_3(\lambda)=\frac{X_{2l}^{(1)}(\lambda)X_{2l}^{(2)}(\lambda)}{X_{1l}^{(1)}(\lambda)X_{2l}^{(2)}(\lambda)-X_{2l}^{(1)}(\lambda)X_{1l}^{(2)}(\lambda)}$$

Expressions (16) and (17) show that the cross-sections are determined by two coupled phases of resonance scattering, and interference phenomena are observed.
From the expression (16) for the elastic resonance scattering the interference phenomena may be observed. The cross-sections for the elastic scattering can be represented in the form.

$$\sigma_{el}=\sigma_{pot}+\sigma_{el}^{nonres}+\sigma_{el}^{nonres}\Re_L(\lambda), \quad (19)$$

where the interference term has the form

$$\Re_L(\lambda)=\left|A_1^{(L)}(\lambda)\right|^2F_1(\lambda)+\left|A_2^{(L)}(\lambda)\right|^2F_2(\lambda)+2\left|A_3^{(L)}(\lambda)\right|^2F_3(\lambda) \quad (20)$$

$$\sigma_{pot}=\frac{4\pi}{k^2}\sum_{l\neq L}(2l+1)Sin^2\delta_l, \qquad \sigma_{nonres}=\frac{4\pi}{k^2}(2L+1)Sin^2\delta_L; \quad (21)$$

σ_{nonres} is the scattering cross-section for angular momentum L in the absence of resonant state.

$$F_j(\lambda)=\frac{\chi_L^2+2\chi_L\overline{z}_j^{(L)}(\lambda)-1}{1+\overline{z}_j^{(L)}(\lambda)^2} \qquad (j=1,2) \quad (22)$$

$$F_3(\lambda) = \frac{\chi_L^2(1+\bar{z}_1^{(L)}(\lambda)\bar{z}_2^{(L)}(\lambda)) + \chi_L \bar{z}_1^{(L)}(\lambda)\bar{z}_2^{(L)}(\lambda)(\bar{z}_1^{(L)}(\lambda)}{(1+\bar{z}_1^{(L)}(\lambda))(1+\bar{z}_2^{(L)}(\lambda))} + $$
$$+\frac{\bar{z}_2^{(L)}(\lambda)) + \bar{z}_1^{(L)}(\lambda)\bar{z}_2^{(L)}(\lambda) - (\bar{z}_1^{(L)}(\lambda) + \bar{z}_2^{(L)}(\lambda)) - 1}{(1+\bar{z}_1^{(L)}(\lambda))(1+\bar{z}_2^{(L)}(\lambda))} \quad (23)$$

with $\chi_L = -ctg\,\delta_L$

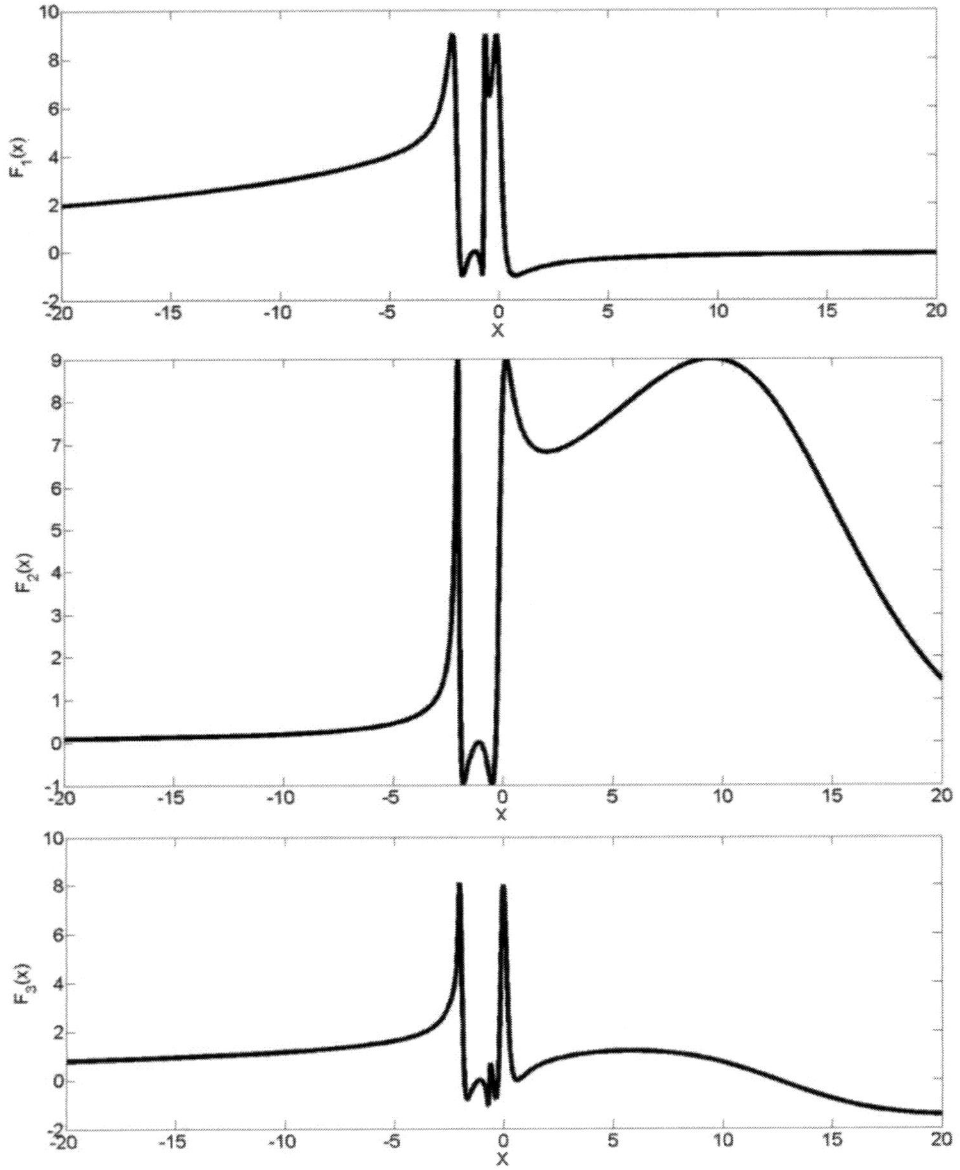

Figure 2. The functions $F_i(x)$ (i=1,2,3) with $x = \frac{\lambda - \tilde{\lambda}_1}{\tilde{\Gamma}_1^{(2)}}$ for $\tilde{\Gamma}_1^{(2)} = 1$, $\tilde{\Gamma}_2^{(1)} = 2$, $\tilde{\Gamma}_1^{(1)} = 1.5$, $\tilde{\Gamma}_2^{(2)} = 0.5$, $\Omega = 1$, $\nu = 0$ (all quantities are in units $\tilde{\Gamma}_1^{(2)}$) at the value $\chi_L = 3$.

The functions $F_i(x)$ (i=1,2,3) (entering the interference term (20)) with $x = \frac{\lambda - \tilde{\lambda}_1}{\tilde{\Gamma}_1^{(2)}}$ are plotted in Fig.2.

3. Conclusion

We obtained cross-sections of resonance scattering with formation of Feshbach resonance. As distinct from the case of two-photon resonance in absence of the decay channel via interaction Ω_E [4] and from the case of elastic scattering at one-photon resonance [5], the cross-sections are determined by two coupled phases of resonance scattering which pass in respective limits to the corresponding expressions reported in [4,5].

The work was supported by Volkswagen foundation grant "Optical information processing, driven by adiabatic interactions between light and matter" and is performed in the scope of the International Associated Laboratory IRMAS and Grant of the State Committee of Science of the Ministry of Education and Science of the Republic of Armenia.

References

[1] Timmermans, E., Tommasini, P., Hussein, M., and Kerman, A., *Phys. Rep.*, 1999, vol. 315, p. 199.
[2] Duine, R.A. and Stoof, H.T.C., *Phys. Rep.*, 2004, vol. 396, p. 115.
[3] Chin, Ch., Grimn, R., Julienne, P., and Tiesinga, E., *Rev. Mod. Phys.*, 2010, vol. 82, p. 1225.
[4] A. Gazazyan, E. Gazazyan and V. Chaltykyan,*Proc.* SPIE 7998, 79980Y, doi:10.1117/12.890508 (2010), Europhysics conference abstracts, *43rd European Group for Atomic System (EGAS)*, University of Fribourg, Fribourg, Swizerland, June 28 - July 2, volume 35C, page 93 (2011).
[5] E. A. Gazazyan, A. D. Gazazyan, and V. O. Chaltykyan, , Journal of Contemporary Physics (Armenian Academy of Sciences), Vol. 46, No. 5, p. 204 (2011).

Journal of Physics: Conference Series **350** (2012) 012010 doi:10.1088/1742-6596/350/1/012010

Electromagnetic Wave Detection Via Interaction With Non-Uniform Distributed Space Charge In Vacuum

H.S. Haroyan, A.H. Makaryan, K.M. Movsisyan, V.R. Tadevosyan

Microwave Radiophysics and Telecommunications Dept., Yerevan State University

E-mail: armenm@ysu.am

Abstract. The detection of electromagnetic radiation of subterahertz (up to 0.14 THz) and IR regions obtained from the interaction with non-uniformly distributed space charge in the interelectrode space of a vacuum triode with plane electrodes is experimentally demonstrated. The dependence of the detected signal from the direction of wave polarization and vacuum triode's current characteristics has been investigated.
A simple theoretical model, according to which, the detection is due to nonlinearity associated with the non-uniform distribution of electrons along the electrostatic field direction, has been proposed to explain the mechanism of detection. The measured detection characteristics reasonably agree with theoretical results, based on low-level signal approximation and assuming that the current-voltage characteristic of the used devices (triodes) obeys the law of 3/2.

Terahertz (THz) range of electromagnetic radiation (0.1-10 THz) bridges the gap between the microwave and optical regions, and offers significant scientific and technological potential in many fields (sensing, imaging and spectroscopy, diagnostics of various facilities, environmental monitoring, imaging, nondestructive testing, security applications and so forth) [1-3], since the rotational transition frequencies of molecules of gases and liquids, oscillation frequency of complex biological molecules and optical phonons of solids are in this range. Non-ionizing nature and weak scattering of THz radiation in the fine-dispersed media makes it very attractive for use in biology and medicine as well[1,2].

Detection at terahertz frequencies obviously differs from the detection at shorter optical wavelengths and longer radio wavelengths. In comparison to shorter wavelengths, due to the low level of photon energies at terahertz frequencies, ambient background thermal noise is dominant than request cryogenic cooling or long-integration-time radiometric techniques.

Besides, the diameter of THz radiation beam is rather large, which makes a mode converter or matched director (antenna) between the signal and detector element necessary. In terahertz field, RF detection shows better performance than optical detection because the crossover frequency at which an ideal thermal noise limited detector (such as a room-temperature Schottky barrier diode) excels the sensitivity of an ideal quantum detector (like a photodiode), falls between 1 and 10 THz [1]. In comparison to longer wavelength radio techniques, lumped electronic components such as resistors, capacitors, and inductors, as well as amplifiers and low-loss transmission media are not available to terahertz detectors.

Thus, fast and sensitive detectors of optical range, operating on the basis of external or internal photoeffect are not suitable for registration of THz radiation. From the other hand the sensitivity of widespread microwave Schottky diodes decreases rapidly in the THz region and sensitive bolometric receivers are inertial, and therefore they are not applicable for registration of fast processes. Therefore, for successful development of terahertz frequency range it is necessary to implement fast, high sensitive and available detectors operating at room temperature [1-4].

In our recent work we have investigated the detection of terahertz waves by using a vacuum diode, as a consequence of nonlinear interaction of electromagnetic waves with the electron flow [3].

In this work we have investigated the detection of electromagnetic radiation of subterahertz and IR regions obtained from the interaction with non-uniformly distributed space charge in the interelectrode space (space between cathode and grid) of a vacuum triode with plane electrodes.

Block diagram of experimental setup is shown in Fig. 1. As a source of electromagnetic radiation in the subterahertz range standard generator G4-161 (maximum output power ~10mW, tuning range 129-142GHz) was used, and in the optical range - a helium-neon laser LG-125 with output power ~12mW (λ = 3.39μ, 1.15μ and 0.63μ). Modulation of radiation was realized by using a mechanical chopper. The modulation frequency was ~1 kHz. As a detector, triode 6S15P was used with the most comfortable design of parallel plate electrodes. To ensure penetration of the electromagnetic wave in the interelectrode space, the lamp was located directly in the path of the electromagnetic wave with polarization vector parallel to the direction of electron flow in the tube, as shown in Fig. 1.

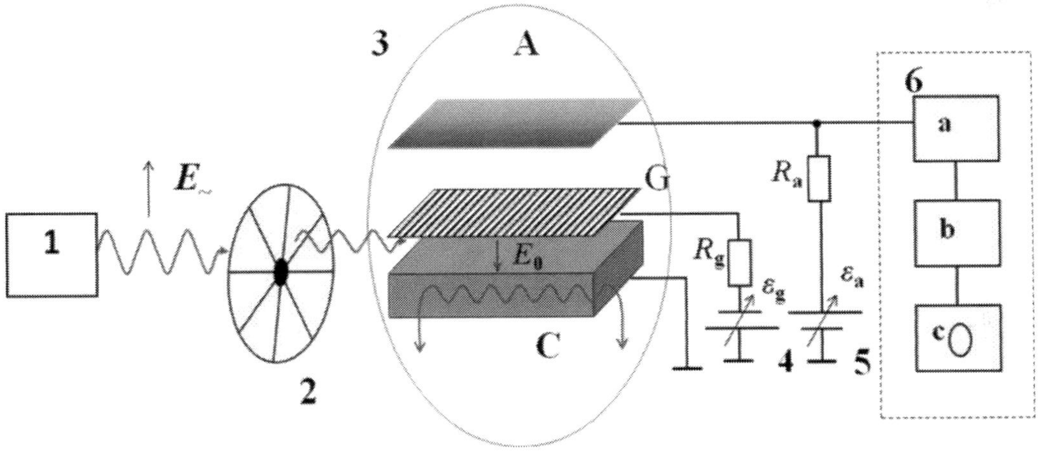

Figure 1. Block diagram of experimental setup:
1 - generator (G4-161 or LG-125), 2 - modulator (F = 1kHz), 3 - vacuum triode, 4- bias voltage source, 5 - anode voltage source, 6 - recording system (a-filter, b-amplifier, c-oscilloscope)

From the regulated DC voltage source ε_a through a ballast resistor R_a anode voltage U_a was applied and the bias voltage on the grid was applied form the source ε_g via a resistor R_g. The detected signal is removed from the anode and after filtering and amplification was registered with an oscilloscope.

The dependence of the detected signal from the direction of incident radiation polarization and current characteristics of vacuum triode has been investigated.

To determine the optimal operating point for obtaining maximum detected signal the anode-anode and anode-grid characteristics of the triode were measured (anode-grid characteristics are shown in Fig. 2.).

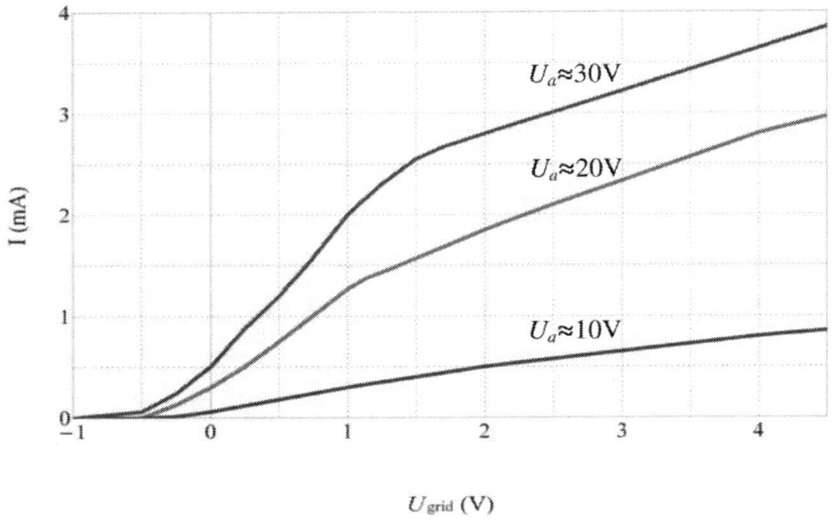

Figure 2. Anode-grid VA characteristics.

In Fig. 3. are shown lamps location in the way of electromagnetic wave propagation and the dependence of the detected signal from the angle θ between the directions of the electric field $E_{\sim}$ and the electrostatic field E_0

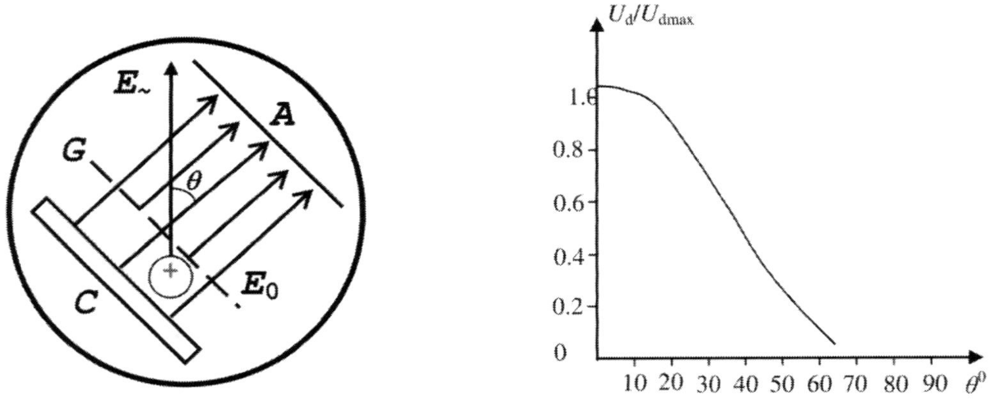

a. b.

Figure 3.- a. lamps location in the way of electromagnetic wave propagation: $E_{\sim}$ - the electric field of electromagnetic wave, E_0 - electrostatic field, θ-angle between the electric fields $E_{\sim}$ and E_0, and " +" -is the direction of electromagnetic wave propagation.
b. the dependence of the detected signal from the angle θ .

In the case of detection of electromagnetic radiation in the subterahertz range, the dependence of the detected signal from the bias voltage on the grid of vacuum triode is shown in Fig. 4. The analogous dependence has been obtained in IR range.

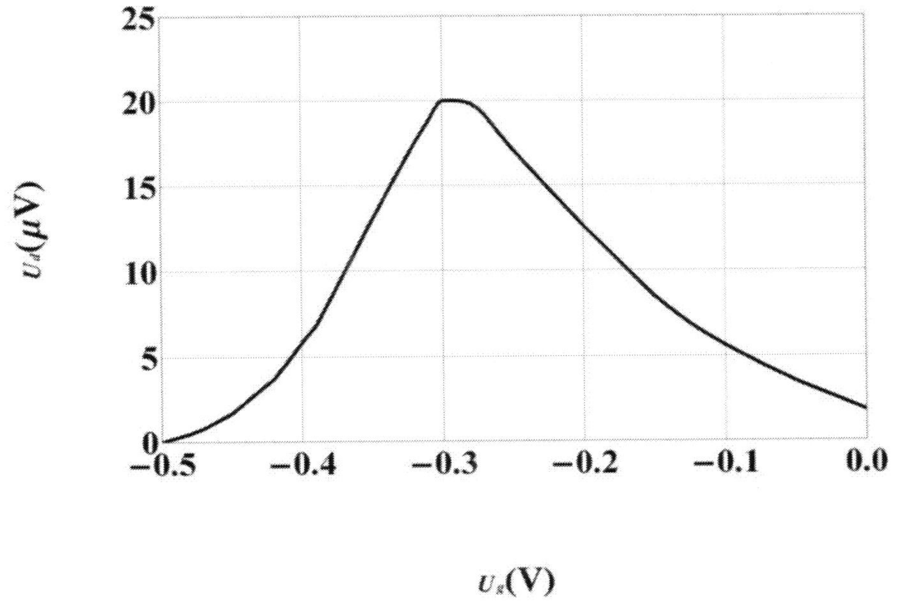

Figure 4. The dependence of the detected signal from the bias voltage on the grid of vacuum triode

As expected, when $\varepsilon_a \approx 60$V (for $R_a \approx 30$kOm, $U_a \approx 30$V) the maximum detected signal was obtained at a bias voltage of -0.3V. The value of maximum detected signal was ~25μV in the optical range with laser power ~1mW and ~20μV in THz range ($F \approx 0.14$THz) with THz power ~10μW.

A simple theoretical model, according to which the detection in vacuum triode is due to the nonlinearity associated with the non-uniform distribution of charge carriers (electrons) along the direction of the electrostatic field, has been proposed to explain the mechanism of detection.

The nonlinear properties of the triode are due to the non-uniform distribution of electrons along the direction of the electrostatic field. We estimate the value of the detected signal, assuming that the current-voltage characteristics obeys the law of 3/2 and does not depend on the frequency of the applied field, when the plane of wave polarization is perpendicular to the electrodes of the triode. If the wave electric field in the space between cathode and grid uniformly, the difference potential grid-cathode, produced by the wave, can be written:

$$U_{g\sim} = E_{\sim}d = E_{0\sim}d\cos\omega t = U_{0\sim}\cos\omega t\,, \tag{1}$$

where $E_{\sim}$ is electric field of the wave, $E_{0\sim}$ its amplitude, $U_{0\sim}$ amplitude of produced voltage, d - is the interelectrode distance, ω- the wave frequency. If the lamp applied also anode constant voltage U_{a0}, and grid constant voltage U_{g0}, in the unsaturated regime for the anode current we have

$$I_a = aU_{eff}^{3/2} = a(U_{g0} + U_{g\sim} + DU_{a0})^{3/2}\,, \tag{2}$$

where D is the permeability of triode. The value of the coefficient a for the used vacuum tubes is $a = 1mA/V^{3/2}$, and $D \approx 0.025$ (gain constant $\mu \approx 40$).

Estimates based on formula (2) show, that by value of electromagnetic wave intensity ~10μW/cm^2, and with U_{g0}≈-0.3V, the maximum of the rectified current value is ~3nA. Current recorded in the experiment was ~1nA. This difference is quite understandable, because in the experiment there was not ensured consistency, and therefore, only a small part of the radiation power penetrated in the interelectrode space.

Because THz waves occupy intermediate region of the spectrum between the IR and subterahertz, the results can be used in future to develop THz detectors based on vacuum devices.

Despite the fact that the sensitivity of the optical range decreased rapidly (~ 100), however, estimates indicate that in solution of the matching problem and an increase of interaction spaces of radiation with an interelectrode space charge, vacuum tubes can be successfully used for detection, and also for frequency conversion of terahertz range radiation.

References

[1]. A.Rostami, H. Rasooli and H. Baghban, 2010, Terahertz Technology, Fundamentals and applications, Springer, Berlin.
[2]. K. Sakai, Terahertz optoelectronics, Springer, Berlin.-2005.
[3]. A. Hakhoumian, A. Makaryan, K. Movsisyan, K. Khachatruan, proc.of Conf. on Laser Physics 2007, 174.
[4]. Lei Hou, Hongkyu Park, Xi-Cheng Zhang, 2011 IEEE Journal of Selected Topics in Quantum Electronics, **17**, no. 1, 177.

Raman - Nath approximation for diffraction of atoms in the laser field taking into account spontaneous emission of atoms for ground and high energy level

L Hovhannisayan and A Zh Muradyan
Department of Physics, Yerevan State University, 1 Alex Manookian, Yerevan 025, Armenia

E-mail: lhovhannisyan@gmail.com

Abstract. The problem of resonant Kapitza – Dirac diffraction is solved in Raman – Nath approximation out of familiar Bessel function approximation (applicable in zero and very large resonance detuning cases). It shows new and promising results for the atom optics and atom interferometry if the atomic momentum state has been prepared in a form of discrete Gaussian distribution. Namely, instead of monotonic broadening within the Bessel function approximation, our formula yields in splitting of initial distribution into two identical peaks, whith preserving form, which symmetrically move away from the distribution center for the interaction time. A table-shaped form for the ultimate momentum distribution also is in frame of new distributions. As to relaxation processes, they have only quantitative influence on the pattern of diffraction.

Introduction

Atom scattering in the field of laser counterpropagating waves[1] attracts a constant attention in atom optics[2] and atom interferometry[3]. The mechanism of scattering is emission of photons from one of the counterpropagating waves into the other counterpropagating wave, when due to each reemitted photon the atom acquires recoil momentum $2\hbar k$, where k - is the wave vector[3]. Outside the Bragg regime of diffraction analytical results for the scattering amplitudes was possible to obtain also for short times of interaction when the kinetic energy operator of the atom may be neglected (Raman-Nath approximation). Under exact resonance and far off-resonance conditions, the probability amplitude of the n-th order diffraction (acquiring of an additional momentum $2n\hbar k$) has the following well-known form[5]:

$$a_n(t) = i^n J_n(\Omega_{Rabi} t), \qquad (1)$$

where the argument of the Bessel function $J_n(\Omega_{\text{Rabi}}t)$ where the argument of the Bessel function $J_n(\Omega_{\text{Rabi}}t)$ is the product of Rabi frequency Ω_{Rabi} and the interaction time t. The evolution of momentum state (1) is shown in Figure 1. The characteristic feature of it is the monotonic unlimited enlargement of the area covered by the momentum states. Therefore, high-order Kapitza-Dirac diffraction appears not useful for atom interferometry. A careful analysis, proceeding from (1), shows that the superpositional nature of the initial state of atomic center of mass translational motion also does not give the desirable breakthrough in the problem[4]. Here, we reexamine the problem deriving an expression for the atom scattering amplitude $a_n(t)$ without putting any restrictions on the value of resonance detuning $\Delta = \omega - \omega_0$, where ω is the laser field frequency and ω_0 is the atomic transition frequency. The amplitude is expressed by means of a definite integral (see (9)) and can be viewed as some type of generalization of the Bessel function. Although it is true, that the difference between (9) and (1) has only a quantitative nature, one has totally diverse behavior if the atom initially is in the superposition of many equidistant momentum states. Evolution of diffraction picture acquires a qualitatively new content, which is very promising for the large-scale beam splitter atom interferometry, which we call here two bunch atom interferometry.

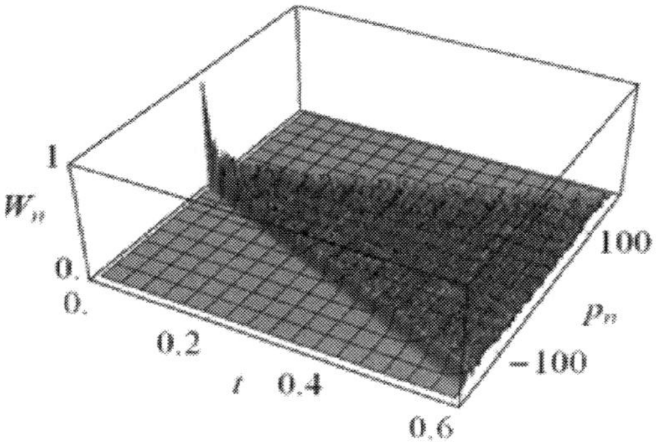

Figure 1. The picture of resonant Kapitza-Dirac diffraction in Raman-Nath approximation ($W_n = |a_n(t)|^2$, $p_n = 2n\hbar k$). in case of a definite value of atomic initial momentum ($a_n(t=0) = \delta_{n0}$).

2. Scattering amplitudes

The problem under consideration is evolution of the momentum distribution of a two-level atom due to interaction with the field of counterpropagating waves. The strength of electric field is

$$E(z,t) = E_1(t)e^{i(kz-\omega t)} + E_2(t)e^{-i(kz-\omega t)} + c.c \quad \textbf{(2)}$$

where $k = \omega / c$ and z is the atomic center of mass coordinate. Wave amplitudes $E_1(t)$ and $E_2(t)$ are assumed of simplest step-wise form, which turn on at the moment $t = 0$. The Hamiltonian in dipole approximation is

$$\hat{H}_0 = \frac{\hbar^2}{2\mathrm{M}} \frac{\partial^2}{\partial z^2} + \hat{H}_0 - d\,E(z,t) \quad \textbf{(3)}$$

where H_0 is the Hamiltonian of free and immovable atom, M is atomic mass and d is the dipole moment operator. The atomic wavefunction can be written in the following form:

$$\psi(z,\mathbf{r},t)=a(z,t)\,\psi_1(\mathbf{r})e^{-i\frac{\varepsilon_1 t}{\hbar}}+b(z,t)\,\psi_2(\mathbf{r})e^{-i\frac{\varepsilon_2 t}{\hbar}-\mathrm{i}\Delta t}, \qquad \textbf{(4)}$$

where $a(z,t)$ and $b(z,t)$ are the target probability amplitudes of atom to reside in ground and excited energy levels respectively, ε_1, ε_2 and $\psi_1(\mathbf{r})$, $\psi_2(\mathbf{r})$ are eigenvalues and eigenfunctions of the free Hamiltonian, $\mathbf{r}$ is the radius-vector of atomic optical electron relative to the center of mass. The Raman-Nath approximation is an approximation of short interaction times when the corresponding atomic displacement is much smaller than the spatial period of the standing wave (constructed by the pair of counterpropagating waves). In some sense it means an immovable atom, giving a right to ignore the kinetic energy operator in the Hamiltonian (3). For illustrating the effect of relaxation of high energy level due to spontaneous emission on probability amplitudes we choose the simplest model: instead of ω_0 we took $\omega_0+i\gamma$. As long as we have a resonance detuning greater than the line broadening, this procedure describes the damping phenomena very well. Then, for $t\succ 0$ we obtain the equations

$$i\hbar\tfrac{\partial}{\partial t}a(z,t)=-d\left(E_1^* e^{-ikz}+E_2^* e^{ikz}\right)b(z,t) \qquad \textbf{(5)}$$

$$\left(i\hbar\tfrac{\partial}{\partial t}-\hbar\Delta+i\hbar\gamma\right)=-d^*\left(E_1e^{ikz}+E_2e^{-ikz}\right)a(z,t). \qquad \textbf{(6)}$$

The right-hand side coefficients do not have any time dependence and the equation for amplitude $a(z,t)$ can be obtained by acting on (5) with operator $i\hbar\frac{\partial}{\partial t}-\hbar\Delta$ and then using Eq.(6). This yields

$$\left(\tfrac{\partial}{\partial t}+i\Delta+\gamma\right)i\hbar\tfrac{\partial}{\partial t}a(z,t)=-\frac{|d|^2}{\hbar^2}\left(E_1^2+E_2^2+2E_1E_2\cos(2kz)\right)a(z,t) \qquad \textbf{(7)}$$

Solution of this equation in momentum space should be the Fourier series

$$a(z,t)=\sum_{n=-\infty}^{\infty}a_n(t)e^{in2kz}, \qquad \textbf{(8)}$$

Unknown amplitude $a_{\mathrm{n}}(t)$ will be sought in the form of a definite integral

$$a_{\mathrm{n}}(t)=\frac{1}{\pi}\int_0^{\pi}e^{i\lambda(\varphi)t}\cos(n\varphi)\mathrm{d}\varphi \qquad \textbf{(9)}$$

with an unknown function $\lambda(\varphi)$. Substitution of (8) and (9) into (7) gives two possible expressions for $\lambda(\varphi)$:

$$\lambda_{1,2}(\phi)=\frac{i\gamma-\Delta}{2}\pm\frac{|\Delta|}{2}\sqrt{1-\frac{\gamma^2}{\Delta^2}+\frac{2i\gamma}{\Delta}+\xi(\phi)}, \qquad \textbf{(10)}$$

where $\xi(\varphi)=\xi_1+\xi_2+2\sqrt{\xi_1\xi_2}\cos(\varphi)$ and $\xi_{1,2}=4d^2E_{1,2}/\hbar^2\Delta^2$.

The general solution is a superposition of two linearly independent solutions:

$$a_n(t)=C_1\int_0^{\pi} exp(i\lambda_1\phi)cos(n\phi)d\phi+C_2\int_0^{\pi} exp(i\lambda_2\phi)cos(n\phi)d\phi, \quad (11)$$

where coefficients C_1 and C_2 have to be determined from initial conditions. If, for example, the atom initially was in rest and resided on the ground state, then

$$C_{1,2}=\frac{1}{2}\pm\frac{sign[\Delta]}{4\sqrt{1-\frac{\gamma^2}{\Delta^2}-2i\frac{\gamma}{\Delta}+\xi)}}F(\left(4\sqrt{\xi_1\xi_2}/1-\tfrac{\gamma^2}{\Delta^2}-2i\tfrac{\gamma}{\Delta}+\xi\right)^{-1})', \quad (12)$$

where $F(x)$ is the elliptic integral of the first kind. Equation (11) with notations (10) and (12) is generalization of familiar (1) for the case of arbitrary resonance detuning.

3. Numerical calculation

Qualitative differences from the Bessel function regularities are obtained when the atomic motional state is prepared in a quantum superposition state. They are especially prominent, when the superposition is of discrete Gaussian form [4]:

$$a(z,t=0)=\sum_{n=-\infty}^{\infty}s_n(t)e^{in2kz}, \quad (13)$$

$$s_n=\frac{1}{\sqrt{\pi}}e^{i\alpha n}\frac{1}{\sqrt[4]{\sigma}}e^{-i\alpha v-\frac{(n\ v)^2}{2\sigma}}, \quad (14)$$

where σ is the FWHM of distribution and α is the initial phase (this case is depicted in Figure 2). Then the diffraction amplitude (11) gets the following form:

$$\begin{aligned}a_n(t)=&\frac{1}{\sqrt{\pi}}e^{i\alpha n}\frac{1}{\sqrt[4]{\sigma}}(\\ &C_1\int\exp(\tfrac{i|\Delta|t}{2}\sqrt{1-\tfrac{\gamma^2}{\Delta^2}+2i\tfrac{\gamma}{\Delta}+\xi(\varphi)})\sum_{v=-\infty}^{\infty}\exp(-i\alpha v-\tfrac{(n-v)^2}{2\sigma})\cos(v\varphi)d\varphi \\ &+C_2\int\exp(-\tfrac{i|\Delta|t}{2}\sqrt{1-\tfrac{\gamma^2}{\Delta^2}+2i\tfrac{\gamma}{\Delta}+\xi(\varphi)})\sum_{v=-\infty}^{\infty}\exp(-i\alpha v-\tfrac{(n-v)^2}{2\sigma})\cos(v\varphi)d\varphi)\end{aligned} \quad (15)$$

To illustrate the new regularities in the diffraction pattern, we will concentrate on more important cases $\alpha=0$ and $\alpha=\pi/2$. Bessel function approximation in case of $\alpha=0$ results in a monotonic increment in the width of prepared Gaussian distribution. Numerical calculations on base of (15) give a qualitatively different result. As is seen in Figure 3, initial distribution splits into two identical parts, which preserving their form, symmetrically move away from each other (or the centre of initial distribution). It should be noted experimentally it has not been possible to increase the distance between the two peaks up to few hundreds of photon momentum yet, while in principle this is realizable under the interaction scheme suggested in the paper. Note, that the relaxation has no role under conditions of Figure 3. The occurrence of relaxation in general affects the picture of evolution through the product parameter γt, where γ is the rate of relaxation. Such a case with $\gamma t=0.3$ is presented in Figure 4. As is seen, the influence on the form of distribution is slightly visible. Main deviation from the reference picture in Figure 3 is the loss of full symmetry of the picture. As would be expected, the symmetry can't be restored totally by means of some new value of γt or remaining parameters. The optimum form of distribution in case of $\gamma t = 0.3$ is depicted in Figure 5.

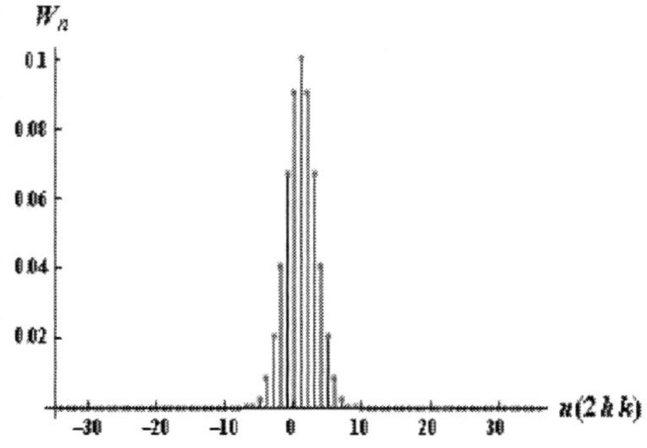

Figure 2. Initial momentum distribution with phase parameter $\alpha = 0$ and half-width $\sigma = 10$.

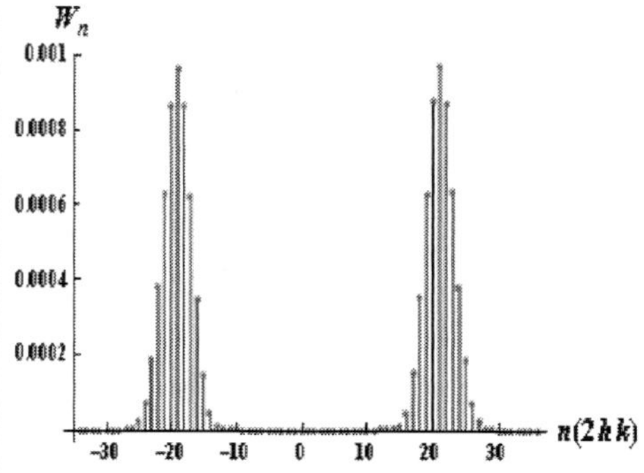

Figure 3. Diffracting in the field of far off-resonance counterpropagating waves in Raman-Nath approximation, atom momentum distribution splits into two symmetrically posed paths. Parameters are the similar to the ones in Figure 2, $\gamma = 0, t = 18\times10^6$ (in units Δ^{-1}).

Not less interesting is the case $\alpha = \pi$, when a table-shaped form of momentum distribution is obtained from the initial Gaussian distribution as is shown in Figure 6. This strongly recommended in high resolution spectroscopy[5] distribution is robust with respect to relaxation processes: affecting the

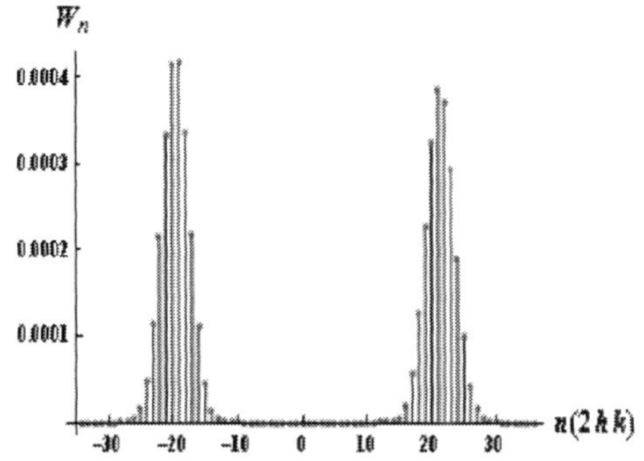

Figure 4. Parameters are the similar to the ones in Figure 3, $g = 0.3, t = 18\times10^6$ (in units Δ^{-1}).

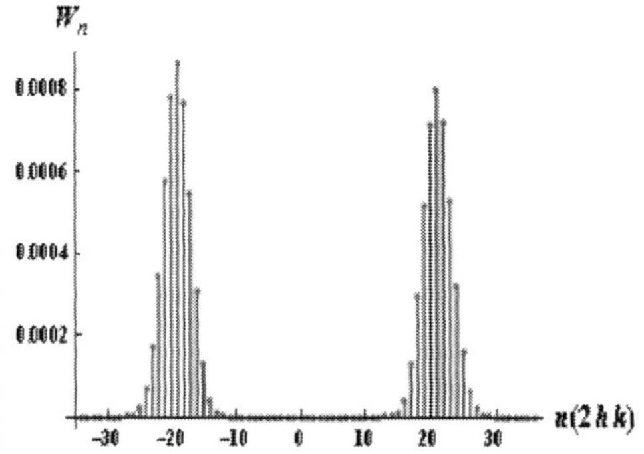

Figure 5. Parameters are the similar to the ones in Figure 2, $g = 0.3$, $t = 17.82\times10^6$ (in units Δ^{-1}).

absolute value of probability amplitudes, the relaxation preserves the table-shaped form of the momentum distribution.

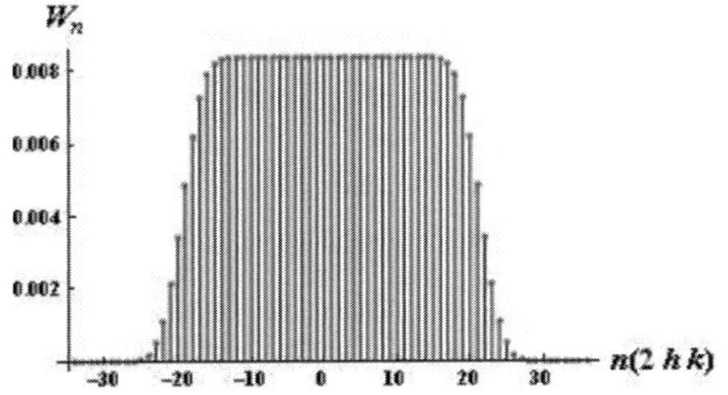

Figure 6. Generation of table-form momentum distribution for a moving atom in the Raman-Nath approximation: $\alpha=\pi$, $\sigma=10$, $t=17.8\times10^6$.

4.Conclusion

The problem of atomic Kapitza-Dirac diffraction is solved in Raman-Nath approximation for arbitrary values of resonance detuning including the relaxation. The obtained formula for the scattering amplitudes generalizes the known expression given by means of the first kind Bessel function. Remarkable point here is the ability to obtain double-peak and table-like forms in momentum distribution, desirable in applied atomic optics and atom interferometry.

This work was supported by the Armenian State Scientific Committee Grant 11-1c215.

5. References

[1] Arutyunyan V M and Muradyan A Zh 1975 *Izvestia AN ARM.* SSR. **60** 275 Cook R J and Bernhardt A F 1978 *Phys. Rev.* A **18** 2533; Martin P J, Gould P L, Oldaker B J, Miklich A H and Pritchard D E1987 *Phys.Rev. A* **36** 2495

[2] Arimodo E, Lew H and Oka T 1979 *Phys. Rev. Lett.* **43** 753; Phillips W D, Prodan J V and Metcalf H J 1985 *J. Opt. Soc. Am.* B **2** 1751; Metcalf H J and van der Straten P 1999 *Laser cooling and trapping* (New York: Springer); Foot Ch J 2005 *Atomic Physics* (New York: Oxford University Press); Castin Y 2001 *Preprint* cond-mat/0105058

[3] Berman P 1997 *Atom Interferometry* (New York :Acad. Press); Cronir A D, Schmiedmayer J and Pritchard D E 2007 *Preprint* 0712.3703v1; Hogen J M, Johnson D M S and Kasevich M A 2008 arXiv(0806:3261)

[4] Ishkhanyan A M 1997 *Laser Physics* **7** 1225-1230 Ishkhanyan A M 2000 *Phys. Rev.* A **61** 063611;

[5] Ye J and Cundiff S. T 2005 *Femtosecond Optical Frequency Comb* Technology (New York: Springer); Jenkins D W and Dow J D 1989 *Phys. Rev.* B **39** 3317

Higher order transmission resonances in above-barrier reflection of ultra-cold atoms

H A Ishkhanyan[1,2], V P Krainov[2], and A M Ishkhanyan[1]

[1]Institute for Physical Research of NAS of Armenia, 0203 Ashtarak-2, Armenia
[2]Moscow Institute of Physics and Technology, 141700 Dolgoprudny, Russia

E-mail: aishkhanyan@gmail.com

Abstract. The reflectionless transmission resonances in above-barrier reflection of Bose-Einstein condensates by the Rosen-Morse potential are considered using the mean field Gross-Pitaevskii approach. Applying an exact third order nonlinear differential equation obeyed by the condensate's density, the exact solution of the problem for the first resonance is derived. It is shown that in the nonlinear case the total transmission is possible for positive potential heights, i.e., for potential barriers. Further, it is shown that an appropriate approximation for higher-order resonances can be constructed using a limit solution of the equation for the density written as a root of a polynomial equation of the third degree. Using this limit function and the solution for the first resonance, a simple approximation for the shift of the nonlinear resonance potential's depth from the corresponding linear resonance's position is constructed for higher order resonances. The result is written as a linear function of the resonance order. This behavior notably differs from the case of the rectangular barrier for which the nonlinear shift is approximately constant for all the resonance orders.

1. Introduction

The basic effects of tunneling through a barrier and above-barrier reflection of particles are well appreciated paradigms of quantum mechanics extensively studied for many decades [1]. The experimental realization of Bose-Einstein condensation in dilute gases of ultracold atoms [2] has stimulated a renewed interest to those effects since the condensates provide a different test of fundamental principles of quantum mechanics due to the essentially nonlinear nature of the many-body dynamics of Bose-condensates [3]. Several physical situations for which the linear analogue is known have recently been discussed including, e.g., the step-potential [4], the rectangular barrier [5], the double-delta-shell and delta-comb configurations [6]. The above-barrier reflection of cold atoms by the squared-sech Rosen-Morse potential has been addressed in [7,8,9]. The exact solution of the Gross-Pitaevskii equation has been reported for a certain depth of the potential [8] and the reflection coefficient has been calculated for the close vicinity of the first linear transmission resonance for the case of small nonlinearity [9]. In the present paper, we consider the higher order transmission resonances viewed in terms of incoming and outgoing waves.

The dynamics of Bose-Einstein condensates in the mean-field approximation is described by a nonlinear Schrödinger equation referred to as the Gross-Pitaevskii equation [3,10]. In the one-dimensional case this equation is written as

$$i\hbar\frac{\partial\Psi}{\partial t}=-\frac{\hbar^2}{2m}\frac{\partial^2\Psi}{\partial x^2}+(V(x)+g|\Psi|^2)\Psi=0, \tag{1}$$

where the nonlinearity parameter g determines the mean-field self-interaction and $V(x)$ is the reflecting potential. Applying the ansatz $\Psi(x,t)=\exp(-i\mu t/\hbar)\psi(x)$, where μ is the chemical potential, Eq. (1) is reduced to the following stationary version

$$-\frac{1}{2}\frac{d^2\psi}{dx^2}+(-\mu+V(x)+g|\psi|^2)\psi=0. \tag{2}$$

The reflecting potential we consider here is the finite-height squared hyperbolic-secant Rosen-Morse potential

$$V(x)=V_0\,\mathrm{sech}^2 x. \tag{3}$$

To meet the conditions for *above*-barrier reflection, we suppose that $\mu>0$ for a potential well with $V_0<0$ and $\mu>V_0$ for a barrier with $V_0>0$. Further, to treat the process in terms of incoming and outgoing waves, we look for a solution of Eq. (2) which at $x\to-\infty$ has the asymptotic behavior of the form $\psi\sim C_1e^{+ikx}+C_2e^{-ikx}$ describing a combination of incident and reflected running waves, and at $x\to+\infty$ has the asymptote $\psi\sim C_3\,e^{+ikx}$ corresponding to a transmitted wave. The reflectionless transmission then corresponds to a specific case when the reflected wave is absent, i.e., $C_2=0$ and, hence, $|C_1|=|C_3|$. Substituting the ansatz $\psi=C_1e^{+ikx}$ into Eq. (2) and noting that the Rosen-Morse potential vanishes at $x\to-\infty$, we readily see that the wave number k is determined as $k=\sqrt{2(\mu-|C_1|^2 g)}$. Hence, since k is supposed to be a real number, it should be $\mu>|C_1|^2 g$. Note finally that the normalization of the wave function can always be incorporated in the definition of the nonlinearity parameter g. For this reason, without loss of generality, we put $|C_1|=1$ and thus adopt the normalization $|\psi(-\infty)|^2=1$.

In the linear case $g=0$ the reflectionless transmission is known to occur for a discrete spectrum of *negative* values of the potential height defined as [1]

$$V_{Ln}=-\frac{1}{2}n(n+1),\quad n=1,2,3,... \tag{4}$$

The corresponding solution of the linear Schrödinger equation for these transmission resonances satisfying the adopted normalization is written in terms of the Gauss hypergeometric function [11]:

$$\psi_{Ln}=e^{ikx}\,{}_2F_1(-n,1+n,1+ik_0,z),\quad k_0=\sqrt{2\mu}, \tag{5}$$

where $z=(1+\tanh x)/2$. Below we consider the transmission resonances for the nonlinear case $g\neq 0$. We adopt the numeration of the successive resonances for which the height of the nth order nonlinear resonance V_{NLn} at $g\to 0$ tends to the height of the nth order linear resonance V_{Ln} given by Eq. (4). Note that, in accordance with this formula, the reflectionless transmission in the linear case is possible only for potential wells. However, we will see below that in the nonlinear case the total transmission is also possible for positive potential heights, i.e., for potential barriers. Note also that the resonance position in the linear case does not depend on the chemical potential μ. We will see that in the nonlinear case the situation is changed: the spectrum of the nonlinear resonances becomes a function of μ.

2. Equation for the probability and the limit solution

An initial observation serving as a starting point for what follows is that the direct linearization of the stationary Gross-Pitaevskii equation (2) using the linear solution (5) as the zero-order approximation (i.e., effectively, a linearization via replacement of the $|\psi|^2$ term in Eq. (2) by $|\psi_{Ln}|^2$) does not produce accurate results. More advanced approaches include the multiple scale analysis [12] (we have applied this method to treat the first nonlinear resonance for the rectangular barrier [5] and the Rosen-Morse potential [9]) and the renormalization technique (this reveals that one should apply the linear solution using $k=\sqrt{2(\mu-g)}$ instead of $k_0=\sqrt{2\mu}$ in Eq. (5)). Here, we suggest an alternative approach based on an exact third order differential equation for the probability $p=|\psi(x)|^2$. This equation is derived by substituting $\psi(x)=\psi_1(x)+i\psi_2(x)$ into Eq. (2). Separating then the real and imaginary parts, we get

$$-\frac{\psi_1''}{2}+[-\mu+V+g(\psi_1^2+\psi_2^2)]\psi_1=0,$$
$$-\frac{\psi_2''}{2}+[-\mu+V+g(\psi_1^2+\psi_2^2)]\psi_2=0, \tag{6}$$

where the prime denotes differentiation with respect to x. Multiplying now the first equation by ψ_1' and the second equation by ψ_2' and summing the two equations leads to the equation

$$-\frac{1}{2}\frac{d}{dx}[(\psi_1')^2+(\psi_2')^2]+(-\mu+V+g\,p)p'=0. \tag{7}$$

Another equation is derived by multiplying the first equation of system (6) by ψ_1 and the second one by ψ_2, summing the equations and applying the identity $\psi_1\psi_1''+\psi_2\psi_2''=-(\psi_1'^2+\psi_2'^2)+p''/2$. The result reads

$$-\frac{p''}{4}+\frac{1}{2}(\psi_1'^2+\psi_2'^2)+(-\mu+V+g\,p)p=0. \tag{8}$$

The last step is straightforward. We differentiate Eq. (8) and use Eq. (7) to arrive at the third order differential equation

$$\frac{d}{dx}\left[-\frac{p''}{4}+(-\mu+V+g\,p)p\right]+(-\mu+V+g\,p)p'=0. \tag{9}$$

Since we are studying the reflectionless transmission through the barrier the boundary conditions we here impose are

$$p(-\infty)=p(+\infty)=1. \tag{10}$$

Note that such boundary conditions define a nonlinear eigenvalue problem for the derived equation (9). Finally, note that, since we look for a solution with traveling-wave-like asymptotic behavior $\psi(-\infty)\sim e^{ikx}$, the function $p(x)$ should additionally obey the conditions

$$p'(-\infty)=0,\quad p''(-\infty)=0. \tag{11}$$

The derived equation (9) turns out to be a productive one since it allows one to linearize the problem using the solution of the linear problem as a starting approximation. A simpler alternative possibility (dealing with elementary functions) suggested by this equation for a relatively large chemical potential μ is to treat the problem starting from a limit solution that is obtained when neglecting the second derivative term in square brackets in Eq. (9). Numerical simulations show that this approach is valid as far as $\mu>0.25$. The development of this approach is useful since it is not based on the solution of the associated linear problem which is known only for a few cases. Instead, it works with a simple approximate solution that is readily written for any potential.

Generally speaking, dropping out the highest derivative term of a differential equation is a singular procedure because the reduced equation becomes of a lower order and, hence, the solution to this equation in general cannot satisfy all the initial conditions (see, e.g., numerous examples discussed in Ref. [12]). However, in our particular case we obtain a solution that incidentally satisfies all the imposed boundary conditions. This is not a unique case. A similar favorable situation is faced, e.g., in treating the Landau-Zener transition in cold molecule formation via photoassociation of an atomic Bose-Einstein condensate [13].

According to aforesaid, the limit solution $p_0(x)$ is a function defined as the solution of the reduced equation

$$\frac{d}{dx}\left[(-\mu+V+g\,p)p\right]+(-\mu+V+g\,p)p'=0. \tag{12}$$

This equation is straightforwardly integrated producing a cubic polynomial equation for p_0:

$$C_0+(-\mu+V+g\,p_0)p_0^2=0, \tag{13}$$

where C_0 is the integration constant. The boundary condition $p(-\infty)=1$ gives $C_0=\mu-g$. The limit function defined by this equation is a symmetric [$p_0(x)=p_0(-x)$], non-oscillatory function of x. The comparison of this function with the exact numerical result for the probability p is shown in Fig. 1 under both resonant (Fig. 1a) and non-resonant (Fig. 1b) conditions (the incoming wave is supposed to move from right to left, the large-amplitude oscillations seen for large positive x in Fig. 1b describe the interference of incident and reflected waves). It is seen that in both cases the limit function satisfactory approximates the behavior of the system in a space region embracing the potential, while it becomes invalid for $x\to+\infty$ if the potential depth is not resonant. Hence, we conclude that this function may be specifically useful for treatment of the reflectionless transmission resonances. We will show below that this is indeed the case.

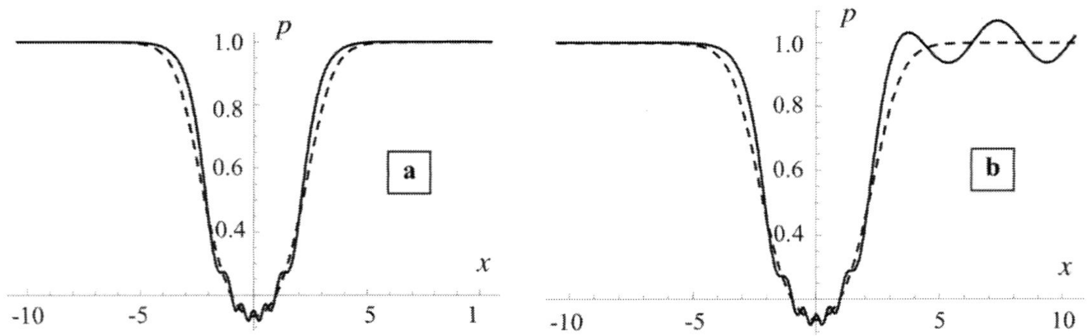

Fig. 1. Comparison of the limit function $p_0(x)$ defined by Eq. (13) (dotted line) with the exact numerical result for p (solid line) for $\mu=0.5$ and $g/\mu=0.25$ (the incoming wave is supposed to move from right to left). a) The depth of the potential is chosen to support the sixth-order transmission resonance: $V_0=V_{NL6}=-20.4914$. b) The depth of the potential here is $V_0=-20$; it does not meet the transmission resonance conditions. The large-amplitude oscillations seen for large positive x describe the interference of incident and reflected waves.

Journal of Physics: Conference Series **350** (2012) 012012 doi:10.1088/1742-6596/350/1/012012

3. Exact solution for the first nonlinear transmission resonance

Equation (9) for probability p allows deriving the exact solution describing the first nonlinear transmission resonance. This solution is obtained when searching of a particular solution of the form

$$p = 1 + a\,\mathrm{sech}^2 x. \tag{14}$$

To simplify the calculations, we rewrite Eq. (9) in the form

$$\frac{d}{dx}\left[-\frac{p''}{4} + \left(-2\mu p + Vp + \frac{3}{2} g p^2\right)\right] + Vp' = 0 \tag{15}$$

and recall that the second derivative of function $f = \mathrm{sech}^2 x$ is written as a quadratic polynomial in f:

$$f'' = 4f\left(1 - \frac{3}{2}f\right). \tag{16}$$

Eq. (15) is then written as

$$\frac{d}{dx}\left[-af\left(1-\frac{3}{2}f\right) + \left((-2\mu + V_0 f)(1+af) + \frac{3}{2}g(1+af)^2\right) + aV_0\frac{f^2}{2}\right] = 0. \tag{17}$$

Since the expression in the square brackets is a quadratic polynomial in f, it is understood that this equation is satisfied if the coefficients of the terms proportional to f and f^2 vanish. We thus compose two algebraic equations:

$$\begin{aligned} &-a + (-2\mu a + V_0 + 3g\,a) = 0, \\ &\frac{3}{2}a(1 + V_0 + g\,a) = 0, \end{aligned} \tag{18}$$

from which we readily obtain

$$a = \frac{-1}{1 + 2(\mu - g)} \tag{19}$$

and

$$V_0 = -1 + \frac{g}{1 + 2(\mu - g)}. \tag{20}$$

Since when passing to the linear limit by tending $g \to 0$ the potential depth defined by the last equation becomes equal to -1 it is clear that this solution describes the nonlinear transmission resonance V_{NL1} corresponding to the first linear resonance $V_{L1} = -1$.

Analyzing now the obtained formula for the first resonance, we note that the shift from the value of the linear resonance's potential depth is negative for attractive interaction ($g < 0$) and is positive for repulsive interaction ($g > 0$). Interestingly, we note that if the nonlinearity is repulsive and strong enough, the positive shift in the dept of the potential may prevail the first term in Eq. (20) and thus produce a positive value of $V_0 = V_{NL1}$. Hence, in contrast to the linear case, in the nonlinear case reflectionless transmission may occur not only for potential wells but also for potential barriers. This happens when $g > (1 + 2\mu)/3$. Note finally that it follows from Eq. (20) that for the whole range of considered variation of parameters, i.e., $g \in (-\infty, \mu)$, $\mu > 0$, the maximal range of variation of the potential height for the first transmission resonance is $V_{NL1} \in (-3/2, -1 + \mu)$. The dependence of the first resonance position on the nonlinearity parameter g for $\mu = 4$ is shown on Fig. 2.

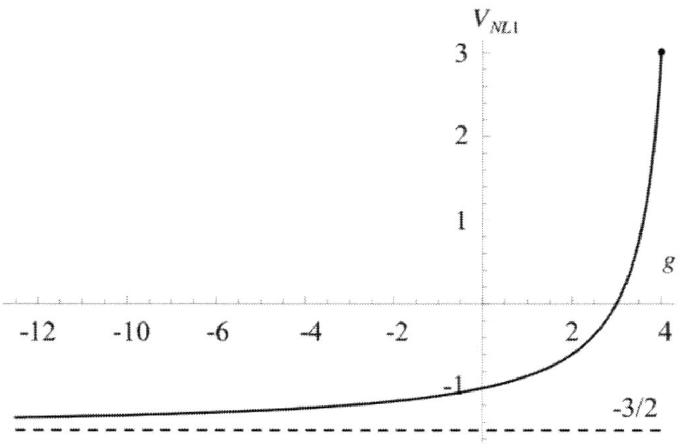

Fig. 2. The position of the first transmission resonance V_{NL1} vs. nonlinearity parameter g for $\mu = 4$. The height of the potential becomes positive starting from $g = (1+2\mu)/3 = 3$.

4. Higher order transmission resonances

To discuss the higher order resonances, we recall that we have previously shown that the position of these resonances in the first-order approximation of the stationary perturbation theory [1] is given as [14]

$$V_{NLn} = V_{Ln} + gF_n / C_n, \tag{21}$$

where

$$F_n = \int_0^1 z^{ik}(1-z)^{-ik} F(u_{Ln}) u_{Ln}\, dz, \quad F(u) = \frac{|u|^2 - 1}{4z(z-1)} u \tag{22}$$

and

$$C_n = \int_0^1 z^{ik}(1-z)^{-ik} u_{Lm} u_{Ln}\, dz, \tag{23}$$

where $u_{Ln}(z)$ is the solution of the linear problem for the n th-order linear resonance given by Eq. (5). It has been numerically proven that these formulas provide a highly accurate description of the nonlinear shift of the position of the reflectionless transmission resonances for all the resonance orders n and all the variation range of the chemical potential μ if the self-interaction parameter g is relatively small, more precisely, the formulas provide a relative error of the order of 10^{-3} if $g/\mu < 0.25$. Notably, the result for the first resonance turns out to be exact, i.e., for $n = 1$ Eqs. (21)-(23) produce the exact result (20).

Now, we use the limit solution (13) to estimate the integrals (22) and (23) for higher order resonances $n > 1$. To do this, we construct an approximate solution of Eq. (13) by means of successive iterations. We rewrite this equation in the following form

$$p_0 = \sqrt{\frac{\mu - g}{\mu - V - g p_0}} \tag{24}$$

and start from the simplest zero-order initial approximation $p_0 = 1$, i.e., we put $p_0 = 1$ in the right-hand-side of this equation to obtain the first approximation for the limit solution as

$$p_0(z) = \sqrt{\frac{\mu - g}{\mu - g - V(z)}}, \quad V(z) = -V_0 4z(z-1). \tag{25}$$

Numerical testing shows that this is a good enough approximation. Notably it corresponds to the limit solution of a linear problem for the effective chemical potential $\mu - g$. For this function, the integrals (22) and (23) are calculated analytically. The result for F_n is simple:

$$F_n = \frac{-V_{Ln}}{\mu - g - V_{Ln}}, \tag{26}$$

and for C_n we obtain

$$C_n = \frac{1}{2}\sqrt{\frac{\mu - g}{-V_{Ln}}} \ln\left(\frac{V_{Ln} + \sqrt{V_{Ln}(\mu - g)}}{-V_{Ln} + \sqrt{V_{Ln}(\mu - g)}} \right). \tag{27}$$

For fixed wave vector $k = \sqrt{2(\mu - g)}$, the asymptotes of F_n and C_n for large $n >> 1$ are $F_n \sim 1$ and $C_n \sim 1/\sqrt{-V_{Ln}} \sim 1/n$, hence, we conclude that

$$V_{NLn} - V_{Ln} \sim gn. \tag{28}$$

Thus, the nonlinear shift of the resonance position is approximately a linear function of the resonance order n. This conclusion is further supported by numerical simulations (Fig. 3). Recalling now the formula (20) for the first transmission resonance, we finally obtain

$$V_{NLn} \approx V_{Ln} + \frac{g\,n}{1 + 2(\mu - g)}. \tag{29}$$

This formula provides a quick estimate of the nonlinear transmission resonance position and hence may be especially useful for qualitative discussions.

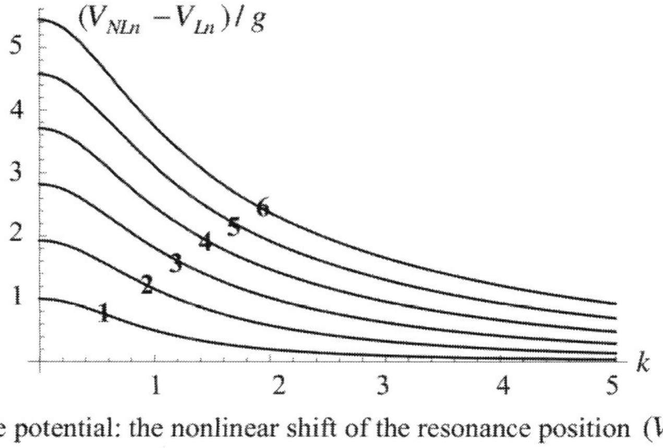

Fig. 3. Rosen-Morse potential: the nonlinear shift of the resonance position $(V_{NLn} - V_{Ln})/g$ as a function of the wave vector k . The resonance orders are indicated by bold-face numbers.

It is interesting to compare the obtained result with the one for the rectangular barrier. The latter problem has recently been addressed in several publications (see, e.g. [5]). The result for the transmission resonances presented in these papers is rather cumbersome. Here we present a much simpler discussion due to application of the approach developed in [14]. This approach rests on the reformulation of the reflectionless transmission problem as an eigenvalue problem for the reflecting potential's height and further application of the Rayleigh-Schrödinger time-independent perturbation theory [1]. To proceed in this way, we recall that in the associated *linear* problem for an *effective* chemical potential $\mu_{eff} = \mu - g$ the transmission resonances for the rectangular barrier

$$V = \begin{cases} 0, & x<0 \quad \text{and} \quad x>1 \\ V_0, & 0 \le x \le 1 \end{cases} \tag{30}$$

occur for [1]

$$V_{Ln} = (\mu - g) - \frac{\pi^2 n^2}{2}, \quad n = 1,2,3,... \tag{31}$$

(compare with Eq. [4]) and the corresponding wave functions are given as $\psi_{Ln} = e^{-ikx}$ if $x<0$, $\psi_{Ln} = (-1)^n e^{-ik} e^{-ikx}$ if $x>1$, and

$$\psi_{Ln} = \cos(\pi n x) - \frac{ik}{\pi n}\sin(\pi n x) \quad \text{if} \quad 0 \le x \le 1. \tag{32}$$

[One should put $g = 0$ in Eqs. (30)-(32) if namely the linear problem is considered. However, for the discussion of the nonlinear case that we consider here one should use the formulas with $g \neq 0$.] The functions ψ_{Ln} defined by Eq. (32) are orthogonal in the interval $x \in [0,1]$:

$$\int_0^1 \psi_{Lm}\psi_{Ln}\, dx = C_n \delta_{mn}, \quad C_n = \frac{1}{2} - \frac{k^2}{2\pi^2 n^2}. \tag{33}$$

Then, according to the procedure developed in [14] the nonlinear shift of the resonance position in the limits of the Rayleigh-Schrödinger perturbation theory is given as

$$V_{NLn} - V_{Ln} = \frac{g}{C_n}\int_0^1 (|\psi_{Ln}|^2 - 1)\psi_{Ln}\psi_{Ln}\, dx. \tag{34}$$

.The integral involved in this formula is readily calculated. The final result for the nonlinear resonance position reads

$$V_{NLn} = V_{Ln} + \frac{g}{4}\left(1 - \frac{3k^2}{\pi^2 n^2}\right). \tag{35}$$

This is a fairly good approximation. The numerical simulations show that the derived formula defines the position of the resonances with accuracy of the order of 10^{-4} (or less) for all the resonance orders and for all the variation range of the chemical potential μ and the nonlinearity parameter g.

The dependence (35) of the resonance position on the wave vector k of the incoming matter-wave is demonstrated in Fig. 4. It is clearly seen that the situation is radically changed compared with the case of the Rosen-Morse potential (Fig. 3). An immediate observation indicated by Eq. (35) is that for the rectangular barrier (30) the nonlinear shift of the resonance position is approximately constant for higher order resonances $n >> 1$:

$$V_{NLn} - V_{Ln}(g=0) \approx -\frac{3}{4}g. \tag{36}$$

Thus, we see that the Rosen-Morse potential suggests an essentially different behavior [Eq. (29)] not indicated by the rectangular barrier [see Eq. (35)]. It is supposed that this is because of the smooth variation of the Rosen-Morse potential's shape. An immediate conjecture following from this observation is that the case of *asymmetric* potentials involving two different scales for the variation of the potential in different space intervals is potent to suggest further effects. We hope to examine this possibility in a future investigation.

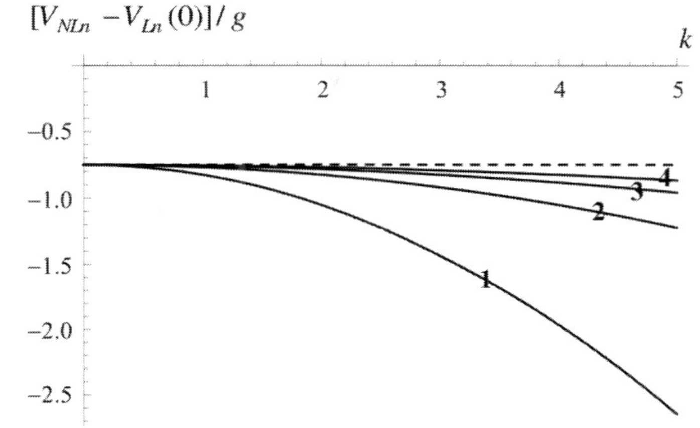

Fig. 4. Rectangular barrier: the nonlinear shift of the resonance position $(V_{NLn} - V_{Ln})/g$ as a function of the wave vector k. The bold-face numbers indicate the resonance orders.

5. Conclusions

Thus, we have discussed, within the mean field Gross-Pitaevskii approach, the reflectionless transmission resonances in the above-barrier reflection of Bose-Einstein condensates by the Rosen-Morse squared hyperbolic-secant potential. Applying an exact third order nonlinear differential equation obeyed by the condensate density, we have derived the exact solution of the problem for the first-order resonance. It follows from this solution that in the nonlinear case the total transmission is also possible for positive potential heights, i.e., for potential barriers, not only for the potential wells as it is the case in the linear case. Further, using a limit solution of the equation for the density (written as a root of a polynomial equation of the third degree) and the solution for the first resonance, we have constructed a simple approximation for the shift of the nonlinear resonance potential's depth from the corresponding linear resonance's position for higher order resonances. The constructed approximation shows that the nonlinear shift of the resonance position is an approximately linear function of the resonance order. This behavior essentially differs from the result for the rectangular barrier for which the nonlinear shift is approximately constant. By noting that this radical difference is probably caused by the smooth variation of the Rosen-Morse potential's shape in contrast to the sharp variation in the case of the rectangular barrier, it is conjectured that the case of *asymmetric* potentials involving two different scales for the variation of the potential in different space intervals is potent to suggest further effects. We intend to address this perspective in a forthcoming discussion.

Acknowledgments

This research has been conducted in the scope of the International Associated Laboratory IRMAS. The work was supported by the Armenian National Science and Education Fund (ANSEF Grants No. 2464 and No. 2591), Armenian State Committee of Science (SCS Grants No. 11A-1c061 and No. 11RB-026), and Russian Foundation for Basic Research (RFBR Grant No. 09-08-00722). H.A. Ishkhanyan acknowledges SPIE for a 2011 Scholarship in Optics and Photonics.

References

[1] Landau L and Lifshitz E 1977 *Quantum Mechanics (Nonrelativistic Theoty)* (New York: Pergamon Press)

[2] Anderson M H, Ensher J R, Matthews M R, Wieman C E, and Cornell E A 1995 *Science* **269** 198; Davis K B, Mewes M-O, Andrews M R, van Druten N J, Durfee D S, Kurn D M, and Ketterle W 1995 *Phys. Rev. Lett.* **75** 3969; Bradley C C, Sackett C A, Tollett J J, and Hulet R

Journal of Physics: Conference Series **350** (2012) 012012 doi:10.1088/1742-6596/350/1/012012

G 1995 *Phys. Rev. Lett.* **75** 1687

[3] Pethick C J and Smith H 2002 *Bose–Einstein Condensation in Dilute Gases* (Cambridge: Cambridge University Press)

[4] Cheng Y S, Gong R Z, and Li H 2007 *Chin. Phys. Lett.* **24** 35; Ishkhanyan H A and Krainov V P 2009 *Laser Physics* **19** 1729

[5] Rapedius K, Witthaut D, and Korsch H J 2006 *Phys. Rev.* A **73** 033608; Ishkhanyan H A and Krainov V P 2009 *Phys. Rev.* A **80** 045601

[6] Rapedius K and Korsch H J 2009 *J. Phys.* B **42** 044005; Witthaut D, Rapedius K, and Korsch H J 2009 *J. Nonlin. Math. Phys.* **16** 207; Seaman B T, Carr L D, and Holland M J 2005 *Phys. Rev.* A **71** 033609

[7] Lee C and Brand J 2006 *Eur. Phys. Lett.* **73** 321

[8] Song J, Hai W, Luo X 2009 *Physics Letters* A **373** 1560; Song J, Hai W, Zhong H, and Luo X 2008 *Commun. Theor. Phys.* **50** 89

[9] Ishkhanyan H A and Krainov V P 2009 *JETP* **109** 585

[10] Gross E P 1961 *Nuovo Cimento* **20** 454; Pitaevskii L P 1961 *JETP* **13**

[11] Abramowitz M, Stegun I A 1965 *Handbook of Mathematical Functions* (New York: Dover)

[12] Nayfeh A H 1981 *Introduction to Perturbation Techniques*, John Wiley (NewYork: John Wiley)

[13] Ishkhanyan A, Joulakian B, and Suominen K-A 2009 *J. Phys.* B **42** 221002; Ishkhanyan A M, Joulakian B, and Suominen K-A 2008 *Eur. Phys. J.* D **48** 397

[14] Ishkhanyan H A, Krainov V P, and Ishkhanyan A M 2010 *J. Phys.* B **43** 085306

International Symposium on Optics and its Applications (OPTICS2011) IOP Publishing
Journal of Physics: Conference Series **350** (2012) 012013 doi:10.1088/1742-6596/350/1/012013

The interference method for measuring small shift of the wavelength of laser radiation

V B Pakhalov[1] and K A Margaryan[1]

[1]Department of Physics, Yerevan State University, 1 Alex Manoogian, 0025 Yerevan, Armenia

E-mail: margaryankarine@gmail.com

Abstract. We experimentally investigate a shift of the interference pattern in dependence from wavelength of coherent light beam propagating through the anisotropic crystal. This method allows us to measure the wavelength of coherent radiation with accurate to 10^{-2} nm.

1. Introduction

The simplest devices for light spectral investigation are prisms and diffraction gratings. But low optical resolution limits their practical applications in problems where more accurate measurements are needed. Higher optical resolution could be achieved with use of Fabry-Perot interferometer, which for instance allows observing of hyperfine structure of green line of mercury lamp [1]. The theoretical and experimental observations of the spatial mode coupling of mode-locked semiconductor lasers of Fabry-Perot were compared [2].

The most popular application of Fabry-Perot interferometer have found as laser resonator. It is well known that in interference experiments the position of interference fringes depends on optical path difference of interfering beams. Fabry-Perot interferometer enables observation of longitudinal modes of He-Ne laser. Measuring the shift of interference fringes it is possible to measure refractive index, object size, mechanical movement.

The semiconductor lasers have opened new prospects in the field of communications and now are getting more powerful and efficient and have found many applications in contemporary science and industry [3]. They are compact, easy integrated and their characteristics are continuously improved.

The output of a semiconductor laser spectrum depends strongly on the temperature and injection current of latter. Mode hopping occurs for specific values of temperature and injection current. This means that mode hopping is eliminated by careful control of these parameters. The stability map is a reliable method of determining the values of parameters resulting in stability [4].

Conoscopic patterns which are the result of interference of ordinary and extraordinary waves, when the plane of oscillations coincides with the external polarizer, are also known for studying anisotropic crystals.

Such method was also used to measure the spatial coherence through the polarization interferometer [5]. However, the above mentioned two methods require fixed wavelength (or frequency) of radiation. In the second case, immersion liquid and the second crystal are also required.

In this paper we propose interferometer based on anisotropic crystal which allows to measure radiation wavelength shift with accuracy ~ 1 Å by measuring interference fringe shift. Such high resolution is obtained by dispersion difference between interfering two beams with orthogonal polarizations. Particularly in calcite crystal the change of 40 nm wavelength leads to 0,002 change of index of refraction. With 1 nm change of wavelength in crystal with 2 cm length the optical path difference changes as much as λ/2 which causes half stripe shifts of interference fringes. The change of shift with 1/10 part we can measure 1 Å wavelength shift. The same result could be obtained with the use of spectrophotometer with diffraction gratings which has 1000 grids on 1 mm. Experimental and theoretical study of this method was also carried out.

2. Experiment

Characteristics of 2 types of lasers are investigated in the following (our) work:

1. HLDPM-650-3 injection red laser (wavelength is 0.65μm, power is 3mWatt)
2. Neodymium laser with the semiconductor pumping (wavelength is 0.9μm) (second harmonic 0.53 μm)

The spectral measurements of these lasers were made by means of OSA spectral analyzer (high and low regimes from generation threshold, figure 1).

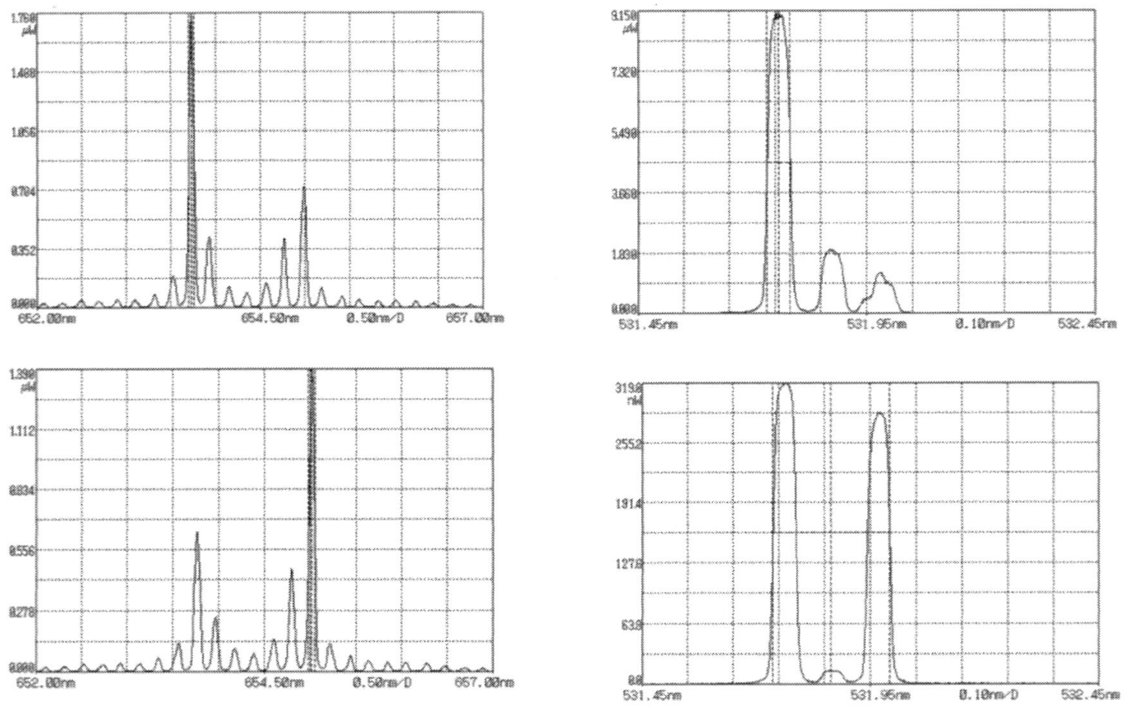

a. b.

Figure 1.The spectrum of radiation of a) "red" laser (2.5; 4.5 mA) and b) "green" laser (140 mA; 230 mA) when increasing the pumping current [6].

After passing through the light emitting diode, the beam gets the analyzer. In figure 1, the vertical axis is the output power. The accuracy of measurements of wavelength is about 0.01 - 0.2nm.

The above mentioned first type lasers almost are constantly and have multimode structure in the low regime below threshold, which width is 10 - 12nm and there are a few longitudinal and transversal modes in the high regime above threshold. Parallel to the current increase strengthening and mode hopping take place from one mode to the other (figure 1).

Separate mode width is 0.05 nm. The distance between modes (nearest modes or in general) is 0.1-1 nm (transversal and longitudinal modes). This distance has the following form

$$\Delta\lambda = \frac{\lambda^2}{2Ln},$$

where L is the thickness of the crystal, n is refractive index. In this experiment with L = 1 mm and n = 3.5, the spectrum is compressed above of threshold 200 times.

The results of detailed measurments, when we used uniaxial crystal placed between crossed polarizers, are presented in the following part of this study (Figure 2).

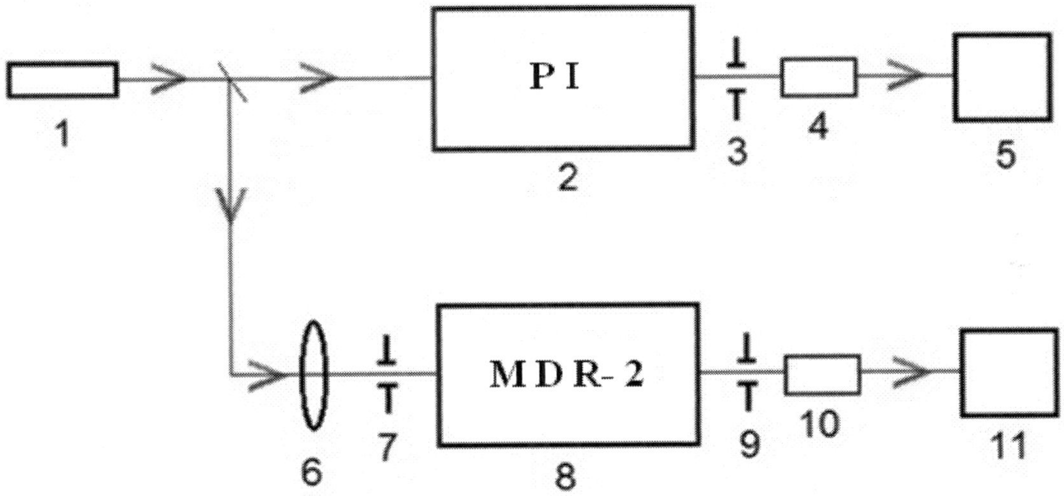

Figure 2.The experimental scheme for measuring the shift of spectral and interference lines
1. semiconductor laser, 2. polarization interferometer, 3,7, 9. diaphragm,
4, 10. CCD cameras, 5, 11. screen, 6. lens, 8. MDR-2 diffraction spectrometer.

The shift of the wavelength of the semiconductor laser occurred when the pumping current or the temperature were changed. We used Iceland spar crystal cut parallel to the optical axis. The polarization of radiation forms 45° with this axis. A big difference of path is achieved while changing the difference between refractive indices of ordinary and extraordinary rays. It results in the shift of the interference lines (figure 3).

Subangstrom measurement accuracy is achieved in the case of 2 cm crystal. The accuracy of these measurements is verified by means of a diffraction spectrometer of MDR-2 series with 1200 lines/mm grating (approximately with the same resolution as for the case of polarization interferometer) and a CCD camera.

The shifts of modes are depicted in figure 4.

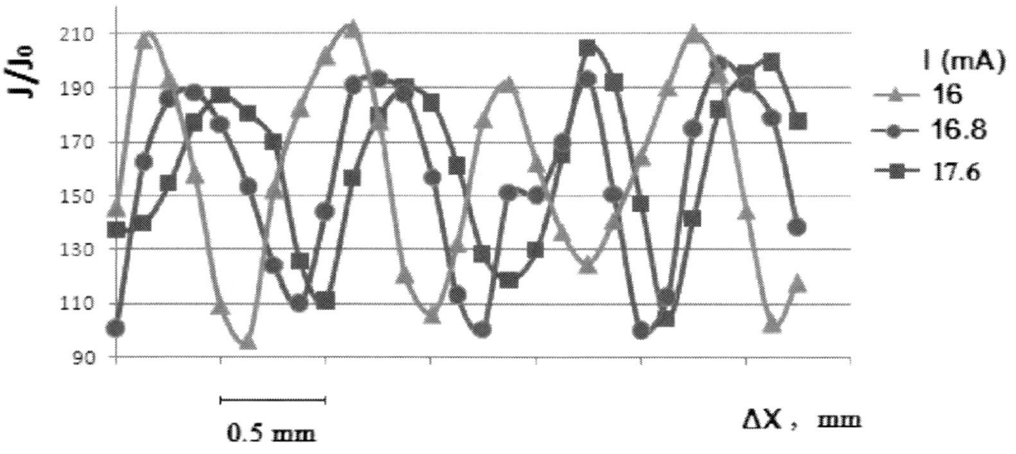

Figure 3.The dependence of shift of the pattern of interference lines on the current of laser.

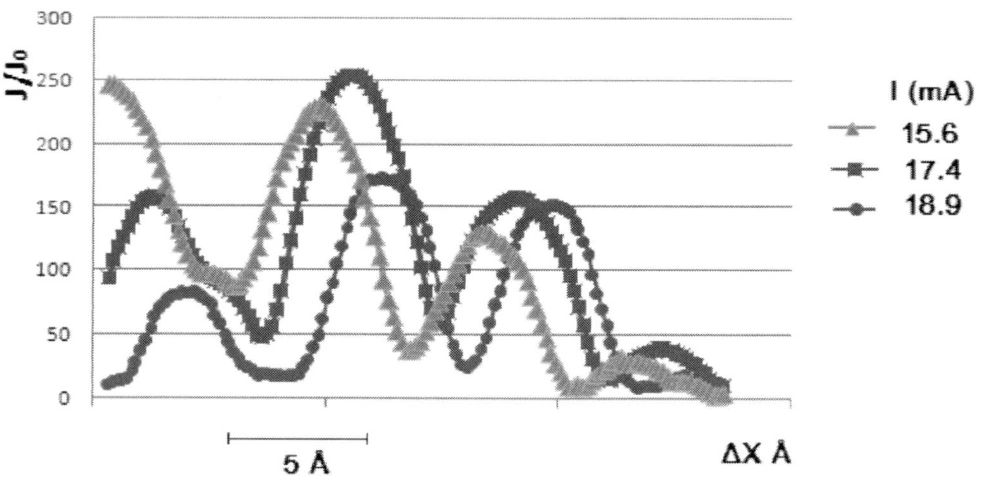

Figure 4. The dependence of shift of the pattern of spectral lines on the current of laser.

Output signal was detected by a digital camera and was analyzed using Lab View software.

Knowing the dependence of the difference $\Delta n(\lambda)$ of refraction indices of ordinary and extraordinary rays, it is possible to define the difference in processes Δx between two rays. The Δx is equal to $\Delta n(\lambda)\cdot d$ and depends on direction of propagation of radiation. The distance of nearest maximum lines of the pattern corresponds to the wavelength ($\Delta x = \lambda$), figure 5.

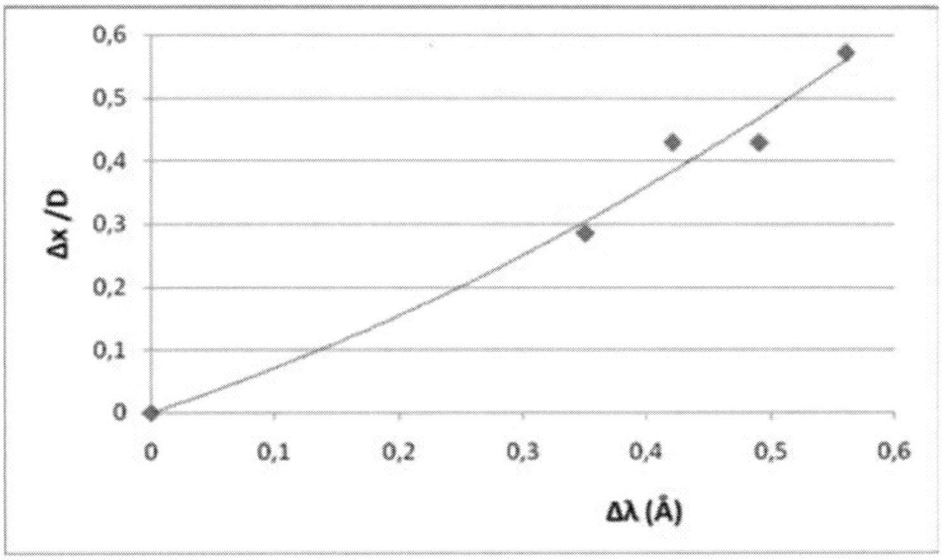

a.

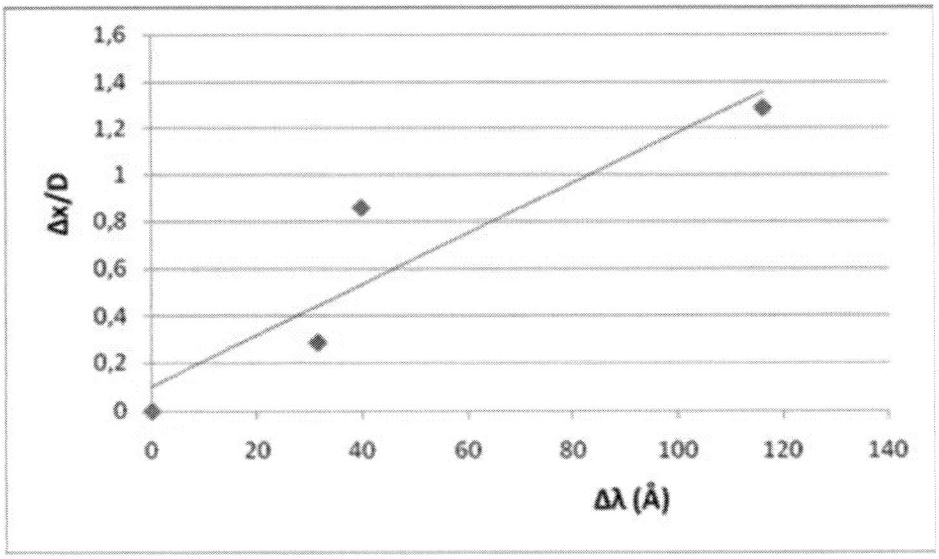

b.

Figure 5. The shift of interference lines depending on the changes of the wavelength of a) "red" and b) "green" lasers [7].

Temperature dependence of shift of semiconductor laser is shown in figure 6.

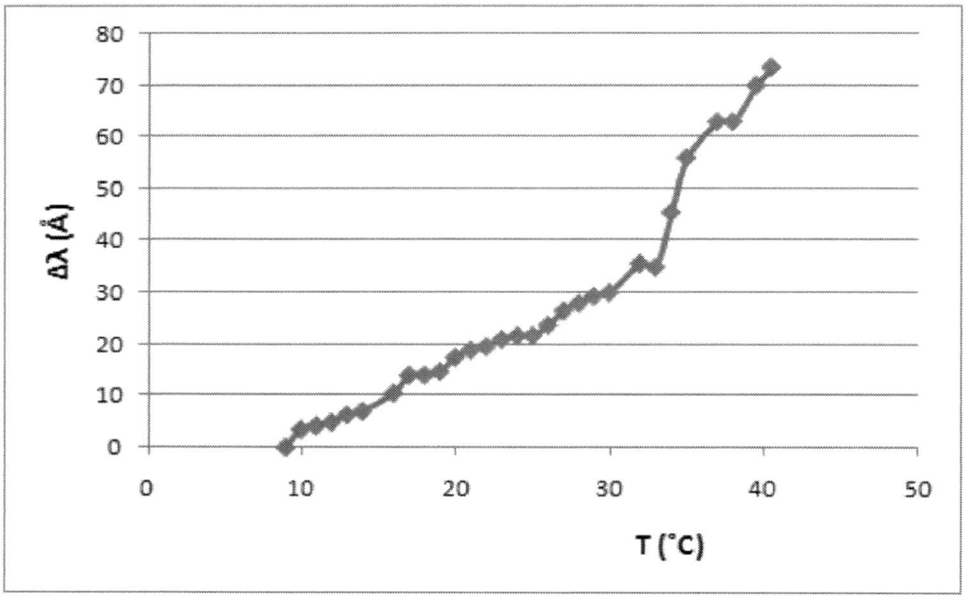

Figure 6. The dependence of displacement of spectral lines on the temperature of "red" laser.

Particularly for 3 cm KDP crystal we get 1nm change of wavelenght. If the minimum displacment of lines is 1/10, the change of wavelenghtis equal to 0.1nm. This accuracy may be achieved by means of a diffraction spectrometer with the 1000 line/mm grating.

3. Conclusion

The measurements involving spectrometer with diffraction grating and optical spectral analyzer (OSA) were measured for establishing the proposed method depending on pumping current and temperature. The accuracy of these measurements is 0.01 nm which corresponds to 1/10 part of the line displacement length. Thus, this method is effective for measuring even a very small displacement of the wavelength of a coherent or partly coherent laser radiation.

4. References

[1] Putilov K A Fabrikant V A 1963 Kurs fiziki *Fizmatgiz* T **3** 97-98.

[2] Yang W 2011 Single-Section Fabry-Perot Mode-Locked Semiconductor Lasers *Review Article* V **2011** 1-11.

[3] Murphy E 2010 The semiconductor laser: Enabling optical communication *Nature Photonics* V **4** 287.

[4] Heumier T A and Carlsten J L 2011 Mode hopping in semiconductor lasers *Photonic Test and Measurement* MT **59717.**

[5] Arutunyan A G, Tunkin V G and Chirkin A S 1973 *Quantum Electronics* V **3** 63.

[6] Pakhalov V B 2010 *Technical Physics Letters* V **36** 354-357.

[7] Pakhalov V B, Margaryan K A 2010 *Proceedings of the SPIE* V **7998** 799811-799811-4.

DIPOLE AND QUADRUPOLE MOMENTS OF ELECTRON IN SPHERICAL NANOLAYER

N.G. Aghekyan[1], E.M. Kazaryan[1], H.A. Sarkisyan[1,2]
1 Russian-Armenian (Slavonic) University, 123 H. Emin, Yerevan, Armenia
2 Yerevan State University, 0025 Al. Manukyan, Yerevan, Armenia
E-mail: shayk@ysu.am

Abstract. In this paper for one electron states dipole and quadrupole moments are studied in spherical impenetrable quantum nanolayer. It is shown that dipole moment is zero for the considered system. An analytical formula for quadrupole moment is derived. Quadrupole moment dependence upon one radius of the nanolayer, when the other one is fixed, is obtained.

1. Introduction

The development of precise methods of the growth of semiconductor nanostructures makes possible the realization of the zero-dimensional systems of different geometrical forms and sizes [1]. As we know quantum dots (QD) are those structures where quantum effects visualize themselves more vividly. Indeed, due to full quantization of the spectrum of charge carriers, QDs have similar properties to real atoms, hence, usually, QDs are called "artificial atoms" [2]. They are considered as very perspective systems, that can be used as a base elements for contemporary nanoelectronics: started from lasers based on QDs and finished with solar cells of new generation [3]-[7]. The physical processes in QDs are being investigated intensively by the specialists, as besides of pure academic interest, the results of investigation can be applicable [8]. Currently spherical, cylindrical, ellipsoidal, pyramidal, lens-shaped and many other QDs are experimentally realized and thoroughly investigated (e. g. [9] - [15]) All of these QDs listed above have specific geometries that imposes their electronic, optical, kinetic and other characteristics. The latest circumstance gives a vast choice of properties of QDs for solving a concrete practical problem.

The implementation of zero-dimensional systems strongly actualized entire classes of quantum mechanical problems, which initially were carrying only academic interest and were considered as models only. In this connection it is enough to mention the problem about size-quantized film with impenetrable walls, when the system is being approximated by quantum well with rectangular infinitely high walls [16]. QD form a family of "artificial atoms" with different symmetries due to their diversity of geometries. The simplest model can be considered the "artificial hydrogen atom" consisting of QD containing one electron. Thereby we can talk about spherical, cylindrical, ellipsoidal, pyramidal "hydrogen atoms". Analogically, when the system consists of QD containing two electrons, then it is a "artificial helium atom" [17].

In order to describe two- and multi-electron states in QD it is possible to use quantum mechanical methods of description of multi-electron atoms such us: perturbation theory for helium atom and Hartree-Fock method (see for example [17] - [22]). Particularly, as it was shown in Ref. [23] for spherical QD it is possible to construct the analog of the Hund rule for occupation of electron shells. The authors have discussed the case when there are forty electrons in the QD. Also for that system the quadrupole moment have been computed. On the other hand, along with multi-electron systems a QD containing one electron can have dipole and quadrupole moments too. It is explained by the fact that the angular distribution of charge density in the QD is spherical symmetric only in limited cases. Particularly, it is well known that even in spherical symmetric QD only S-state of the electron is spherical symmetric. Thereby, spherical symmetric QD which contains one electron can have non-zero dipole and quadrupole moments. In this connection it is interesting, along with multi-electron systems, calculate these moments for one electron in a spherical symmetric QD. It is important to note that during last years the interest regarding to physical properties of layered QDs risen strongly, particularly for spherical and cylindrical nanolayers (see for example [24] - [27]). The localization of electron in spherical nanolayer can be controlled by changing both the inner radius and the outer radius. In other words in layered structures one gains additional levers to control the wave functions and energy spectrum of charge carriers. For this reason it becomes interesting to study dipole and quadrupole moments of electron in spherical nanolayer, as well as to derive the dependence of mentioned quantity on the values of inner and outer radii.

In current work the average values of dipole moment and tensor of quadrupole moment is obtained in layered spherical QD with impenetrable walls. The spin of electron is taken into account as well and hence the problem is discussed in z-representation of spin operators.

2. Theory

As far as we are intended to calculate quantum mechanical average values of dipole and quadrupole moments, we, definitely, need to discuss one electron states in spherical quantum nanolayer with the following confinement potential:

$$V_{conf}(r) = \begin{cases} 0, \ R_1 \le r \le R_2 \\ \infty, \ [r < R_1, \ r > R_2]. \end{cases} \tag{1}$$

With the consideration of electron spin we will find the solution of Schrödinger equation with the confinement potential (1) in z-representation by the matrices of Pauli, requiring the wave function to be the eigenfunction of the square of total angular momentum operator $\hat{J}^2 = \left(\hat{\vec{L}} + \hat{\vec{s}}\right)^2$ and of its z-projection $\hat{J}_z = \hat{L}_z + \hat{s}_z \left(\hat{s}_z = \frac{\hbar}{2}\sigma_z\right)$. In the above mentioned representation for $\hat{J}_z$ and $\hat{J}^2$ we have [28]

$$\hat{J}_z = \hbar \begin{pmatrix} -i\frac{\partial}{\partial\varphi} + \frac{1}{2} & 0 \\ 0 & -i\frac{\partial}{\partial\varphi} - \frac{1}{2} \end{pmatrix}, \tag{2}$$

$$\hat{J}^2 = \begin{pmatrix} \hat{L}^2 + \frac{3}{4}\hbar^2 + \hbar\hat{L}_z & \hbar\hat{L}_+ \\ \hbar\hat{L}_- & \hat{L}^2 + \frac{3}{4}\hbar^2 - \hbar\hat{L}_z \end{pmatrix}, \tag{3}$$

where $\hat{L}_\pm = \hat{L}_x \pm i\hat{L}_y$. Considering the geometry of QD, it is more convenient to solve the problem in spherical coordinates. As far as the considered centrosymmetrical field does not depend on the spin, then the radial and angular parts can be separated and presented in the following form:

$$\Psi_n^{j=\ell+1/2}(r, \theta, \varphi) = \frac{F_{n,\ell}(r)}{\sqrt{2\ell+1}} \begin{pmatrix} \sqrt{\ell + m_j + 1/2} \cdot Y_{\ell, m_j - 1/2}(\theta, \varphi) \\ -\sqrt{\ell - m_j + 1/2} \cdot Y_{\ell, m_j + 1/2}(\theta, \varphi) \end{pmatrix}, \tag{4}$$

and

$$\Psi_n^{j=\ell-1/2}(r,\theta,\varphi)=\frac{F_{n,\ell}(r)}{\sqrt{2\ell+1}}\begin{pmatrix}\sqrt{\ell-m_j+1/2}\cdot Y_{\ell,m_j-1/2}(\theta,\varphi)\\ \sqrt{\ell+m_j+1/2}\cdot Y_{\ell,m_j+1/2}(\theta,\varphi)\end{pmatrix}, \tag{5}$$

where $Y_{\ell,m_j\pm1/2}$ are spherical harmonics.
By presenting the radial function in the following form:

$$F_{n,\ell}(r)=\frac{1}{r}\chi_{n,\ell}(r), \tag{6}$$

for the function $\chi_{n,\ell}(r)$ we come to the following equation:

$$-\frac{\hbar^2}{2\mu}\left(\frac{d^2\chi_{n,\ell}}{dr^2}-\frac{\ell(\ell+1)}{r^2}\chi_{n,\ell}\right)+V_{conf}(r)\chi_{n,\ell}=E_{n,\ell}\chi_{n,\ell}. \tag{7}$$

The presence of impenetrable walls imposes boundary conditions as follows:

$$\chi_{n,\ell}(R_1)=\chi_{n,\ell}(R_2)=0. \tag{8}$$

After some transformations, for the radial functions we obtain:

$$F_{n,\ell}(r)=\sqrt{\frac{\pi k}{2r}}\left(C_1J_{\ell+1/2}(kr)+C_2J_{-(\ell+1/2)}(kr)\right), \tag{9}$$

where $J_{\ell\pm1/2}(x)$ is the half-integer order Bessel function of the first kind, $k=\sqrt{\frac{2\mu E_{n,\ell}}{\hbar^2}}$. Energy spectrum of the electron, in its turn, will be defined by the circumstance of equality to zero of the following determinant:

$$\det \mathrm{J}=\begin{vmatrix}J_{\ell+1/2}(kR_1) & J_{-(\ell+1/2)}(kR_1)\\ J_{\ell+1/2}(kR_2) & J_{-(\ell+1/2)}(kR_2)\end{vmatrix}=0. \tag{10}$$

Thus, for obtaining the final form of the wave functions corresponding to quantum numbers $j=\ell+1/2$ and $j=\ell-1/2$, it is necessary to substitute (9) into (4) and (5).

Now, when we have obtained the wave functions of the system, let us turn to the dipole moment of electron in the layered QD. Dipole moment of electron equals to:

$$P_i=ex_i, \tag{11}$$

where e is electron charge and x_i is the corresponding coordinate in Cartesian coordinate system. Let us calculate quantum mechanical average values of P_i:

$$<P_i>=\int\left|\Psi_n^{\ell\pm1/2}(r,\theta,\varphi)\right|^2P_idV. \tag{12}$$

The square of electron wave function module has the following form:

$$\left|\Psi_n^{j=\ell+1/2}(r,\theta,\varphi)\right|^2=\frac{|F_{n,\ell}(r)|^2}{2\ell+1}f^{j=\ell+1/2}(\theta), \tag{13}$$

where

$$f^{j=\ell+1/2}(\theta)=\left\{(\ell+1/2+m_j)\left|Y_{\ell,m_j-1/2}\right|^2+(\ell+1/2-m_j)\left|Y_{\ell,m_j+1/2}\right|^2\right\} \tag{14}$$

and

$$\left|\Psi_n^{j=\ell-1/2}(r,\theta,\varphi)\right|^2 = \frac{|F_{n,\ell}(r)|^2}{2\ell+1} f^{j=\ell-1/2}(\theta), \tag{15}$$

where

$$f^{j=\ell-1/2}(\theta) = \left\{(\ell+1/2-m_j)\left|Y_{\ell,m_j-1/2}\right|^2 + (\ell+1/2+m_j)\left|Y_{\ell,m_j+1/2}\right|^2\right\}. \tag{16}$$

In notations (14) and (16) we have have considered that the angular part of square modulus of the wave function depends only on θ variable.

It is easy to show that $< P_x >$ and $< P_y >$ equal to zero. Indeed, by substituting wave functions (4) and (5) into (12), and using well known expressions in spherical coordinate system for x, y and z we can assure that integrals by azimuthal angle φ equal to zero. $< P_z >$ is zero too. In order to proof that we should consider the following expressions:

$$Y_{\ell,m}\cos\theta = b_{\ell,m}Y_{\ell+1,m} + b_{\ell-1,m}Y_{\ell-1,m}, \quad b_{l,m} = \sqrt{\frac{(\ell+m+1)(\ell-m+1)}{(2\ell+1)(2\ell+3)}} \tag{17}$$

$$\oint Y^*_{\ell,m}Y_{\ell',m'}d\Omega = \delta_{\ell,\ell'}\delta_{m,m'}. \tag{18}$$

Hence we conclude that dipole moment is zero. Thereby, it is important to discuss quadrupole moment of electron as we have shown that dipole moment is zero. The tensor of quadrupole moment is defined by the following expression[29]:

$$Q_{ik} = 3x_ix_k - r^2\delta_{ik}. \tag{19}$$

It is a symmetric tensor of second rank $Q_{ik} = Q_{ki}$ and the trace equals to zero:

$$Sp\{Q_{ik}\} = 0. \tag{20}$$

Let us now turn to the calculation of quadrupole moment of electron in the spherical coordinate system. Based on the definition of the quantum mechanical average for $< Q_{ik} >$, in general, the following can be written:

$$< Q_{ik} >= \int \left|\Psi_n^{j\pm\ell/2}(r,\theta,\varphi)\right|^2 Q_{ik}dV. \tag{21}$$

Therefore, in order to calculate $< Q_{ik} >$ we come to the following integral:

$$< Q_{ik} >= \int_{R_1}^{R_2} \left\{\frac{|F_{n,\ell}(r)|^2}{2\ell+1} r^2 \oint Q_{ik} f^{j=\ell\pm1/2}(\theta)d\Omega\right\} dr, \tag{22}$$

where $d\Omega$ is the element of solid angle.

Based on the definition of tensor Q_{ik} (19) with direct calculations can be shown, that the average values of nondiagonal components are equal to zero:

$$< Q_{xy} >=< Q_{xz} >=< Q_{yz} >= 0. \tag{23}$$

For diagonal components in its turn can be shown that the following equations takes place: [28]

$$< Q_{xx} >=< Q_{yy} >= -\frac{1}{2} < Q_{zz} > . \tag{24}$$

Therefore, the problem reduces to calculating the integral:

$$< Q_{zz} >= \int_{R_1}^{R_2} \frac{|F_{n,\ell}(r)|^2}{2\ell+1} r^4 dr \oint \left(3\cos^2\theta - 1\right) f^{j=\ell\pm1/2}(\theta) d\Omega. \tag{25}$$

The integration by angles can be achieved using the relation

$$\oint \left(3\cos^2\theta - 1\right) |Y_{\ell,m}(\theta,\varphi)|^2 (\theta) d\Omega = \frac{2\ell(\ell+1) - 6m^2}{(2\ell-1)(2\ell+3)}. \tag{26}$$

After the calculations, the details of which can be found in [28], the expression (25) will be written in the following form:

$$< Q_{zz} >= \frac{1}{2}\left(1 - \frac{3m_j^2}{j(j+1)}\right) \mathrm{I}_{n,\ell}(R_1, R_2), \tag{27}$$

where the following notation is introduced

$$\mathrm{I}_{n,\ell}(R_1, R_2) = \int_{R_1}^{R_2} |F_{n,\ell}(r)|^2 r^4 dr. \tag{28}$$

Let us note, that the formula (27) is correct both for the value of quantum number $j = \ell + 1/2$ and for $j = \ell - 1/2$.

Finally, for our case we can write the electrostatic field of the system is as follows:

$$\varphi(r) = \frac{q}{\varepsilon r} + \frac{1}{2\varepsilon r^5}\left\{\sum_{ik} < Q_{ik} > x_i x_k\right\}, \tag{29}$$

where $i, k = x, y, z$ and ε is the average dielectric constant of surrounding medium. Here we have considered that dipole moment of this system is zero. As far as $< Q_{ik} >$ is equal to

$$< Q_{ik} >= \frac{1}{2}\left(1 - \frac{3m_j^2}{j(j+1)}\right) \mathrm{I}_{n,\ell}(R_1, R_2) \begin{pmatrix} -\frac{1}{2} & 0 & 0 \\ 0 & -\frac{1}{2} & 0 \\ 0 & 0 & 1 \end{pmatrix}, \tag{30}$$

therefore, for electrostatic field can be written:

$$\varphi(r) = \frac{q}{\varepsilon r} + \frac{< Q_{zz} >}{4\varepsilon r^3}\left(3\cos^2\vartheta - 1\right) = \frac{q}{\varepsilon r} + \frac{1}{8\varepsilon r^3}\left(1 - \frac{3m_j^2}{j(j+1)}\right) \mathrm{I}_{n,\ell}(R_1, R_2)\left(3\cos^2\vartheta - 1\right), \tag{31}$$

where ϑ is the angle between $\vec{r}$ radius vector of observation point and OZ axis.

3. Discussion

Quadrupole moment of electron in spherical symmetric field, regardless of the form of its potential, possesses an important characterictics connected with the presence of the spin. By direct calculation it can be shown, that the relation $< Q_{zz} > / \mathrm{I}_{n,\ell}(R_1, R_2)$ equals to zero at a value $j = 1/2$ both for the states $S_{1/2}$ and $P_{1/2}$. Thus, spherical symmetry can possess not only the S state but also the P state. If we consider a similar task without spin, then we can see that a spherical symmetry occurs only for the state S. On figures 1, 2, 3 and 4 it is represented the dependence of $\mathrm{I}_{n,\ell}(R_1, R_2)$ quantity on one radius of the nanolayer when the other one is fixed for different orbital and radial quantum numbers n and ℓ. From these figures, obtained by numerical calculations, we can conclude that:

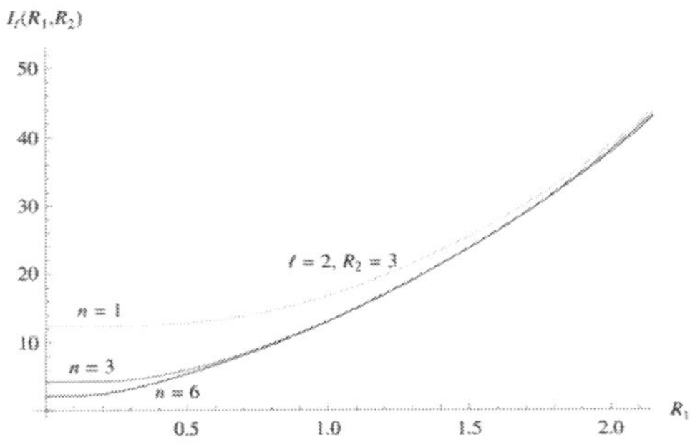

Figure 1. Dependence of $\mathrm{I}_{n,\ell}(R_1, R_2)$ (in units of Bohr radius square) on R_1 inner radius of nanolayer when $R_2 = 3$ (in units of Bohr radius) for different radial quantum numbers n when orbital quantum number ℓ is fixed ($\ell = 2$).

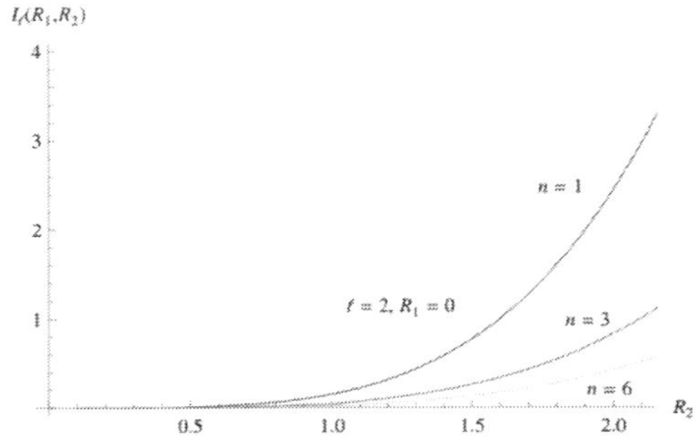

Figure 2. Dependence of $\mathrm{I}_{n,\ell}(R_1, R_2)$ (in units of Bohr radius square) on R_2 outer radius (in units of Bohr radius) of nanolayer when $R_1 = 0$ for different radial quantum numbers n when orbital quantum number ℓ is fixed ($\ell = 2$).

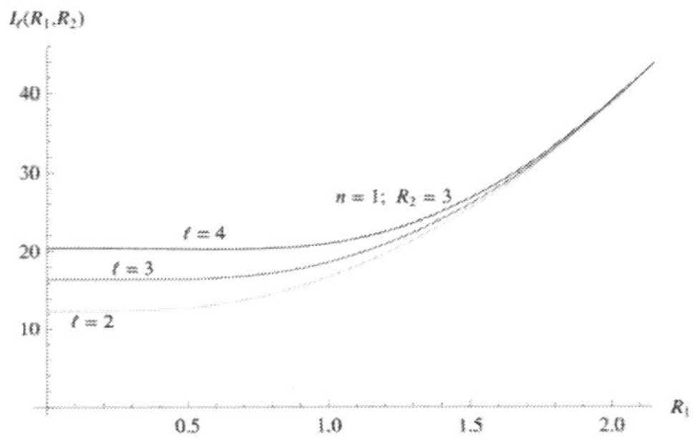

Figure 3. Dependence of $\mathrm{I}_{n,\ell}(R_1, R_2)$ (in units of Bohr radius square) on R_1 inner radius (in units of Bohr radius) of nanolayer when $R_2 = 3$ (in units of Bohr radius) for different orbital quantum numbers ℓ when radial quantum number n is fixed ($n = 1$).

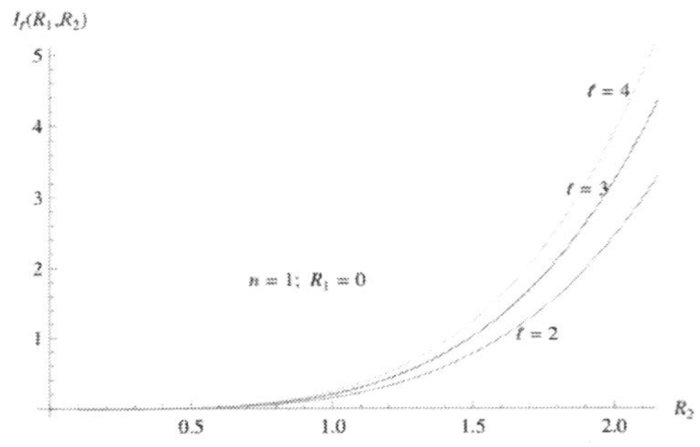

Figure 4. Dependence of $\mathrm{I}_{n,\ell}(R_1, R_2)$ (in units of Bohr radius square) on R_2 outer radius (in units of Bohr radius) of nanolayer when $R_1 = 0$ for different orbital quantum numbers ℓ when orbital quantum number n is fixed ($n = 1$).

1) Quadrupole moment increases for both cases when inner radius increases and outer is fixed and when outer radius increases and inner one is fixed. Moreover, from corresponding curves we can see that quadrupole moment increases more rapidly for the case when inner radius increases than for the case when outer radius increases. Hence, we can proof that spherical symmetry violation takes place for both cases but for the case when inner radius increases the spherical symmetry violation takes place more rapidly. Also we should note that for the case of spherical quantum dot (see figure 2 or 4) quadrupole moment tends to zero when outer radius R_2 tends to zero. This is because spherical quantum dot becomes sufficient small so that we could neglected the non-spherical distribution of electron in the space. In other words, by decreasing the radius of spherical qunatum dot we restore spherical symmetry, hence quadrupole moment becomes closer to zero.

2) When radial quantum number increases (n) and all other parameters are constant (ℓ, R_1, R_2) quadrupole moment decreases. Furthermore, when we increase n it goes to saturation.

We should also note that for the case of spherical QD (see figure 2) when R_2 is sufficiently small then the dependence of quadrupole moment from n weakens. Particularly, we can say there is no dependence on n for sufficiently small outer radii. Whereas, for the case of spherical nanolayer the dependence is weaker as bigger the inner radius R_1 is. Particularly, we can say there is no dependence on n for sufficiently big inner radii (see figure 1).

3) When orbital quantum number increases (ℓ) and all other parameters (n, R_1, R_2) are constant quadrupole moment increases in contrast with the radial quantum number case. Here also quadrupole moment value has a saturation when we increase ℓ (see figure 3 and 4). We should also note that for the case of spherical QD (see figure 4) when R_2 is sufficiently small then the dependence of quadrupole moment from ℓ weakens as it was for the previous case. Particularly, we can say there is no dependence on ℓ for sufficiently small outer radii. Whereas, for the case of spherical nanolayer the dependence is weaker as bigger the inner radius R_1 is. Particularly, we can say there is no dependence on ℓ for sufficiently big inner radii (see figure 3).

As we already noted, in case when the quantum number of orbital moment is fixed ($\ell = 2$) and we discuss different values for radial quantum number n ($n = 1; 3; 6$) with increase of R_1 the curves of $\mathrm{I}_{n,\ell}(R_1, R_2) =< r^2 >$ come closer to each other. This is an expected result as far as $< r^2 >$ for very small thickness of the nanolayer for $n = 1; 3; 6$ approximately will have same value. Therefore, quadrupole moment will be defined by angular distribution which, in its turn, is defined by the following coefficient: $\left(1 - \frac{3m_j^2}{j(j+1)}\right)$ which is constant for fixed angular and spin quantum numbers. We should also note that the same situation takes place for the case when n is fixed and $\ell = 2; 3; 4$:

$$< r^2 >_{n,2} =< r^2 >_{n,3} =< r^2 >_{n,4} . \tag{32}$$

This interpretation also explains the saturation effects discussed above with increase of n when ℓ, R_1, R_2 are fixed and with the increase of ℓ when n, R_1, R_2 are fixed. These facts are being obvious from the curves of square of modulus of radial function for different n and ℓ from R_1 to R_2. Moreover, those curves demonstrate that with the increase of n saturation tends to a small values of $\mathrm{I}_{n,\ell}(R_1, R_2)$ while with the increase of ℓ the values of $\mathrm{I}_{n,\ell}(R_1, R_2)$ increase and saturation takes place for bigger values.

4. Conclusion

In this paper for one electron states dipole and quadrupole moments are studied in spherical impenetrable quantum nanolayer. It is shown that dipole moment is zero for the considered system. An analytical formula for quadrupole moment is derived. Quadrupole moment dependence on one radius of the nanolayer, when the other one is fixed, is obtained for different radial (n) and orbital (ℓ) quantum numbers. The dependence of quadrupole moment on orbital and radial quantum numbers when other parameters are fixed is discussed in details.

References

[1] Bimberg D, Grundmann M and Ledentsov 1999 Quantum Dot Heterostructures, Wiley, New York
[2] Maksym P A and Chakraborty T 1990 *Physical Review Letters* **65** 108
[3] Asryan L V and Suris R A 1996 *Semiconductor Science and Technology* **11** 554
[4] Asryan L V and Luryi S 2001 *IEEE Journal of Quantum Electronics* **37** 905
[5] Nozik A J 2002 *Physica E* **14** 115
[6] Aroutiounian V, Petrosyan S, Khachatryan A and Touryan K 2001 *Journal of Applied Physics* **89** 2268
[7] Cao Q, Yoon S F, Liu C Y and Ngo C Y 2007 *Nanoscale Research Letters* **2** 303
[8] Alferov Zh I 1998 *Semiconductors* **32** 1
[9] Chuu D S, Hsiao C M and Mei W N 1992 *Physical Review B (Condensed Matter)* **46** 3898
[10] Zhu Jia-Lin 1989 *Physical Review B (Condensed Matter)* **39** 8780
[11] Bayer M, Schilling O, Forchel A, Reinecke T L, Knipp P A, Pagnod-Rossiaux Ph and Goldstein L 1996 *Phys. Rev.* B **53** 15810

[12] Atayan A K, Kazaryan E M, Meliksetyan A V and Sarkisyan H A 2008 *European Physical Journal* B **63** 485
[13] Dvoyan K G, Hayrapetyan D B, Kazaryan E M and Tshantshapanyan A A 2007 *Nanoscale Research Letters* **2** 601
[14] Gambaryan K M, Aroutiounian V M, Boeck T, Schulze M and Soukiassian P 2008 *Journal of Physics D: Applied Physics* **41** 162004
[15] Grundmann M, Stier O and Bimberg D 1995 *Physical Review B (Condensed Matter)* **52** 11969
[16] Bastard G 1981 *Physical Review B (Condensed Matter)* **24** 4714
[17] Pfannkuche D, Gerhardts R R, Maksym P A and Gudmundsson V 1993 *Physica B: Physics of Condensed Matter* **189** 6
[18] Hawrylak P and Pfannkuche D 1993 *Physical Review Letters* **70** 485
[19] Pino R 1998 *Physical Review B (Condensed Matter and Materials Physics)* **58** 4644
[20] Lozovik Y 2003 *Solid State Communications* **126** 269
[21] Hanson R, Kouwenhoven L P, Petta J R, Tarucha S and Vandersypen L M K 2007 *Reviews of Modern Physics* **79** 1217
[22] Wunsch B 2009 *Physical Review* B **79** 235408
[23] Blümel R and Vorrath T 2003 *European Physical Journal* B **32** 227
[24] Lorke A, Luyken R J, Govorov A O, Kotthaus J P, Garcia J M and Petroff P M 2000 *Physical Review Letters* **84** 2223
[25] Chakraborty T and Pietilainen P 1994 *Physical Review B* **50** 8460
[26] Harutyunyan V A 2009 *Physica E* **41** 695
[27] Aghekyan N G, Kazaryan E M, Petrosyan L S and Sarkisyan H A 2010 *Journal of Physics: Conference Series* **248** 012048
[28] Flugge S 1971 Practical Quantum Mechanics Part 1, Springer, Germany
[29] Bredow M M, Rumyantsev V V and Toptygin I N 1985 Classical Electrodynamics, Moscow, Nauka

Screened Coulomb bound states in a finite confining potential semiconductor quantum well

K H Aharonyan

Physics Department, State Engineering University of Armenia, 105 Teryan Street, Yerevan 375009, Armenia

Abstract. The model to investigate the screened Coulomb interaction in realistic EuS/PbS/EuS finite confining potential quantum well is theoretically developed. A strong contrast of existing material parameters across the well interfaces and typical band-nonparabolicity effect of lead salt semiconductor are taken into account while a variational approach has been used. In presence of inhomogeneous dielectric polarization a quasi two-dimensional screened Coulomb form factor analysis is carried out by using random phase approximation. Initial deviations from the inhomogeneous polarization absent case are established and in particular quantum well width dependent screened logarithmical-type interaction potential is obtained. A screened binding energy has been investigated and an explicit logarithmical-type expression is used for the first time. Depending on the realistic nanostructure specifics a monotonic decrease of the enhanced binding energy is received when in-plane carrier density /temperature ratio parameter increases.

1. Introduction

A great deal of attention to a quasi-two-dimensional (Q2D) electron gas (EG) in a variety of semiconductor nanosystems (quantum wells (QW), double heterojunction structures (DHS), superlattices) is caused both by fabrication advances and purely scientific interest as well as by realization of these structures for the optoelectronic needs [1]. Carriers in a Q2D EG, that could be generated optically by intense illumination or by modulation doping, are confined within an active semiconductor layer with compressed wave functions (quantum confinement (QC)) and characterized by continuously varying carrier density probability distribution in the QW growth direction [2,3]. In these structures the advanced selective doping makes possible to investigate a charged electron (hole) system rather than by a neutral electron-hole plasma, as in the 3D samples [4]. The filling of the QW by free charge carriers results to the changes of the Q2D optical properties [5-8], which are caused by many-body effects such as a screening of the carriers Coulomb interaction. The way that the free carriers screen the bare Coulomb potential depends on the dimensionality of the system. In 3D bulk samples the screened potential decays exponentially in a real space and effectively becomes a short-range potential.

Conversely, in pure 2D system the real–space statically screened potential in long-wavelength limit is

$$V_S(\rho) = -(e^2/\varepsilon_w)[\rho^{-1} - q_S \int_0^{\infty} dq J_0(q\rho)(q+q_S)] . \quad (1)$$

In Eq. (1) $J_0(x)$ is the Bessel function, $\rho(x,y)$ is the in-plane 2D distance between the charges, ε_w is the QW material static dielectric constant, q_S is the 2D screening parameter [5] and q is the 2D in-plane wave vector. The screened potential falls off in accordance with 2D pure Coulomb law at intermediate distances and then decays by power law at large distances [9-11]. There is no screening length in 2D system. Thus, the 2D screening effects decrease because the free carriers can only screen the Coulomb field between any two charges of the 2D system, whereas the part of the field extending outside the confinement region remains unscreened. So the field lines that abandon the system are not directly affected by those charges. Note, that the screening parameter dependence from the free carrier density is different in 2D and 3D cases [2,10]. The screening parameter is saturated for small temperature or large carrier density n_S in a 2D EG (due to the constant density of states), while being continued to increase with volume density n_V in a 3D EG. As a result of the 2D screening saturation a Coulomb center cannot be screened out at any 2D carrier density.

The realistic QW has a nonzero width d and usually holds finite confining barrier potential height. A theoretical study of screened Q2D Coulomb interaction with arbitrary dielectric constants of the QW and barrier ($\varepsilon_w, \varepsilon_b$) have been started in the works of Ritova (Refs. 9, 10) by using diagram technique and Thomas-Fermi approximation (TFA), respectively. In Ref. 10 the author discussed the all possible relations between ε_w and ε_b ($\varepsilon_w \neq \varepsilon_b$, $\varepsilon_w \sim \varepsilon_b$). In particular, for the strong contrast relation case such as $\varepsilon_r = \varepsilon_w / \varepsilon_b >> 1$ (dielectric confinement (DC)), the statically screened Q2D potential has been established for the distances $\rho >> d$ (the long-wavelength limit) in the form

$$V_S(\rho) = -(2e^2/\varepsilon_w d) K_0\left(\sqrt{2q_S/d}\,\rho\right), \tag{2}$$

where K_0 is the modified Bessel function of the second kind. Eq. (2) both qualitatively and quantitatively differs from the widely discussed pure 2D potential function (1). The latter corresponds to the weak or negligible DC and is independent from the QW width. Screened potential strongly depends on the QW width and exponentially decays for large in-plane distances outside the circle with the radius of $\rho_0 = (d/2q_S)^{1/2}$. Hence, DC leads to the recovering of Q2D screening length. On the other hand, because of asymptotic behavior of $K_0(x)$ function for intermediate in-plane distances $d < \rho < \rho_0$, the Q2D potential function (2) becomes pure logarithmically spatially dependent and increases when QW width decreases. Thus, in this case the hydrogen-like properties of the Coulomb Q2D screened centers would be strongly enhanced compared with the systems where the DC is insufficient. This is confirmed by recent results which have been reported with logarithmical potential in EuS/PbS/EuS QW, where an effective enhancement of Coulomb unscreened bound states is revealed due to the DC [12, 13]. The purpose of present work is the study of screened Coulomb hydrogen-like bound states in the realistic EuS/PbS/EuS QW system. This is done in view of the finite height of the potential barrier and with strong material parameter contrast in the QW heteroboundaries (MPCH), namely the dielectric constants and effective mass.

2. Theoretical framework and model

Bound states of the screened Coulomb center influenced by spatial QW confinement have been the subject of intense investigations in the past time. The experiments in this field are based on the III-V group semiconductors [4, 14-16]. From standpoint of the theoretical interest the various calculation techniques within the effective-mass approximation (EMA) are performed. The binding energy studies for both infinite and finite confining potential QW models are carried out by perturbation theory [17, 18], numerical resolution [19] and variational procedure [2, 3, 21-25]. In this field the most studied systems are $GaAs/Al_xGa_{1-x}As$, Si/SiO_2 DHS structures concerning to the cases of the screened impurities [2, 3, 20, 22] and excitons [17- 19, 21, 23-25] by use of the random phase approximation (RPA) and TFA. All this binding energy issues refer to the situation, when the dielectric constants difference between QW and barrier ($\varepsilon_W, \varepsilon_b$) is small or neglected. For the case of $\varepsilon_W >> \varepsilon_b$ the interaction turns out to be significantly larger [10] and the hydrogen-like center binding energy would greatly enhanced with decreasing of well width [26, 27]. Due to the large ε_W of the material that especially would take place in the lead salt based QW's [12, 13]. Narrow-direct gap lead salt solid semiconductor solutions are very well lattice-matched materials with the europium salts (0.5 %) and can be epitaxially grown with the high quality in a controlled way. They possess a multivalley band structure with nearly mirror valence (v) and conduction (c) strong nonparabolic bands (NPB). There is no heavy hole band in the NaCl structure. Electrons and holes have small and almost the same effective masses m_w. Due to the large dielectric constant the bound states in the bulk sample have tiny binding energies. In the europium salts (as a wide gap barrier material) the Bohr radius is smaller than the lattice parameter (for EuS material it is estimated to be about 0.25 nm with dielectric constant ε_b =5 and effective mass $m_b \sim m_0$ [12]). Therefore in EuS/PbS/EuS DHS screened hydrogen-like bound states could enhance strongly with the DC. Firstly, it would concern to the binding energy in the long wave limit. The model for the Q2D EG embedded in a QW is based on the nearly rectangular band alignment with the finite c- band offset Δ_c. The Hamiltonian of Coulomb center in QW is

$$\hat{H} = H_{kin} + H_{eff} + V_S, \tag{3}$$

where $H_{eff} = V_{fin}(z) + V_{self-im}(z)$ - is the unperturbed part and V_S is the screened Coulomb potential. $V_{fin} = 0$ when $0 < z < d$ and $V_{fin} = \Delta_c$ when $z \leq 0, z \geq d$, $V_{self-im}$ is the one-electron self-image interaction potential, which, as well as the size-quantized potential, modifies one-particle states. This contribution has the same order as the quasiparticle binding energy [12, 13]. Since EuS/PbS/EuS QW system band offsets ($\Delta_{c(v)}$= 1.2 eV (0.15 eV) [12]) are much larger than the ground state binding energies. In this case the self-image interaction changes the rectangular localizing potential into the smoothly shaped potential. Then a confinement feature does not change distinctly when the perturbing potential varies smoothly on the scale of k_F^{-1} ($k_F = \sqrt{2\pi n_s}$ is the 2D Fermi vector) and is so weak, that can be treated in a first-order perturbation theory [2]. In doped QW's the presence of both free carriers and charged impurities induces band bending [3]. We restrict ourselves to the n-type QW with the electrons occupying only states in the first ε_1^0 subband (electric size quantum limit (SQL)). Within the accepted approximation the band bending could be treated perturbatively for thin QW's which leaves eigenfunctions unchanged but shifts ε_1^0. In present paper we are interested in this eigenfunctions and consider the finite rectangular symmetrical model. Our study involve two steps: 1. calculation of the screened Coulomb potential using RPA in presence of the MPCH, 2. carrying out the Coulomb impurity binding energy's variational calculation in EuS/PbS/EuS QW with the nondegenerate Q2D EG.

3. The Screened Potential and the Binding Energy

Following to the strong confinement regime $a_0 >> d$, the distance between quantized energy levels is larger than interaction energy. Thus the appropriate ground state wave function has a separable form such as $\phi = \phi(\rho, z) = \psi(\rho)\varphi(z)$. Here $\psi(\rho)$ is an in-plane wave function and $\varphi(z)$ is an one-particle 1D Schrödinger-like equation EMA solution along z axis. Because of strong confinement regime, such as in Refs. [12, 13], a long wave limit is used here. For the latter case there are effective wave vectors $q < d^{-1}$ measured from the L point, wave function spatial change is determined by d . In the variational calculations the functional $E[\phi] = \int \phi^*[H_{kin} + H_{eff} + V_S]\,\phi dV$ is elaborated. So we have to determine the average value of the screened potential V_S over $(\varphi(z))^2$, which we will deal using RPA method. For the latter the Fourier expanded screened Coulomb interaction potential $V_S(q,z)$ is related to the bare potential Fourier component $V(q,z)$ as $V_S(q,z) = V(q,z)/\varepsilon^{Q2D}_{finit}(q)$. Here $\varepsilon^{Q2D}_{finite}(q)$ is the Q2D EG dielectric function and has the form [28]

$$\varepsilon^{Q2D}_{finit}(\omega, q) = 1 - [2\pi e^2/Sq] \cdot \gamma \cdot F_\Sigma(q). \tag{4}$$

In Eq. (4) γ is the irreducible polarizability of 2D EG, $F_\Sigma(q)$ is the Coulomb form factor which takes into account the deviation from the pure two-dimensionality, S is the interface area. As follows from Eq. (4), only the wave vector dependent dielectric function of the Q2D EG and averages of bare potential $V(q,z)$ over $(\varphi(z))^2$ are required. We have calculated the quantities $\varepsilon^{Q2D}_{finite}(q)$ and $< V_S(q) >$ in the recent works [13, 28]. In this case the lowest eigenfunction $\varphi(z)$ used for the square-well potential is

$$\varphi_w = N\cos(k_w z),\ \varphi_b = B\exp(-k_b z) \ \text{ with } \ k_w = \sqrt{2m_w \varepsilon^c_{1,\ell}/\hbar^2},\ k_b = \sqrt{2m_b(\Delta^c - \varepsilon^c_{1,\ell})/\hbar^2}, \tag{5}$$

where $N = (2/d)^{1/2} N^{-1}_{fg}$, $N_{fg} = 1 + (2f/gk_b d)$, $f = 1 + \alpha n_\ell (k_b/k_w)^2$, $g = 1 + (\alpha n_\ell k_b/k_w)^2$, $B = N/\sqrt{g}$. The subband energies have the form of the dispersion equation

$$tg(k_w d/2) = \alpha n_\ell k_b / k_w, \tag{6}$$

where $\alpha = (m^*_\ell / m_b)$, $m_w = m^*_\ell n_\ell$, m^*_ℓ is the longitudinal effective mass, $n_\ell = \sqrt{1 + (2\hbar k_w)^2 / m^*_\ell E^w_g})$ is the NPB term. To take into account the latter we have used "energy-dependent effective mass" approach [29, 13].

In Ref. 28 using the self-consistent field approach or RPA in presence of the MPCH we revealed the Coulomb form factor implicit expression in the long - wavelength limit $q < d^{-1}$. With the inequality $k_w > k$ (k^{-1}, k_w^{-1} are the typical spatial changes of the potential energy's Fourier-component and wave function in z direction, respectively) it gives the condition $k_w d \sim 1$ and form factor takes $F_\Sigma(q) = 2/[qd + (2/\varepsilon_r)]$ form. With this and for the static limit of irreducible polarization γ [22], the Eq. (4) finally goes to the expression

$$\varepsilon^{Q2D}_{finit}(q) = 1 + [2q_S / q(qd + (2/\varepsilon_r)]. \tag{7}$$

Recently [13], we have evaluated the averaged bare image potential-mediated interaction energy operator matrix element $< \alpha' | V(q) | \alpha >$. And according to Ref. [28] we get the Q2D form factor expressing a deviation from the two-dimensionality with the MPCH for the finite confining barrier potential as $F_C(q)=[4N^2/(kq+2\varepsilon_r^{-1})]\,[((2k_b+q)g)^{-1}+(d/4)+(\alpha_\ell n_\ell k_b/2k_w^2 g)]$. Now if combining Eq. (7) with the $F_C(q)$ for the screened potential averaged value we have

$$<V_S(q)>=\frac{4\pi e^2}{\varepsilon_w}\frac{1}{q}\frac{1}{[1+(2q_S/q(qd+(2/\varepsilon_r))]}\frac{4N^2}{qd+2\varepsilon_r^{-1}}\left(\frac{g^{-1}}{2k_b+q}+\frac{d}{4}+\frac{\alpha_\ell n_\ell k_b}{2k_w^2 g}\right). \quad (8)$$

Eq. (8) for the 2D vectors q << $2k_b$ is identical to the analogous result of Ref. 10 received by usage of TFA and this is quite expected because in long-wavelength limit the RPA and TFA coincide. In that case Eq. (8) goes to the following form

$$<V_S^i(q)>=-(4\pi e^2/\varepsilon_w d)\{q[q+(2/\varepsilon_r d)]+(2q_S/d)\}^{-1}. \quad (9)$$

In view of the long wave ($qd<<1$) and the DC ($\varepsilon_r^{-1}<<1$) conditions there are two distinct 2D wave vector ranges, such as $k_b d > k_w d > 1 > kd >> 2\varepsilon_r^{-1}$ and $k_b d > k_w d > 1 > 2\varepsilon_r^{-1} >> kd$, hence Eq. (8), (9) possesses the following analytical features :

(i) For the moderate small wave vectors with $k_b d > k_w d > 1 > kd >> 2\varepsilon_r^{-1}$:

$$<V_S^i(q)>=-(4\pi e^2/\varepsilon_w d)[q^2+(2q_S/d)]^{-1}. \quad (10)$$

(ii) For the small enough wave vectors with $k_b d > k_w d > 1 >> 2\varepsilon_r^{-1} >> kd$:

$$<V_S^{ii}(q)>=-(4\pi e^2/\varepsilon_b)(q+\varepsilon_r q_S)^{-1}. \quad (11)$$

Note, that when neglecting the linear term in denominator of Eq. (9), the condition $2q_S d >> \varepsilon_r^{-2}$ have been used. Only in the case of moderate long wave related wave vectors Eq. (10) leads to the potential function from Eq. (2). Using $K_0(x)$ Bessel function properties for the screened potential it can be written

$$V_S(\rho)=\begin{cases} -(2e^2/\varepsilon_w d)\sqrt{\pi\rho_0/2\rho}\exp(-\rho/\rho_0) & \rho>\rho_0\ , \quad a)\\ -(2e^2/\varepsilon_w d)\left[\ln\left(\sqrt{q_S\rho/2d}\right)+\gamma\right] & \rho_0>\rho>>d\ . \quad b)\end{cases} \quad (12)$$

According to the Eq. (12a) the screening potential exponentially decays outside the circle with the radius $\rho_0=(d/2q_S)^{1/2}$ and a strong DC effect leads to the recovering of Q2D screening length [10]. At the same time, as shown in Eq. (12b), for the intermediate in-plane distances such as $d<\rho<\rho_0$ due to the DC Q2D screened potential function becomes logarithmically spatially dependent, deviates particularly from the pure 2D as well as 3D forms, and increases with decreasing of the d. The latter enhancement of screened potential is quite natural, because the DC overbalances in relation of screening effect. In Eq. (11) the revealed new features of the screened potential with DC are the followings: a) a screened potential's Fourier-image (therefore real–space statically screened potential) is more enhanced because instead of large QW dielectric constant it only depends on the barrier small analogous, b) the appropriate real –space statically screened potential with $<V_S^{ii}(q)>$ repeats the spatial dependence (1) only at very long wave related wave vectors, c) from Eq. (11) follows, that the screening length at very long wave related wave vectors depends from the dielectric contrast between the QW and barrier.

In this paper we are interested in the screened potential from Eq. (12b). For the ground 1s bound state level we choose normalized one parameter trial wave function in the form $\psi(\rho)=\sqrt{2/\pi}(\lambda e^{-\lambda\rho})$

and minimize the functional versus the variational parameter λ. The binding energy E_b^S is obtained as the difference of the first subband energy ε_1^0 and expectation energy E_{1s}. For the 2D kinetic energy operator's matrix element from [13] $K(\lambda)=\hbar^2\lambda^2 m_r/2m_\perp$, where $m_r=\alpha_\ell n_\ell \gamma_r$ is the effective mass renormalization term with $\gamma_r=[2(1+(k_b/k_w)^2)+gk_b d/\alpha_\ell n_\ell]/(2f+gk_b d)$. The screened interaction energy operator matrix element with Eq. (12b) is

$$V_S(\lambda)=-(2e^2/\varepsilon_w d)\iint[\ln(\rho/2\rho_0)+\gamma]\left|\psi(\rho)\right|^2 d\vec{\rho}, \tag{13}$$

Finally, for the Coulomb center ground state screened binding energy in finite confining QW we have

$$E_b^S(d)=-(e^2/\varepsilon_w d)\,[\ln(16n_\ell/a_0\, q_S\, m_{r\Lambda})-3], \tag{14}$$

$m_{r\Lambda}=m_r-4n_\ell\Lambda(d/a_0)$, $\Lambda=k_b d(2-g)/(k_w d)^2(2f+gk_b d)$. As follows, with decreasing of the QW width the screened binding energy mainly grows in terms of $E_b^S \sim d^{-1}$, which differs from the corresponding dependence $E_b \sim d^{-1}\ln d$ for the unscreened case with the DC effect [12,13]. We note, that the dependence of $E_b^S(d)$ is much more complex, because the quantities n_ℓ and $m_{r\Lambda}$ weakly depend on QW width. The relevance of the Eq. (14) firstly depends on the feasibility of the condition $d << \rho < \rho_0$ and for that from an extra condition an under logarithm term holds. The latter is related to the abovementioned condition $2q_S d >> \varepsilon_r^{-2}$. If combining both conditions we get $d^{-1} >> q_S >> \varepsilon_r^{-2} d^{-1}$, which determines allowed interval for the key parameter n_S/T. In this work we pay an attention to the case of nondegenerate EG. For that the high temperatures and low densities are required and strong condition $m_w k_B T/\pi n_S \hbar^2 >> 1$ is held. For this case the screening parameter expression in SQL $q_S=(2/a_0)[1-\exp(-\pi\hbar^2 n_S/m_\perp k_B T)]$ [5] goes to $q_S=4\pi n_S e^2/\varepsilon_w k_B T$ and eventually with the allowed interval condition leads to the relation as

$$\varepsilon_w k_B/8\pi e^2 d >> n_S/T >> \varepsilon_w k_B/8\pi e^2 d\,\varepsilon_r^2\,. \tag{15}$$

Now let's carry out the Q2D screened binding energy on-center donor numerical calculations for EuS/PbS/EuS realistic DHS to illustrate the offered theoretical model.

Table 1. The numerical results of one-particle parameters in EuS/PbS/EuS SQW for c-band offset.

d nm	$k_w^e d$	$k_b^e d$	f_e	g_e	n_{el}	$m_{r\Lambda}$
1.5	1.83	8.21	6.82	2.68	2.74	0.96
5	2.17	29.13	26.41	4.61	1.35	0.98

In Table 1 numerical results of the one-particle electron state parameters are given (with the parameter ratios ε_r=34, α_ℓ = 0.105) for the strongly nonparabolic c-band offset case (1.2eV). The model is solved for PbS QW grown with [111] orientation, where the four valleys are split into the one longitudinal and three equivalent oblique ones [29]. For PbS sample they are characterized by longitudinal (along [111]) and transversal (normal to [111]) band edge effective masses respectively: m_ℓ^*=0.105 m_0, $m_\perp^*$=0.08 m_0 (m_0 is the free electron mass)). For bulk PbS sample a static dielectric constant value is ε_w =170 with impurity Bohr radius $a_0 \approx$ 102.5nm. At this moment let's display on-center donor binding energy numerical result (Eq. (14)) in presence of nondegenerate EG.

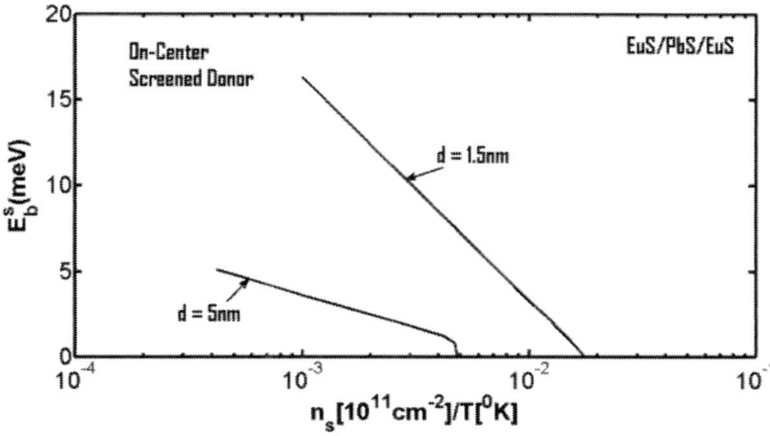

Fig. 1. The binding energy of the screened donor impurity as a function of the ratio n_s/T in EuS / PbS / EuS DHS. The top (low) curve corresponds to the QW width 1.5 nm (5 nm) calculation result.

In Fig. (1) the binding energy dependence versus n_S/T parameter is plotted for the QW width values d = 1.5 nm and d = 5 nm, respectively. The top (low) curve in the graph corresponds to the d = 1.5 nm (5 nm) calculation result. As follows from the graph the enhanced screened binding energy logarithmical dependence for the nondegenerate EG extends approximately two decades with the ratio of n_s/T (Eq. (15)). Except that the characteristic curve decreases almost linearly with increasing of n_s/T. As shown in Fig. 1, an influence of DC on the screening in the EuS/PbS/EuS QW is appreciably for thin enough QW's in accordance with Eq. (14).

8. Conclusions

The theoretical study of screened Coulomb interaction in realistic EuS/PbS/EuS finite confining potential QW is carried out in the presence of QW and barrier material parameter's strong contrast across the heteroboundaries. A characteristic band-nonparabolicity effect of the lead salt semiconductor material is taken into account. In the presence of inhomogeneous dielectric polarization of the QW system Q2D screened Coulomb form factor is evaluated using a randon phase approximation. The enhanced 2D Coulomb-type (Eq. (11)) and logarithmical-type (Eq. (12 b)) screened Q2D potentials are revealed for the first time. A screened binding energy problem has been investigated variationally and an analytical logarithmical-type result is used for the first time. Depending of realistic EuS/PbS/EuS QW specifics, the numerical analysis is provided. A very strong enhancement of the binding energy $E_b^S(d)$ in relation to tiny unscreened bulk value (~ 0.1 meW) is received especially for the small QW widths. As follows, the screened binding energy decreases monotonically with increasing of parameter

n_S / T for the nondegenerate Q2D electron gas. The results indicate to the favorable conditions for detecting of screened Coulomb bound states in EuS/PbS/EuS finite confining potential QW system.

Acknowledgment
The author is thankful to Professor E.M. Kazaryan for the helpful discussions.

References

[1] Davies J H 1998 *The Physics of Low Dimensional Semiconductors* (Cambrige : Cambrige University Press)
[2] Stern F and Howard W E 1967 *Phys. Rev.* **163** 816
[3] Brum J A , Bastard G and Guillemot C 1984 *Phys. Rev B* **30** 905
[4] Gubarev S I , Kukushkin I V, Tovstonog S V, Akimov M Yu, Smet J, von Klitzing K and Wegschelder W 2000 *JETP Lett.* **72** 324
[5] Spector H N , Lee J and Melman 1985 *J. Appl. Phys.* **58** 1893
[6] Aharonyan K H and Kazaryan E M 1983 *Thin Solid Films* **105** 149
[7] Scmitt-Rink S , Ell C and Haug H 1986 *Phys. Rev .B* **33** 1185
[8] Bauer G E W 1992 *Phys. Rev .B* **45** 9153
[9] Rytova N S 1965 *Sov.Phys. Doklady* **10** 754
[10] Rytova N S 1967 *Vestnik Moscow Univ.* **3** 30
[11] Stern F 1967 *Phys. Rev. Lett.* **18** 546
[12] Aharonyan K H 2010 *Physica E* **43** 111 ,
[13] Aharonyan K H 2011 *Physica E* **44** 487 ; 2011 *Proc.8-th ICSMN-2011* 192
[14] Chemla D S , Miller D A B , Smith P W , Gossard A C and Wiegmann W 1984 *IEEE J Quant. Electron.* QE-**20** 265
[15] Finkelstein G, Strikman H and Bar-Josef 1995 *Phys. Rev. Lett.* **74** 976
[16] Gubarev S I, Volkov O V, Koval'skiĭ V A, Kulakovskiĭ D V and Kukushkin I V 2002 *JETP Lett.* **76** 575
[17] Ben Brahim Aouani N, Mandhour L , Bennaceur R , Jaziri S , Amand T and Marie X 1998 *Solid State Commun.* **108** 109
[18] Ping E X and Jiang H X 1993 *Phys. Rev B* **47** 2101
[19] Edelstein W S and Spector H N 1989 *Surface Science* **224** 581
[20] Hipolito O and Campos V B 1978 *Phys. Rev B* **19** 3083
[21] Kleinman D A 1985 *Phys. Rev B* **32** 3766
[22] Hawrylak P and Quinn J J 1986 *Phys. Rev B* **33** 8264
[23] Averkiev, Pikus G E and Shmatov 1989 *Sov. Phys. Solid State* **30** 1884
[24] Henriques A B 1991 *Phys. Rev B* **44** 3340
[25] Kulakovskiĭ D V , Gubarev S I and Lozovik Yu E 2002 *JETP* **94** 785
[26] Chaplik A V and Entin M V 1971 *JETP* **61** 2469
[27] Keldish L V 1979 *JETP Lett.* **29** 658
[28] Aharonyan K H 2011 *Physica E* **43** 1618 ; 2011 *Proc.SPIE* 7998 79981D1
[29] de Andrada e Silva E A 1999 *Phys. Rev. B* **60** 8859

Effects of hydrostatic pressure, temperature, electric field and aluminum concentration on the electronic states in $GaAs/Ga_{1-x}Al_xAs$ concentric double quantum rings

H M Baghramyan[1], M G Barseghyan[1,2,4], C A Duque[3] and A A Kirakosyan[1]

[1] Department of Solid State Physics, Yerevan State University, 1 Alex Manoogian,Yerevan 0025, Armenia
[2] Base College of Yerevan State University of Architecture and Construction, 105 Vahan Teryan, Yerevan 0009, Armenia
[3] Instituto de Física, Universidad de Antioquia, AA 1226, Medellín, Colombia

Email: mbarsegh@ysu.am

Abstract. Using the effective mass approximation and transfer matrix formalism the ground state energy of an electron in $GaAs/Ga_{1-x}Al_xAs$ concentric double quantum rings is calculated taking into account the combined effects of electric field, hydrostatic pressure, temperature and aluminum concentration. The ground state energy dependences on the mentioned factors are presented for different values of the sizes of rings and electric field strength. It is found that the considered effects mainly depend on the dimensions of quantum rings.

1. Introduction

Nowadays it is possible to make high-quality semiconductor quantum rings using the self-assembly technique [1,2]. Semiconductor quantum rings have attracted intensive interest due to their unique topological geometry and energy spectrum. Very recently lattice-matched GaAs/GaAlAs concentric double quantum rings (CDQRs) were fabricated by means of the droplet epitaxy technique with high uniformity and excellent rotational symmetry. The making of these concentric structures is clearly demonstrated by images of atomic force microscopy [3].

Recently there has been considerable interest in the properties of single [4-6] and few electron states [7-9], exciton states [3,10], the spin-orbit interaction [11,12] as well as the impurity effects [13] in vertically and laterally coupled double quantum rings. The theoretical investigation of CDQRs under the magnetic field applied perpendicularly to the ring plane is reported in [4,5]. Electron-hole transition energies are calculated as a function of the system geometry confinement, within a single-particle picture, neglecting interaction effects. The effect of hydrostatic pressure on electronic states in $GaAs/Ga_{0.7}Al_{0.3}As$ CDQRs under axial magnetic field is investigated in [6]. For both symmetric and asymmetric CDQRs it is found that the electron-heavy hole transition energies augment with the applied pressure is mainly due to the pressure-induced increase in the GaAs gap.

In this work the effects of growth-direction applied electric field, hydrostatic pressure, temperature and aluminum concentration on the ground state energy in $GaAs/Ga_{1-x}Al_xAs$ CDQRs are studied. The

[4] M G Barseghyan

paper is organized as follows: in Section 2 we describe the theoretical framework, Section 3 is dedicated to the results and discussion and, finally, our conclusions are given in Section 4.

2. Theoretical framework

In figure 1 the schematic view of the CDQRs is presented. The dimensions of heterostructure (radii, widths and the thickness of rings) and the direction of the applied electric field are depicted.

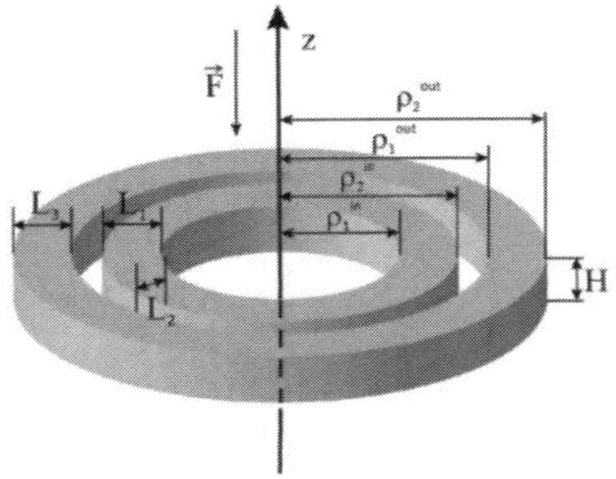

Figure 1. The CDQR heterostructure: L_1 is the width of the inner ring, L_2 is the width of the coupling barrier, L_3 is the width of the outer ring, ρ_1^{in} and ρ_2^{in} are the inner and outer radii of the inner ring, ρ_1^{out} and ρ_2^{out} are the inner and outer radii of the outer ring, H is the thickness of rings, and $\vec{F}$ is the applied electric field strength.

The sizes of the structure depend on the pressure (P) according to the expressions [6, 14]

$$H(P)=H(0)\left[1-\left(S_{11}+2S_{12}\right)P\right],\ \rho(P)=\rho(0)\left[1-2\left(S_{11}+2S_{12}\right)P\right]^{1/2}, \tag{1}$$

where S_{11} and S_{12} are the components of the compliance tensor in GaAs ($S_{11}=1.16\times10^{-3}\,\mathrm{kbar}^{-1}$ and $S_{12}=-3.7\times10^{-4}\,\mathrm{kbar}^{-1}$).

The Hamiltonian of the electron in CDQRs in the effective mass and parabolic band approximations under the combined effects of pressure, temperature and applied electric field is given by the expression

$$\hat{H}=\hat{H}_{\rho\varphi}+\hat{H}_z, \tag{2}$$

where

$$\hat{H}_{\rho\varphi}=-\frac{\hbar^2}{2}\left(\frac{1}{\rho}\frac{\partial}{\partial\rho}\left(\frac{\rho}{m(x,P,T)}\frac{\partial}{\partial\rho}\right)+\frac{1}{m(x,P,T)}\frac{1}{\rho^2}\frac{\partial^2}{\partial\varphi^2}\right)+V_\rho(\rho,x,P,T), \tag{3}$$

$$\hat{H}_z=-\frac{\hbar^2}{2\,m(0,P,T)}\frac{\partial^2}{\partial z^2}+V_z(z,P). \tag{4}$$

In the expression (3) $m(x,P,T)$ is the aluminum concentration (x), temperature (T) and pressure dependent electron effective mass and it is given by [15]

$$m(x,P,T)=m_0\left[1+\frac{\Pi^2(x)}{3}\left(\frac{2}{E_g^\Gamma(x,P,T)}+\frac{1}{E_g^\Gamma(x,P,T)+\Delta_0(x)}\right)+\delta(x)\right]^{-1}, \tag{5}$$

where m_0 is the free electron mass, $\Pi(x)$ is the interband matrix element [$\Pi^2(x)=28900-6290\,x$], and $\Delta_0(x)$ is the valence-band spin-orbit splitting [$\Delta_0(x)=341-66\,x$]. The remote-band effects are taken into account via the $\delta(x)$ parameter, which is given by $\delta(x)=-3.935+0.488\,x+4.938\,x^2$. The energy gap at the i- point ($i=\Gamma$, X) of the conduction band is given by the expression

$$E_g^i(x,P,T) = a_i + b_i x + c_i x^2 + \alpha_i P - \beta_i T^2 (\gamma_i + T)^{-1}, \tag{6}$$

where the values of parameters $a_i, b_i, c_i, \alpha_i, \beta_i$, and γ_i are given in the table 1.

The confining potentials are defined taking into account the crossover between the Γ and X minima of the conduction band [15-17]. There are two critical values of pressure at which the crossover can be observed: 1) at $P_1(x,T)$ there is a crossover between the Γ and X points at the $Ga_{1-x}Al_xAs$ barrier with energies $E_g^{\Gamma}(x,P,T)$ and $E_g^{X}(x,P,T)$ respectively and 2) at $P_2(x,T)$ the crossover is between the Γ point at the GaAs well and the X point at the $Ga_{1-x}Al_xAs$ barrier. In the direct gap regime $(P < P_1)$ the confining potential height is independent on pressure and obtained as a fraction r of the difference between the $Ga_{1-x}Al_xAs$ and GaAs energy gaps at the Γ point. If $P_1 < P < P_2$ the position of the X minimum is lower than the Γ one and the height of the confining potential decreases with pressure and can be defined as a fraction r of the energy $E_g^{X}(x,P,T) - E_g^{\Gamma}(0,P,T) + S_{\Gamma X}(x,P,T)$, where $S_{\Gamma X}(x,P,T) = S_0 (P - P_1(x,T)) x/P$ is the Γ-X mixing strength coefficient and $S_0 = 250\,\text{meV}$ is the adjustable parameter which fits the experimental measurements [16]. If $P > P_2(x,T)$ a type I-type II semiconductor heterostructure transition occurs. It is necessary to note that Γ-X crossover can be observed with variation of aluminum concentration. If $x_c \approx 0.38$ (at $P = 0$ and $T = 4\text{K}$) crossover takes place at the barriers and there is no observed crossover with variation of temperature. Below the pressure values considered are between 0 and $P_2(x,T)$, and x will be taken to lie between 0 and x_c. With the inclusion of the crossover effect the confining potential in radial direction is defined by

$$V_\rho(\rho,x,P,T) = \begin{cases} 0, & \text{if } \rho_1^{\text{in}} \le \rho \le \rho_2^{\text{in}} \text{ and } \rho_1^{\text{out}} \le \rho \le \rho_2^{\text{out}}, \\ V_\rho(x,P,T), & \text{if } \rho < \rho_1^{\text{in}},\ \rho_2^{\text{in}} < \rho < \rho_1^{\text{out}} \text{ and } \rho > \rho_2^{\text{out}}, \end{cases} \tag{7}$$

where

$$V_\rho(x,P,T) = r \begin{cases} E_g^{\Gamma}(x,P,T) - E_g^{\Gamma}(0,P,T), & \text{if } P \le P_1(x,T), \\ E_g^{X}(x,P,T) - E_g^{\Gamma}(0,P,T) + S_{\Gamma X}(x,P,T), & \text{if } P_1(x,T) < P \le P_2(x,T), \end{cases} \tag{8}$$

r being the fraction of band gap discontinuity. The confining potential in z-direction

$$V_z(z,P) = \begin{cases} -eFz, & \text{if } |z| \le H(P)/2, \\ \infty, & \text{if } |z| > H(P)/2. \end{cases} \tag{9}$$

Table 1. Parameters for energy gap [15]

Γ-minimum	$a_\Gamma = 1519.4$ meV	$b_\Gamma = 1360$ meV	$c_\Gamma = 220$ meV
	$\alpha_\Gamma = 10.7$ meV/kbar	$\beta_\Gamma = 0.5405$ meV/K	$\gamma_\Gamma = 204$K
X-minimum	$a_X = 1981$ meV [18]	$b_X = 207$ meV	$c_X = 55$ meV
	$\alpha_X = -1.35$ meV/kbar	$\beta_X = 0.46$ meV/K	$\gamma_X = 204$K

The eigenfunctions of the Hamiltonian (2) can be presented in the form $\Phi(\rho,\varphi,z) = (2\pi)^{-1/2} e^{i\mu\varphi} f(\rho)\, g(z)$, where $\mu = 0, \pm1, \pm2, \dots$ is the magnetic quantum number. In the next only the ground state of electron is taken into account and, consequently, only the value $\mu = 0$ is needed. Then $f(\rho)$ and $g(z)$ will be the eigenfunctions of the Hamiltonians (3) and (4) respectively with the corresponding eigenvalues ε_ρ and ε_z. $f(\rho)$ is a linear combination of Bessel functions. In order to find ε_ρ, the transfer matrix formalism is used [19], whereas in z-direction the Schrödinger equation can be modified to the Airy equation [20].

3. Results and discussion

In the calculations we have taken $r = 0.6$. At $T = 4\text{K}$ the first crossover pressure occurs at $P_1 = 8.37\text{kbar}$. In figure 2 the ground state energy as a function of pressure (a), (b) and the probability density in radial direction for several values of pressure (c) are presented (in all figures of this section the fixed values of CDQRs' sizes are given for $P = 0$).

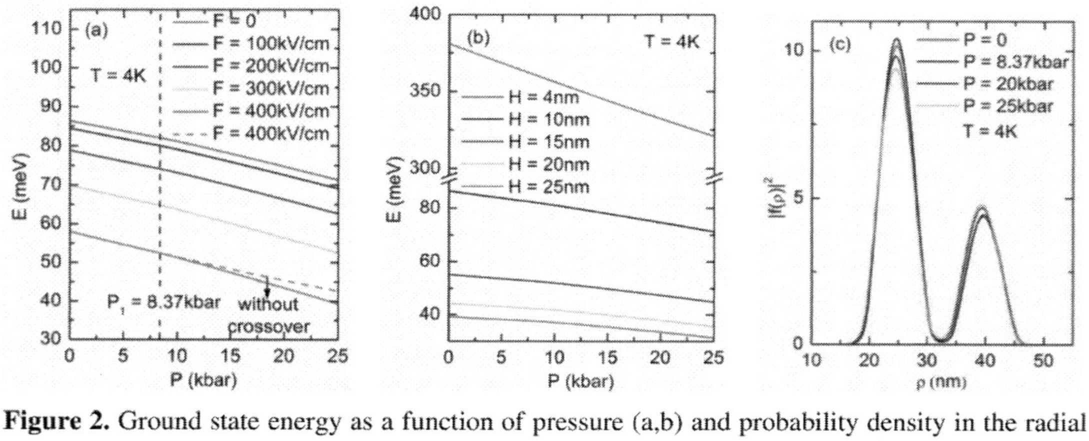

Figure 2. Ground state energy as a function of pressure (a,b) and probability density in the radial direction for different values of pressure (c). In (a) the results are for $x = 0.3$, $\rho_1^{in} = 20$ nm, $L_1 = 10$ nm, $L_2 = 5$ nm, $L_3 = 10$ nm, $H = 10$nm; in (b) – $F = 0$, $x = 0.3$, $\rho_1^{in} = 20$ nm, $L_1 = 10$ nm, $L_2 = 5$ nm, $L_3 = 10$ nm; in (c) – $F = 0$, $x = 0.3$, $L_1 = 10$ nm, $L_2 = 5$ nm, $L_3 = 10$ nm, $H = 10$nm.

As it is shown in figure 2 (a) the energy is a decreasing function of the electric field strength and pressure. For $P \le P_1$ the radial-confining potential height is constant and the energy decrease is due only to the increasing of the electron effective mass. If $P > P_1$ the decrease in the electron energy with pressure is faster, because besides the decrease associated with the augment of the electron effective mass, now there is a reduction associated with the fall in the radial-potential barrier height. For electric field strength of 400 kV/cm it is also shown the energy dependence without including the Γ-X crossover. It should be noted that the energy variations due to the changes of the structure sizes with pressure (see equation (1)) are very small compared with the variations associated with pressure-related changes in the effective mass and height of the confining potential. In figure 2 (b) the energy is a decreasing function of the ring thickness. It is clear that the stronger size-quantization effect and pressure influences are larger for smaller values of ring thickness. This last effect is related to the fact that in z-direction the electron energy changes with pressure are only due to the variation of the effective mass with pressure. The effects associated with the effective mass are stronger for smaller values of ring thickness. From figure 2 (c) it is seen that the probability density is greater in the inner ring (although the widths of the rings are equal) because of the unique topological geometry of the rings. Also we observe that if $P \le P_1$ the maximum value of probability density increases within the inner ring and decreases in outer ring. If $P > P_1$, the opposite situation is observed. This is because in the region $P \le P_1$ there will be only an increase of the electron effective mass with the pressure, with the augment of the probability density in the inner ring and a decrease of it in the outer ring. When $P > P_1$ the radial potential height is also reduced, causing the increase of the tunneling probability from inner to outer ring.

The ground state energy as a function of temperature and the probability density in radial direction for several values of temperature are presented in figure 3 (a), (b) and (c), respectively. In figures 3 (a)

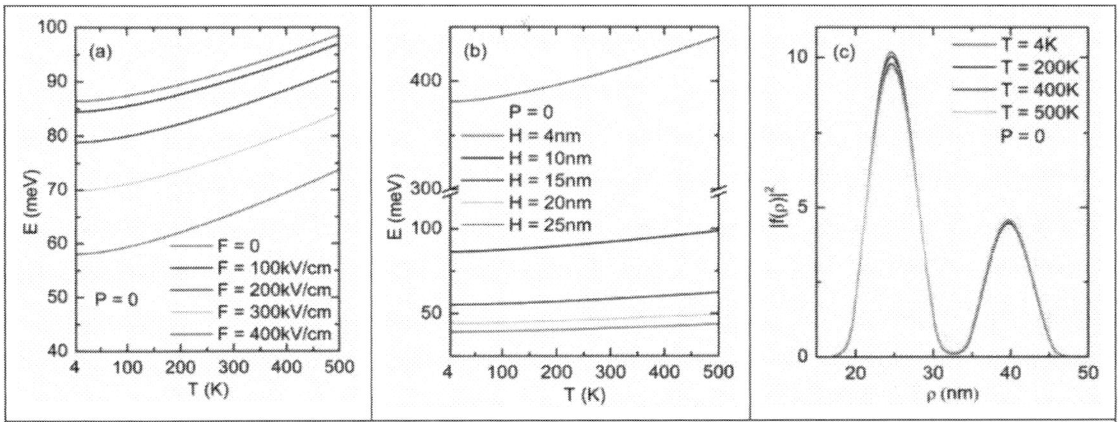

Figure 3. Ground state energy as a function of temperature (a,b) and probability density in radial direction for different values of temperature (c). In (a) the results are for $x = 0.3$, $\rho_1^{in} = 20$ nm, $L_1 = 10$ nm, $L_2 = 5$ nm, $L_3 = 10$ nm, $H = 10$ nm; in (b) – $F = 0$, $x = 0.3$, $\rho_1^{in} = 20$ nm, $L_1 = 10$ nm, $L_2 = 5$ nm, $L_3 = 10$ nm; in (c) – $F = 0$, $x = 0.3$, $L_1 = 10$ nm, $L_2 = 5$ nm, $L_3 = 10$ nm, $H = 10$nm.

and (b) the energy increase with temperature can be explained by the effective mass reduction. Also the energy is a decreasing function of the electric field strength and the quantum rings thickness and the influence of temperature is larger for smaller values of ring thickness. From figure 3 (c) it is observed the expected decrease of the probability density within the inner ring (because of the effective mass decrease with temperature).

In figure 4 are presented the ground state energy as a function of aluminum concentration (a) and the probability density in radial direction for several values of x (b). The energy is an increasing function of x given the strengthening of the radial-confining potential for bigger barrier heights (although the effective mass in the barrier region is larger as well, leading to the decrease of energy,

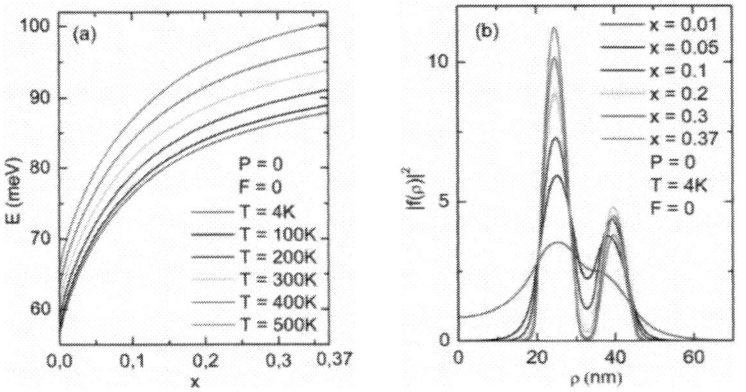

Figure 4. Ground state energy as a function of the aluminum concentration (a) and probability density in radial direction for different values of aluminum concentration (b). In (a) the results are for $\rho_1^{in} = 20$ nm, $L_1 = 10$ nm, $L_2 = 5$ nm, $L_3 = 10$ nm, $H = 10$nm; in (b) – $L_1 = 10$ nm, $L_2 = 5$ nm, $L_3 = 10$ nm, $H = 10$nm.

the influence of variation of the confining potential height is stronger). It can be also seen an augment of the distance between lines with temperature, which is because of the non-linear behavior of energy as a function of temperature (see figure 3). The probability density behavior is strongly dependent on the range of variation of the concentration: (1) the increase of x in the range from 0.2 to 0.37 is reflected in the augment of the probability density in the inner ring while there will be a decrease of the probability density in the outer ring due to the increase of the confining potential height; (2) in the range from 0.1 to 0.01 the probability density both in the inner and outer rings decrease, because now the confining potential in radial direction become very small and the curve of probability density goes down as whole. Besides that, the energies don't start from 0 values because in vertical direction the potential is infinite, although if $x = 0$ there is no potential in radial direction.

4. Conclusions

In this work the combined effects of pressure, temperature, electric field and aluminum concentration on electronic states in GaAs/$Ga_{1-x}Al_xAs$ CDQRs are investigated using the effective mass and parabolic band approximations and the transfer matrix formalism. It is shown that the ground state energy is a decreasing function of electric field strength and pressure and an increasing function of temperature and aluminum concentration. We have shown that the effect of pressure and temperature is larger for smaller values of thickness of the rings. It is also found that the behavior of probability density is strongly dependent on the range of variation of the pressure and aluminum concentration. As a final conclusion it can be mentioned that our findings confirm that the pressure, the temperature, the electric field and the aluminum concentration could be suitable tools for tuning the electronic and optical properties of quantum rings.

Acknowledgments: This research was partially supported by Colombian Agencies: COLCIENCIAS, CODI-Universidad de Antioquia (Estrategia de Sostenibilidad Grupo de Materia Condensada-UdeA, 2011-2012), Facultad de Ciencias Exactas y Naturales-Universidad de Antioquia (CAD-exclusive dedication project 2011-2012) and "El Patrimonio Autónomo Fondo Nacional de Financiamiento para la Ciencia, la Tecnología y la Innovación Francisco José de Caldas" Contract RC – No. 275-2011. The work was developed with the help of CENAPAD-SP, Brazil. This work was partially supported by Armenian State Committee of Science (Project No.11B-1c039).

References

[1] Lorke A, Luyken R J, Govorov A O, Kotthaus J P, Garcia J M and Petroff P M 2000 *Phys. Rev. Lett.* **84** 2223–6
[2] Lee J H, Wang Zh M, Abuwaar Z Y, Strom N W and Salamo G J 2006 *Nanotechnology* **17** 3973–6
[3] Mano T, Kuroda T, Sanguinetti S, Ochiai T, Tateno T, Kim J, Noda T, Kawabe M, Sakoda K, Kido G and Koguchi N 2005 *Nano Lett.* **5** 425–8
[4] Culchac F J, Porras-Montenegro N, Granada J C and Latgé A 2008 *Microelectron. J.* **39** 402–6
[5] Culchac F J, Porras-Montenegro N and Latgé A 2008 *J. Phys.: Condens. Matter* **20** 285215–20
[6] Culchac F J, Porras-Montenegro N, Granada J C and Latgé A 2009 *J. Appl. Phys.* **105**, 094324–8
[7] Castelano L K, Hai G -Q, Partoens B and Peeters F M 2006 *Phys. Rev.* B **74** 045313–7
[8] Liu Y M, Huang G M and Shi T Y 2008 *Phys. Rev.* B **77** 115311–4
[9] Szafran B and Peeters F M 2005 *Phys. Rev.* B **72** 155316–24
[10] Dias da Silva L G G V, Villas-Bôas J M and Ulloa S E 2007 *Phys. Rev.* B **76** 155306–10
[11] Kuan Wen-Hsuan, Tang Chi-Shung and Chang Cheng-Hung 2007 *Phys. Rev.* B **75** 155326–34
[12] Alves F M, Trallero-Giner C, Lopez-Richard V and Marques G E 2008 *Phys. Rev.* B **77** 035434–40
[13] Farias G A, Degani M H, Freire J A K, Costa e Silva J and Ferreira R 2008 *Phys. Rev.* B **77** 085316–24
[14] Elabsy A M 1993 *Physica Scripta* **48** 376–8
[15] Reyes-Gómez E, Raigoza N and Oliveira L E 2008 *Phys. Rev.* B **77** 115308–13
[16] Elabsy A M 1994 *J. Phys.: Condens. Matter* **6** 10025–30
[17] Raigoza N, Morales A L, Montes A, Porras-Montenegro N and Duque C A 2004 *Phys. Rev.* B **69**, 045323–30
[18] Aspnes D E 1976 *Phys. Rev.* B **14** 5331–43
[19] Wojs A, Hawrylak P, Farard S and Jacak L 1996 *Phys. Rev.* B **54** 5604–8
[20] Abramowitz M and Stegun I A 1964 *Handbook of mathematical functions with formulas, graphs and mathematical tables* (Courier Dover Publications: Mineola, New York) pp 446–52

Effects of hydrostatic pressure and temperature on interband optical transitions in InAs/GaAs vertically coupled double quantum dots

H M Baghramyan[1,3], M G Barseghyan[1,2,4] and A A Kirakosyan[1]

[1] Department of Solid State Physics, Yerevan State University, 1 Alex Manoogian,Yerevan 0025, Armenia
[2] Base College of Yerevan State University of Architecture and Construction, 105 Vahan Teryan, Yerevan 0009, Armenia

Email: henrikh.baghramyan@gmail.com

Abstract. We consider the effect of hydrostatic pressure, temperature and the variations of structure's sizes on interband transition energy and absorption coefficient in *InAs/GaAs* vertically coupled double quantum dots. The threshold energy of interband optical transitions is examined as a function of hydrostatic pressure and temperature for the different geometries of the structure. We also investigated the dependencies of the interband light absorption coefficient on the incident photon energy.

1. Introduction

Coupled quantum structures, consisting of two smaller band-gap nanomaterials, which are different in band-gap energies and are separated by a thin embedded barrier with a larger band gap, have been receiving considerable attention because of the interest in both investigations of fundamental physics [1-3] and for potential applications in electronic and optoelectronic devices [4-6]. Among the many coupled double quantum structures *InAs/GaAs* coupled double quantum dots (DQDs) are of current interest because of their promising applications in quantum information storage devices [7] and detectors [8]. M. Bayer, et al. [9] demonstrate coupling and entangling of quantum states in a pair of vertically aligned, self-assembled quantum dots (QDs) by studying the emission of an interacting electron-hole pair in a single dot molecule as a function of the distance between dots. From optical studies of excitons and corresponding calculations authors demonstrated that an electric field on vertically coupled pairs of $In_{0.6}Ga_{0.4}As/GaAs$ QDs controls the mixing of the excitonic states on the two dots and also provides controllable coupling between carriers in the dots [10].

Many works related to the theoretical investigation of electronic and excitonic states of semiconductor DQD have been reported recently [11-15]. In the adiabatic approximation and using an exact diagonalization technique the energy spectrum in vertically coupled DQD is calculated [11]. The effect of the electric field on electronic states in vertically coupled DQD [12] and the effect of the magnetic field on exciton ground-state energy in double and triple vertically coupled QDs have been

[3] H M Baghramyan
[4] Work was partially financed by Armenian State Committee of Science (Project No.11B-1c039)

investigated [13,14]. Using the three-dimensional finite-difference method the interband transition energies of *InAs/GaAs* DQD have been calculated [15]. In this work the photoluminescence (PL) experimental transition energies for *InAs/GaAs* DQDs, as determined from the temperature-dependent PL spectra, were in reasonable agreement with those of the theoretical interband transition energies calculated by using a finite-difference method taking into account shape-based strain and nonparabolic effects.

Hydrostatic pressure is a thermodynamic variable for the solid state, which provides a powerful tool to control and investigate the electronic states and optical properties of semiconductor materials. In [16] the effect of hydrostatic pressure on the interband optical transitions in *InAs/GaAs* QDs with lens shape geometry has been investigated.

In this work the effect of hydrostatic pressure and temperature on interband transition energy is investigated in *InAs/GaAs* vertically coupled DQDs. The threshold energy of interband optical transitions is examined as a function of hydrostatic pressure and temperature for the different geometries of the structure. We also investigate the dependencies of the interband light absorption coefficient on the incident photon energy. The paper is organized as follows: in Section 2 we describe the theoretical framework, Section 3 is dedicated to the results and discussion, and finally, our conclusions are given in Section 4.

2. Theoretical framework

In figure 1 we present the schematic view of vertically coupled DQDs.

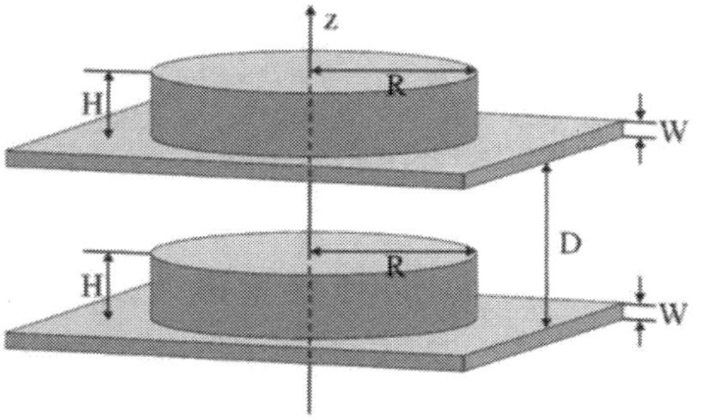

Figure 1. Schematic image of vertically coupled DQDs. H is the height of QD, R is the radius of QD, W is the thickness of wetting layers and D is the spacer thickness.

In the effective-mass approximation, the Hamiltonian in cylindrical coordinates for electron (heavy hole) in *InAs/GaAs* vertically coupled DQDs under the influence of hydrostatic pressure (P) and temperature (T) is given by

$$\hat{H}_{e,hh} = -\frac{\hbar^2}{2m_{e(hh)}(P,T)}\left(\frac{\partial^2}{\partial\rho^2}+\frac{1}{\rho}\frac{\partial}{\partial\rho}+\frac{\partial^2}{\partial z^2}+\frac{1}{\rho^2}\frac{\partial^2}{\partial\varphi^2}\right)+V_{e(hh)}(\rho,z,P,T), \quad (1)$$

where $m_{e(hh)}(P,T)$ and $V_{e(hh)}(\rho,z,P,T)$ are, respectively, the pressure- and temperature- dependent electron (heavy hole) effective mass and the confining potential. For *InAs* $m_{e(hh)}(P,T)$ are given by [16,17]:

$$m_e(P,T) = m_0\left(1+3.8\cdot 10^4\mathrm{meV}\left(\frac{3E_g(P,T)+760\mathrm{meV}}{3E_g(P,T)\left(E_g(P,T)+380\mathrm{meV}\right)}\right)\right)^{-1}, \quad m_{hh} = m_0(\gamma_1-2\gamma_2)^{-1}, \quad (2)$$

where $\gamma_1 = 20.4$ and $\gamma_2 = 8.37$ [16,17], m_0 is the free electron mass, $E_g(P,T)$ is the bulk *InAs*(*GaAs*) bandgap

$$E_g(P,T) = E_g^0 + \alpha P - \beta T^2 (T+c)^{-1}, \tag{3}$$

$E_g^0 = 533\text{meV}$ ($E_g^0 = 1519\text{meV}$) is the energy gap at $P = 0$ kbar and $T = 4\text{K}$, $\alpha = 7.7\text{meV/kbar}$ ($\alpha = 10.8\text{meV/kbar}$) is the pressure coefficient and $\beta = 0.276\text{meV/K}$ ($\beta = 0.5405\text{meV/K}$) and $c = 83$ K ($c = 204\text{K}$) are temperature coefficients [16].

In equation (1) $V_{e(hh)}(\rho, z, P, T)$ is the pressure and temperature dependent confining potential for electron (heavy hole):

$$V_{e(hh)}(\rho,z,P,T) = \begin{cases} G_{c(v)}\left(E_g^{\text{GaAs}}(P,T) - E_g^{\text{InAs}}(P,T)\right) & \text{outside QD}, \\ 0 \quad \text{inside QD}, \end{cases} \tag{4}$$

where G_c and G_v are band offsets for conductance (0.54) and valence (0.46) bands respectively [20].

Since the height of each QD is much smaller than their radii the motion along the z axis is strongly confined. Taking this into account the Schrödinger equation with Hamiltonian (1) can be solved using the adiabatic approximation [11, 12, 19]. So, the wave functions of electron and heavy hole can be given by $\Psi(\rho,\varphi,z) = (2\pi)^{-1/2} e^{i\mu\varphi} g_\rho^\nu(z) f^{\nu,\mu}(\rho)$, where $g_\rho^\nu(z)$ is the wave function in z direction, $f^{\nu,\mu}(\rho)$ is the radial wave function [11], $\mu = 0, \pm1, \pm2, ...$ is the magnetic quantum number and ν is the subband number.

For calculating the absorption coefficient we employ the known expression [20]

$$\alpha(\omega,P,T) = \frac{4\pi^2 e^2}{ncm_0^2 V_0 \omega} \sum_{(i,f)} \left|\vec{e}\vec{p}_{if}\right|^2 \delta\left(E_f(P,T) - E_i(P,T) - \hbar\omega\right)\left[f\left(E_i(P,T)\right) - f\left(E_f(P,T)\right)\right] \tag{5}$$

where e is the electron charge, c is the speed of light, V_0 is the normalized volume of the sample, n is the refraction index, ω is the frequency of light, $\vec{e}$ is the light wave polarization vector, summation is carried out over all initial (i) and final (f) states, and $f(E)$ is the Fermi - Dirac distribution function.

For interband transitions in equation (5) the momentum matrix element is $\vec{P}_{if} = \vec{P}_{cv} Q_{if}$, where $\vec{P}_{cv} = \langle u_c \,|\, \hat{\vec{p}} \,|\, u_v \rangle$, u_c and u_v are the Bloch amplitudes at the centre of the Brillouin zone, $Q_{if} = \left\langle \psi_f^* \middle| \psi_i \right\rangle$ is the overlap integral, and ψ_f and ψ_i are the envelope functions of final and initial states, respectively. For interband transitions, we obtain selection rule $\mu = \mu'$. Under the condition of the conduction band being practically empty (low temperatures), for the absorption coefficient we obtain:

$$\alpha(\omega,P,T) = \frac{4\pi^2 e^2}{ncm_0^2 V_0 \omega} \sum_{\mu,\nu,\nu'} \left| I_z^{\text{I}} \int_0^{R(P)} \langle f^{\nu',\mu}(\rho) \,|\, f^{\nu,\mu}(\rho) \rangle \rho d\rho + I_z^{\text{II}} \int_{R(P)}^{\infty} \langle f^{\nu',\mu}(\rho) \,|\, f^{\nu,\mu}(\rho) \rangle \rho d\rho \right|^2 \times$$

$$\frac{\Gamma}{\left(\hbar\omega - \Delta_{\mu\nu\nu'}\right)^2 + \Gamma^2}, \tag{6}$$

where $I_z^{(1)}$ and $I_z^{(2)}$ are the overlap integrals of wave functions in z direction when $0 < \rho < R(P)$ and $\rho > R(P)$ respectively, $R(P)$ is the pressure dependent radius of QD, Γ ($= 4meV$) is the Lorentzian parameter and $\Delta_{\mu\nu\nu'} = E_e^{\nu',\mu}(P,T) + E_h^{\nu,\mu}(P,T) + E_g^{InAs}(P,T)$ is the threshold energy.

3. Results and discussion

The calculations are performed only for the first allowed interband transitions from the heavy hole ground (symmetric) states with $\nu = 1, \mu = 0$ to the electron ground states with $\nu' = 1,\ \mu' = 0$. In all figures of this section we have used 5.4Å value for wetting layer thickness $W(0)$.

In figures 2 (a) - (c) we present the threshold energy of interband transitions in *InAs/GaAs* vertically coupled DQDs as a function of hydrostatic pressure. Several values of the spacer thickness (D) (figure 2 (a)), height (H) (figure 2 (b)) and radius (R) of QDs have been considered. For comp-

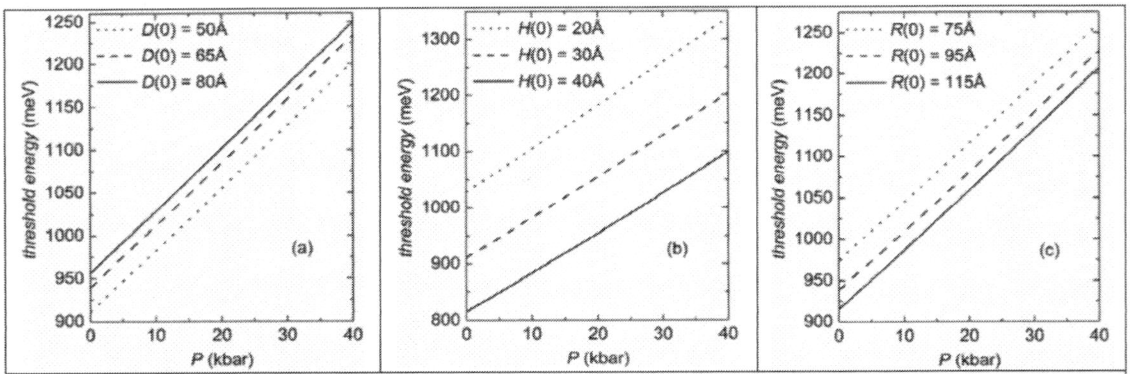

Figure 2. The threshold energy of the interband optical transitions in vertically coupled *InAs/GaAs* DQDs as a function of hydrostatic pressure. Several values of spacer thickness $D(0)$ ($H(0)$ = 30Å, $R(0)$ = 120Å – (a)), height $H(0)$ (D(0) = 50Å, $R(0)$ = 120Å – (b)) and radius $R(0)$ ($D(0)$ = 50Å, $H(0)$ = 30Å – (c)) of QDs have been considered at $T = 4K$.

lete range of hydrostatic pressure considered, the following effects are observed: 1) the height of confining potential is an increasing function of hydrostatic pressure (because of the difference of pressure coefficients of *InAs* and *GaAs*), which itself increases the threshold energy; 2) the energy gap of *InAs* material is an increasing function of hydrostatic pressure, which itself increases the threshold energy; 3) the dimensions of the structure are decreasing functions of hydrostatic pressure and they force the threshold energy to increase; 4) the electron-effective mass is an increasing function of hydrostatic pressure, which itself decreases the threshold energy. Since, the first three factors are stronger than the last one, the threshold energy is an increasing function of hydrostatic pressure. From

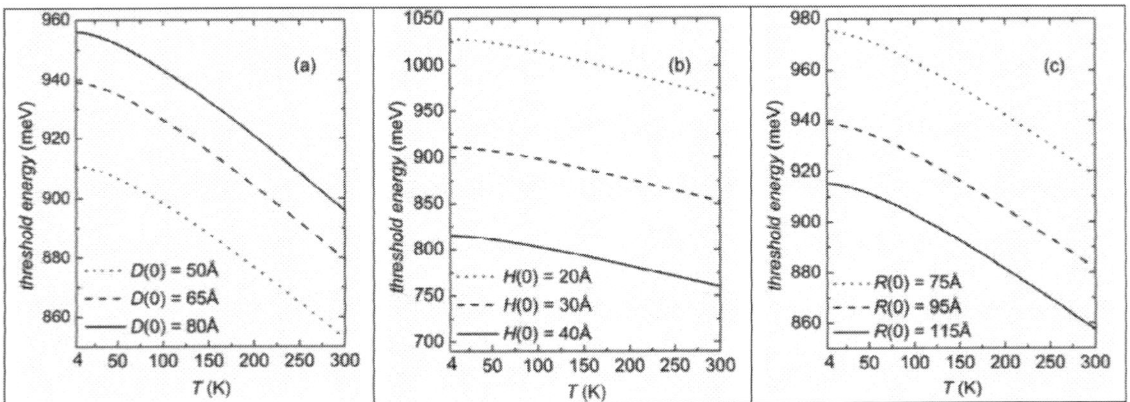

Figure 3. The threshold energy of the interband optical transitions in vertically coupled *InAs/GaAs* DQDs as a function of temperature. Several values of spacer thickness $D(0)$ ($H(0)$ = 30Å, $R(0)$ = 120Å – (a)), height $H(0)$ ($D(0)$ = 50Å, $R(0)$ = 120Å – (b)) and radius $R(0)$ (D(0) = 50Å, $H(0)$ = 30Å – (c)) of QDs have been considered for $P = 0$.

Journal of Physics: Conference Series **350** (2012) 012017 doi:10.1088/1742-6596/350/1/012017

figure 2 (a) it is clear that with the increase of the spacer thickness the threshold energy increases, which is caused by the increase of distance between the energy levels of symmetric states of electron and hole. With the increase of the heights (figure 2 (b)) and radii (figure 2 (c)) of QDs the size quantization of electron and hole weakens and threshold energy decreases.

The effect of temperature on threshold energy of interband transitions in *InAs/GaAs* vertically coupled DQDs is presented in figure 3. The effect of temperature leads to the following variations of parameters: 1) the height of confining potential is a decreasing function of temperature (because of the difference of temperature coefficients of *InAs* and *GaAs*), which itself decreases the threshold energy; 2) the energy gap of *InAs* material is an decreasing function of temperature, which itself decreases the threshold energy; 3) the effective mass of electron is a decreasing function of temperature, which itself increases the threshold energy. Since the first two factors are stronger than the last one, the threshold energy is a decreasing function of temperature. The effect of variations of structure's dimensions on threshold energy is the same as in previous case.

In figure 4, the dependences of the absorption coefficient caused by interband transitions from the heavy hole ground states with $\nu = 1, \mu = 0$ to the electron ground states with $\nu' = 1, \mu' = 0$ on incident photon energy for different values of the spacer thickness (figure 4 (a)), height (figure 4 (b)), radius

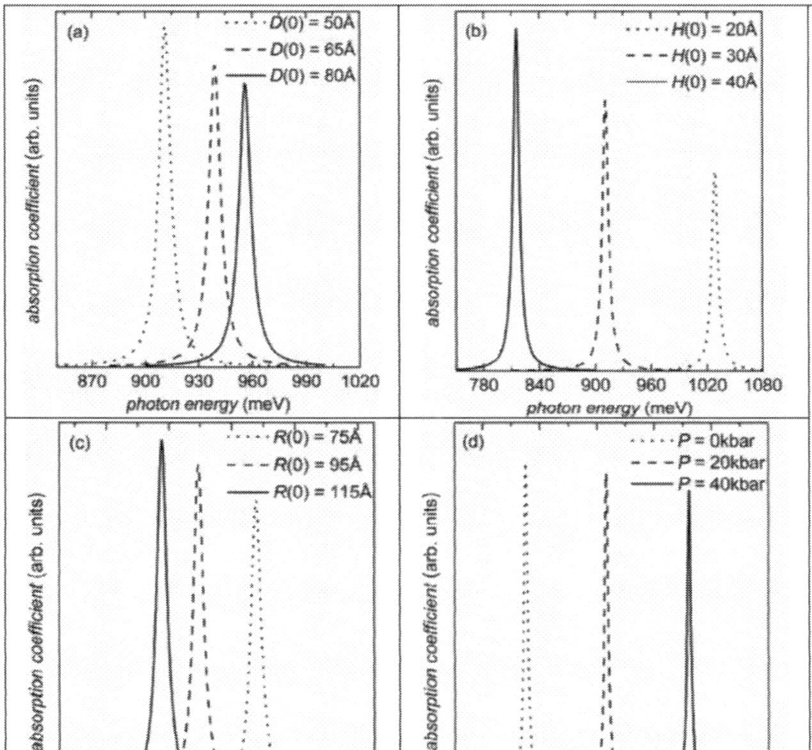

Figure 4. The absorption coefficient of the interband optical transitions in vertically coupled *InAs/GaAs* DQDs as a function of incident photon energy. Several values of spacer thickness $D(0)$ ($H(0)$ = 30Å, $R(0)$ = 120Å – (a)), height $H(0)$ ($D(0)$ = 50Å, $R(0)$ = 120Å – (b)) and radius $R(0)$ ($D(0)$ = 50Å, $H(0)$ = 30Å – (c)) of QDs for P = 0 and hydrostatic pressure P ($D(0)$ = 50Å, $H(0)$ = 30Å, $R(0)$ = 75Å – (d)) have been considered at T = 4K.

(figure 4 (c)) of QDs and hydrostatic pressure (figure 4 (d)) have been considered. Figure 4 (a) and (d) show that with the increase of the spacer thickness and hydrostatic pressure respectively, the peak position of the absorption coefficient is shifted to the high energy region ("blue shift"). Such a behavior caused by the increase of the threshold energy between initial and final states. From figure 4 (b) and (c) it is clear that with the increase of heights and radii of QDs the peak position of the absorption coefficient is shifted to the low energy region ("red shift"). In all cases the maximum value of the absorption coefficient is greater for the smaller values of incident photon energy. The reason of this result is the decrease of the energy distance between initial and final states which leads to the increase of overlap integrals.

4. Conclusions

The effects of hydrostatic pressure, temperature and variation of the structure's sizes on interband transition energy and absorption coefficient are investigated in *InAs/GaAs* vertically coupled DQDs. The threshold energy of interband optical transitions is examined as a function of hydrostatic pressure and temperature for the different geometries of the structure. We also investigated the dependencies of the interband light absorption coefficient on the incident photon energy. The main findings can be summarized as follows: 1) the threshold energy of interband transition is an increasing function of hydrostatic pressure, 2) the threshold energy of the interband transitions is a decreasing function of temperature, 3) the "blue shift" is observed in the interband absorption spectrum with the increase of spacer thickness of DQDs and hydrostatic pressure, 4) the "red shift" is observed in the interband absorption spectrum with the increase of heights and radii of DQDs.

References

[1] Gerardot B D, Strauf S, de Dood M J A, Bychkov A M, Badolato A, Hennessy K, Hu E L, Bouwmeester D and Petroff P M 2005 *Phys. Rev. Lett.* **95** 137403–6
[2] Kim T W, Lee H S and Park H L 2006 *J. Appl. Phys.* **100** 113505–8
[3] Robledo L, Elzerman J, Jundt G, Atature M, Hogele A, Falt S and Imamoglu A 2008 *Science* **320** 772–5
[4] Erwin S C, Zu L, Haftel M I, Efros A L, Kennedy T A and Norris D J 2005 *Nature* **436** 91–4
[5] Yang T, Tatebayashi J, Aoki K, Nishioka M and Arakawa Y 2007 *Appl. Phys. Lett.* **90** 111912–4
[6] Fushman I, Englund D, Faraon A, Stoltz N, Petroff P and Vucković J 2008 *Science* **320** 769–72
[7] Stinaff E A, Scheibner M, Bracker A S, Ponomarev I V, Korenev V L, Ware M E, Doty M F, Reinecke T L and Gammon D *Science* 2006 **311** 636–9
[8] Gustavsson S, Studer M, Leturcq R, Ihn T and Ensslin K 2007 *Phys. Rev. Lett.* **99** 206804–7
[9] Bayer M, Hawrylak P, Hinzer K, Fafard S, Korkusiński M, Wasilewski Z R, Stern O and Forchel A 2001 *Science* **291** 451–3
[10] Ortner G, Bayer M, Lyanda-Geller Y, Reinecke T L, Kress A, Reithmaier J P and Forchel A 2005 *Phys. Rev. Lett.* **94** 157401–4
[11] Korkusiński M and Hawrylak P 2001 *Phys. Rev.* B **63** 195311–7
[12] Liu Jin-Long, Li Shu-Shen, Niu Zhi-Chuan, Yang Fu-Hua and Feng Song-Lin 2003 *Superlatt. and Microstruct.* **33** 29–40
[13] Janssens K L, Partoens B and Peeters F M 2002 *Phys. Rev.* B **66** 075314–22
[14] Janssens K L, Peeters F M, Schweigert V A and Partoens B 2001 *Physica* B **298** 277–81
[15] Kwona H Y, Woo J T, Lee D U, Kima T W and Park Y J 2009 *Solid State Communication* **149** 52–55
[16] Rodríguez A H, Trallero-Ginner C, Duque C A and Vázquez G J 2009 *J. Appl. Phys.* **105** 044308–13
[17] Trallero-Giner C, Kunc K and Syassen K 2006 *Phys. Rev.* B **73** 205202–7
[18] Raigoza N, Reyes-Gómez E, Duque C A and Oliveira L E 2007 *J. Phys.: Condens. Matter* **19** 256202
[19] Kazaryan E M, Meliksetyan A V, Petrosyan L S and Sarkisyan H A 2006 *Physica* E **31** 228–31
[20] Basu P K 2002 *Theory of Optical Processes in Semiconductors* (Oxford University Press: New York) pp 28–44

Direct interband absorption of light in a strongly oblate truncated ellipsoidal quantum dot's ensemble

D.B. Hayrapetyan[a,b1], **K.G. Dvoyan**[a]
[a]Dept. of Physics and Technology, Russian-Armenian (Slavic) University, 123 Hovsep Emin Str., 0051 Yerevan, Armenia, Phone: (+37493) 289-451, Fax: (+37410) 229-244;
[b]Dept. of Physics, State Engineering University of Armenia,
105 Terian Str., 0009 Yerevan, Armenia, Phone: (+37493) 690-662; Fax: (+37410) 567-991;

E-mail: dhayrap82@gmail.com

Abstract. Within the framework of adiabatic approximation the energy levels and direct interband absorption of light in a strongly oblate truncated ellipsoidal quantum dot's ensemble are studied. Analytical expressions for the particle energy spectrum and absorption threshold frequencies in the regime of strong size quantization are obtained. Selection rules for quantum transitions are revealed. To facilitate the comparison of obtained results with the probable experimental data, size dispersion distribution of growing quantum dots by the minor semiaxe by two experimentally realizing distribution functions have been taken into account. Distribution functions of Lifshits-Slezov and Gaussian have been considered.

1. Introduction

Development of the novel growth techniques, such as the Stranski–Krastanov epitaxial method etc., makes possible to grow semiconductor quantum dots (QDs) of various shapes and sizes [1-3]. As it is known, the energy spectrum of charge carriers in QDs is completely quantized and resembles the energy spectrum of atoms ("artificial atoms"). A key factor to control the energy level structure of semiconductor QDs is the external geometrical shape. It is shown that even the small change in external shape of QD strongly influences the energy spectrum and other characteristics of such semiconductor structures ([4] and Ref. therein). From the theoretical point of view, spherical QDs are the easier to investigate taking into account their symmetry, which allows to obtain analytical solutions for the energy spectrum, coefficient of absorption, charge carriers mobility, etc. [8,9]. However, modern growth techniques make possible to obtain QDs of different geometrical shapes and sizes. Spherical, cylindrical, pyramidal, lens shaped, ellipsoidal QDs are considered in many works [10-15]. The confinement potential of QD can be approximated by a parabolic potential with a high accuracy in most cases. However, an effective parabolic potential may arise due to QD external shape [5,6].

Theoretical investigations of optical properties of QDs of various shapes are important task for modelling and constructions of new generation semiconductor devices [7]. Investigations of the optical absorption spectrum of various semiconductor structures are a powerful tool for determination

[1] To whom any correspondence should be addressed.

of many characteristics of these systems: forbidden band gaps, effective masses of electrons and holes, their mobilities, dielectric permittivities, etc. There are many works devoted to the theoretical and experimental study of the optical absorption both in bulk semiconductors and in low-dimentional systems. The presence of size quantization (SQ) essentially influences the absorption mechanism. In fact, the formation of new energy levels of the SQ makes possible new interlevel transitions.

Special interest represents consideration of semiellipsoidal QDs, which two geometrical parameters (semiaxes) allows one to control energy spectrum inside QD.

In this paper the electronic states and direct interband absorption of light in a strongly oblate truncated ellipsoidal QD (SOTEQD) are discussed. Absorption edge and absorption coefficient are obtained in a strong SQ regime. To facilitate the comparison of obtained results with the probable experimental data size dispersion distribution of growing QDs by the minor semiaxe has been taken into account for two experimentally realizing distribution functions. Distribution function of Lifshits-Slezov has been considered in the first model and distribution function of Gauss has been considered in the second case.

2. Theory

Let us consider an opaque strongly oblate truncated ellipsoidal QD (fig. 1 a)). The potential energy of a particle (electron, hole, exciton) in the cylindrical coordinates can be presented in the form

$$U(X,Y,Z)=\begin{cases}0, \dfrac{X^2+Y^2}{a_1^2}+\dfrac{Z^2}{c_1^2}\leq 1, \cap Z\geq -b_1 \\ \infty, \dfrac{X^2+Y^2}{a_1^2}+\dfrac{Z^2}{c_1^2}>1, \cap Z<-b_1\end{cases}, \quad c_1 \ll a_1 , \tag{1}$$

where c_1 and a_1 are the minor and major semiaxes of the SOTEQD, respectively.

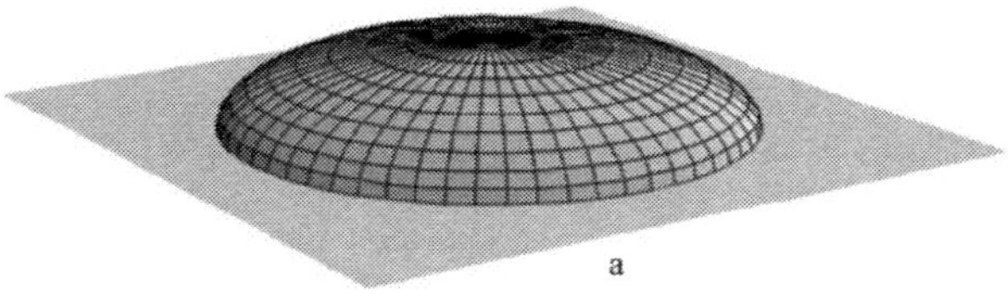
a

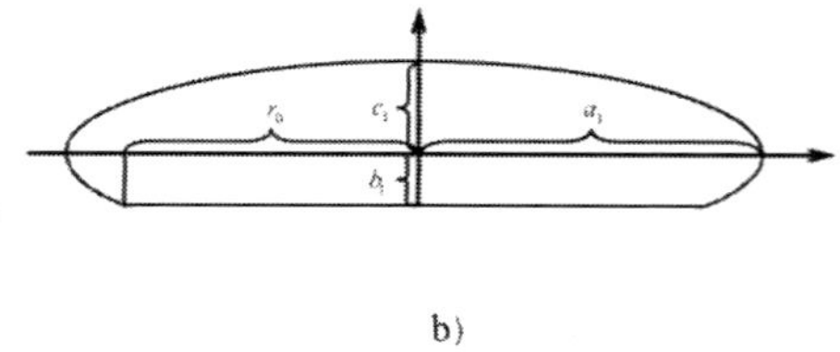

b)

Figure 1. Strongly oblate truncated ellipsoidal QD.
a) 3D picture, b) cross-section.

In the regime of strong SQ the energy of the Coulomb interaction between an electron and hole is much less than the energy determined by the walls of the SOTEQD. In this approximation the Coulomb interaction can be neglected. Then the problem is reduced to the determination of the energy states of an electron and hole separately. From the geometrical form of a QD it follows that the particle motion along the Z-direction is faster than in the plane perpendicular to it. This allows one to use the adiabatic approximation [16]. The Hamiltonian of this system in the cylindrical coordinates has the form

$$\hat{H}=-\frac{\hbar^2}{2m^*}\left\{\frac{\partial^2}{\partial\rho^2}+\frac{1}{\rho}\frac{\partial}{\partial\rho}+\frac{1}{\rho^2}\frac{\partial^2}{\partial\varphi^2}\right\}-\frac{\hbar^2}{2m^*}\frac{\partial^2}{\partial Z^2}+U . \tag{2}$$

It can be represented as a sum of the "fast" ($\hat{H}_1$) and "slow" ($\hat{H}_2$) subsystems Hamiltonians in dimensionless quantities:

$$\hat{H} = \hat{H}_1 + \hat{H}_2 + \hat{U}(r,\varphi,z), \tag{3}$$

where

$$\hat{H}_1 = -\frac{\partial^2}{\partial z^2}, \tag{4}$$

$$\hat{H}_2 = -\left[\frac{\partial^2}{\partial r^2} + \frac{1}{r}\frac{\partial}{\partial r} + \frac{1}{r^2}\frac{\partial^2}{\partial \varphi^2}\right]. \tag{5}$$

Here $\hat{H} = \frac{\hat{\mathrm{H}}}{E_R}$, $r = \frac{\rho}{a_B}$, $z = \frac{Z}{a_B}$, e and μ_p are the charge and effective mass of a particle (electron, hole, exciton), correspondingly, $E_R = \frac{\hbar^2}{2m^* a_B^2}$ is the effective Rydberg energy, $a_B = \frac{\kappa\hbar^2}{m^* e^2}$ is the effective Bohr radius of a particle, κ is the dielectric permittivity. The wave function is sought in the form

$$\psi(r,\varphi,z) = e^{im\varphi}\chi(z;r)R(r). \tag{6}$$

At a fixed value of the coordinate r of the slow subsystem the particle motion is localized in a one-dimensional potential well with the effective width (fig. 1 b))

$$z(r) = \begin{cases} 2c\sqrt{1-\frac{r^2}{a^2}}, r \geq r_0 \\ c\sqrt{1-\frac{r^2}{a^2}} + b, r < r_0 \end{cases}, \tag{7}$$

where $a = \frac{a_1}{a_B}$, $c = \frac{c_1}{a_B}$, $b = \frac{b_1}{a_B}$ and $r_0 = a\sqrt{1-\frac{b^2}{c^2}}$. Solving the Schrödinger equation for the first subsystem, we obtain for the energy spectrum of a particle

$$\varepsilon_1(r) = \frac{\pi^2 n^2}{\left(c\sqrt{1-\frac{r^2}{a^2}} + b\right)^2}, \quad n = 1, 2, \ldots. \tag{8}$$

Here n is a quantum number (QN) of the "fast" subsystem. For the low levels of the spectrum a particle is localized mainly in the region $r \ll a$. Based on this one can expand $\varepsilon_1(r)$ into a series

$$\varepsilon_1(r) \simeq \alpha_n + \beta_n^2 r^2, \tag{9}$$

where $\alpha_n = \frac{\pi^2 n^2}{(b+c)^2}$, $\beta_n = \frac{\pi n}{a}\sqrt{\frac{c}{(b+c)^3}}$. Expression (9) enters the Schrödinger equation of the slow subsystem as an effective potential. For the total energy and the wave functions of the system finally we get

$$\varepsilon = \alpha_n + 2\beta_n(N+1), \quad N = 0, 1, 2, \ldots, \tag{10}$$

$$\psi(r,\varphi,z)=\frac{e^{im\varphi}}{\sqrt{2\pi}}\sqrt{\frac{2}{c\sqrt{1-\frac{r^2}{a^2}}+b}}\sin\left(\frac{\pi n}{c\sqrt{1-\frac{r^2}{a^2}}+b}z+\frac{\pi nb}{c\sqrt{1-\frac{r^2}{a^2}}+b}\right)\times$$
$$\times\sqrt{2\beta_n}\frac{\sqrt{n_r!}}{\Gamma^{1/2}(n_r+|m|+1)}e^{-\frac{\beta_n r^2}{2}}\left(\beta_n r^2\right)^{\frac{|m|}{2}}L_{n_r}^{|m|}\left(\beta_n r^2\right) \quad (11)$$

Here N is oscillatory QN, m is magnetic QN, n_r is radial QN, $L_{n_r}^{|m|}$ is Laguerre polynomials.

3. Direct interband absorption of light

We proceed to the consideration of the direct interband absorption of light in a SOTEQD in the regime of strong SQ, when the Coulomb interaction between an electron and hole can be neglected. WE consider the case of a heavy hole, when $m_e^* \ll m_h^*$. The absorption coefficient is defined by the expression

$$K=A\sum_{\nu,\nu'}\left|\int\Psi_\nu^e\Psi_{\nu'}^h d\vec{r}\right|^2\delta\left(\hbar\Omega-E_g-E_\nu^e-E_{\nu'}^h\right), \quad (12)$$

where ν and ν' are the sets of QNs corresponding to an electron and heavy hole, E_g is the forbidden band width of a bulk semiconductor, Ω is the frequency of the incident light, A is a quantity proportional to the square of the matrix element, taken over the Bloch functions [17]. Finally, in the regime of strong SQ, for the quantity K and absorption edge (AE) we obtain

$$K=A\sum_{n,m,N}\delta\left(\hbar\Omega-E_g-E_e-E_h\right), \quad (13)$$

$$W_{100}=1+\frac{\pi^2 d^2}{(b_1+c_1)^2}+\frac{2\pi d^2}{a_1}\sqrt{\frac{c_1}{(b_1+c_1)^3}}, \quad (14)$$

where $W_{100}=\frac{\hbar\Omega_{100}}{E_g}$ and $d=\frac{\hbar}{\sqrt{2\mu E_g}}$. Expression (14) characterizes the dependence of the effective forbidden band width on the semiaxes a_1 and c_1. The AE shifts to the short-wave region with decrease in the both semiaxes, but the dependence on the minor semiaxis is stronger. Consider now the selection rules for QNs. For the magnetic QN the transitions between the levels with $m=-m'$ are allowed, and for the QN of the first subsystem – the transitions with $n=n'$. For the oscillatory QN the transitions for the levels with $N=N'$ are allowed. Note that the analytic form of expression (13) is presented with allowance for the mentioned selection rules of QNs.

So far we have studied the absorption of a system consisting of semiconductor QDs having identical dimensions. For comparison of the obtained results with experimental data, one has to take into account the random character of SOTEQD dimensions (or semiaxis) obtained in the growth process. The absorption coefficient should be multiplied by concentration of QDs. Instead of distinct absorption lines, account of size dispersion will give a series of maximums. In the first model we use the Lifshits-Slezov distribution function [9]:

$$P(u)=\begin{cases}\frac{3^4 eu^2\exp\left(-1/(1-2u/3)\right)}{2^{5/3}(u+3)^{7/3}(3/2-u)^{11/3}}, u<3/2\\ 0, u>3/2\end{cases}, \quad u=\frac{b+c}{\bar{b}+\bar{c}}, \quad (15)$$

where $\bar{c}$ is some average value of the semiaxis. In the second model the Gaussian distribution function is used (see e.g. [18]):

$$P(u)=Ce^{\frac{(u-1)^2}{\sigma/\bar{c}}}. \tag{16}$$

Two types of QDs ensembles with symmetric and asymmetric distributions can be formed using these two distribution functions specifying the growth process. Gaussian distribution describes the symmetric case and the Lifshits-Slezov distribution describes the asymmetric case. With a consideration of size distribution function the absorption coefficient is defined by the following expression

$$K=A\sum_{m,n,N}\int_0^{3/2}P(u)du\delta\left(\hbar\Omega-E_g-E_\nu^e-E_\nu^h\right). \tag{17}$$

In the strong SQ regime, with account of general size distribution function $P(u)$, we obtain for the absorption coefficient corresponding formula:

$$K=\frac{A}{E_g}\sum_{m,n,N}\frac{1}{\sqrt{\lambda_3^2+4\lambda_1\lambda_2}}\left(\frac{2\lambda_2}{\sqrt{\lambda_3^2+4\lambda_1\lambda_2}-\lambda_3}\right)^2 P\left(\frac{2\lambda_2}{\sqrt{\lambda_3^2+4\lambda_1\lambda_2}-\lambda_3}\right) \tag{18}$$

where $\lambda_1=\frac{\hbar\Omega-E_g}{E_g}$, $\lambda_2=\frac{\pi^2n^2}{4}\left(\frac{d}{\bar{b}_1+\bar{c}_1}\right)^2$, $\lambda_3=\frac{\pi n d^2(N+1)}{a_1}\sqrt{\frac{\bar{c}_1}{\left(\bar{b}_1+\bar{c}_1\right)^3}}$.

4. Discussion

As it is seen from formula (10), the energy spectrum of CCs in a SOTEQD has equidistant character. These results is true only for the low levels of the spectrum (for small values of QNs), when the adiabatic approximation is applicable. Numerical calculations for the regime of strong SQ were carried out for a *GaAs* QD with the following parameters: $m_e^*=0.067m_e$, $m_e^*=0.12m_h^*$, $\kappa=13.8$, $E_R=5.275\,\text{meV}$, $a_e=104\,\text{Å}$ and $a_h=15\,\text{Å}$ are the effective Bohr radii of an electron and hole, $E_g=1.43\,\text{eV}$ is the forbidden band width of a bulk semiconductor. In the regime of strong SQ the frequency corresponding to the transition between the equidistant levels (for the value $n=0$), for fixed values $a=2.5a_e$, $c=0.5a_e$ and $b=0.2a_e$ is $\omega_{10}=1.85\times10^{14}\text{s}^{-1}$, which corresponds to the infrared region of the spectrum. At the same value of the QN but with $a=2a_e$, $c=0.4a_e$ and

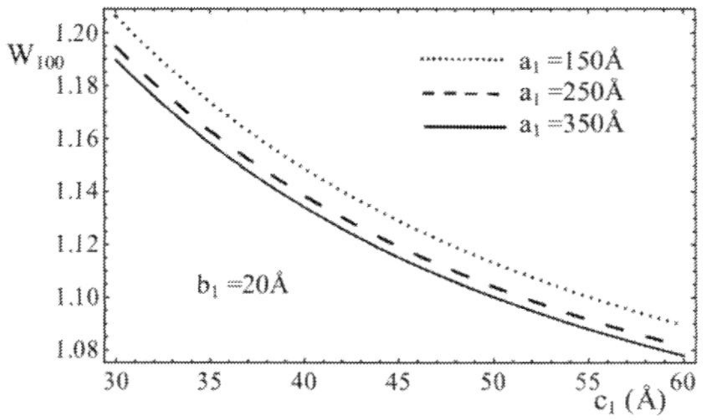

Figure 2. Dependences of the AE on the minor semiaxis of the SOTEQD at a fixed value of the major semiaxis.

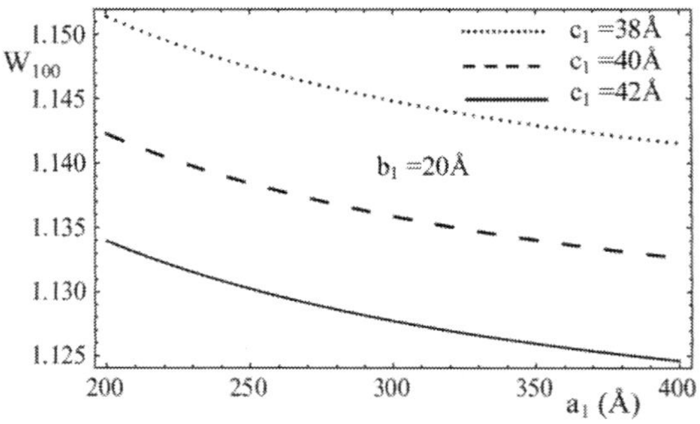

Figure 3. Dependences of the AE on the major semiaxis of the SOTEQD at a fixed value of the minor semiaxis.

$b = 0.2a_e$ we have $\omega_{10} = 2.5\times10^{14}\,s^{-1}$, which is 1.35 times as much as before. As it is seen from formula (10), with increasing semiaxes the particle energy decreases, and it is more "sensitive" to the variations of the minor semiaxis due to SQ contribution. It should be noted that with increasing semiaxes the energy levels become closer together, but remain equidistant.

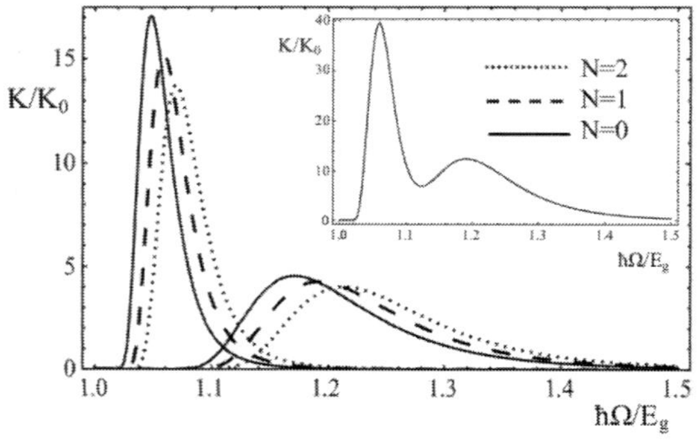

Figure 4. Dependencies of absorption coefficient on the frequency of incident light for the first two equidistant families in the case of Gaussian distribution (The sum of absorption coefficient for all transitions).

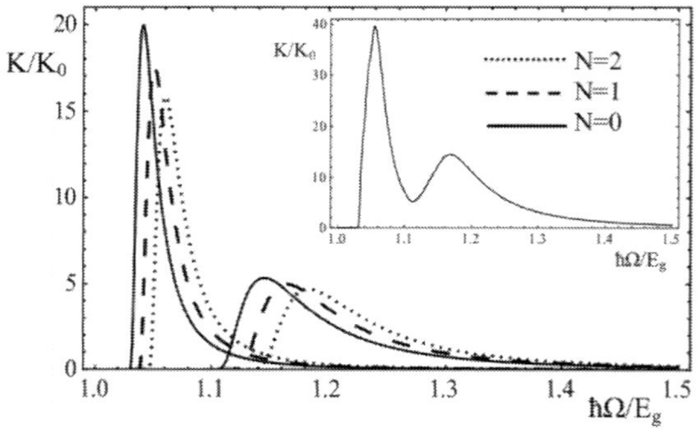

Figure 5. Dependencies of the absorption coefficient on the frequency of incident light for the first two equidistant families in the case of Lifshits-Slezov distribution (The sum of absorption coefficient for all transitions).

Figures 2 and 3 present the dependences of the AE on the minor and major semiaxes of a SOTEQD, respectively. With decreasing semiaxes the absorption edge increases, which is a consequence of the SQ increase (the "effective" forbidden band width increases). As can be seen from the plots, the change in the AE manifests itself clearly in the dependence on the minor semiaxis. For the same reason the curves corresponding to different values of the minor semiaxis (fig. 2) provide a larger shift than in the opposite case (fig. 3).

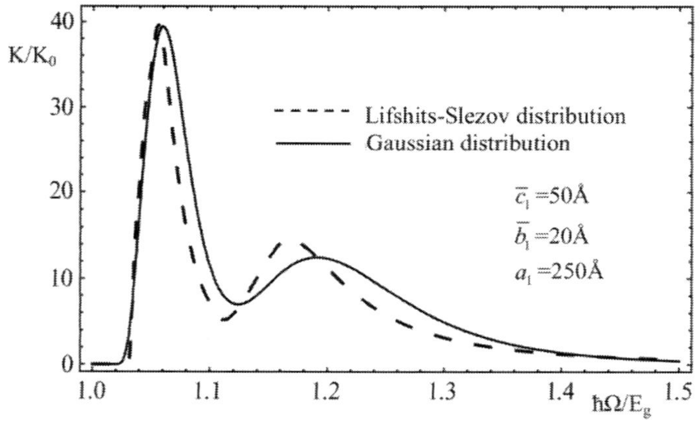

Figure 6. Dependences of the absorption coefficient on the frequency of incident light.

Fig.4 illustrates the dependence of absorption coefficient K on the frequency of incident light for the first two equidistant families in the case of Gaussian distribution, when $\bar{c}_1 = 50$ Å and $a_1 = 250$ Å. The sum of the absorption coefficients corresponding to the separate transitions are also presented on the top of the right side of the figure.

As it is mentioned above, instead of distinct absorption lines, consideration of size dispersion will give a series of maximums. Fig.5 illustrates the same dependence for the Lifshits-Slezov distribution case. Note that both in the model of Gaussian distribution and in the model of Lifshits-Slezov QDs distribution a single distinctly expressed maximum of absorption is observed. When the frequency of light is increased, the second weakly expressed maximum can be observed. Further increase of the incident light frequency results in a fall of absorption coefficient. Fig.6 illustrates the dependencies of the absorption coefficient on the frequency of incident light for both cases.

Schematic diagrams for appropriate transitions where the absorption of light is present are depicted on the fig.7 to understand in detail the process of absorption. From the comparison of diagram and fig.4 (fig.5), it is obvious that first clear expressed maximum corresponds to the $n = n' = 1$ transition family and weak expressed picks are the result of the transitions between quidistant levels. The second weaker maximum corresponds to the $n = n' = 2$ transition family. It is also obvious from fig.4 (fig.5) that the intensity corresponding to the above mentioned family is weaker than the first maximum. This is a result of the small value of the electron and hole wave functions overlapping integral, which means that the transition probability decreases.

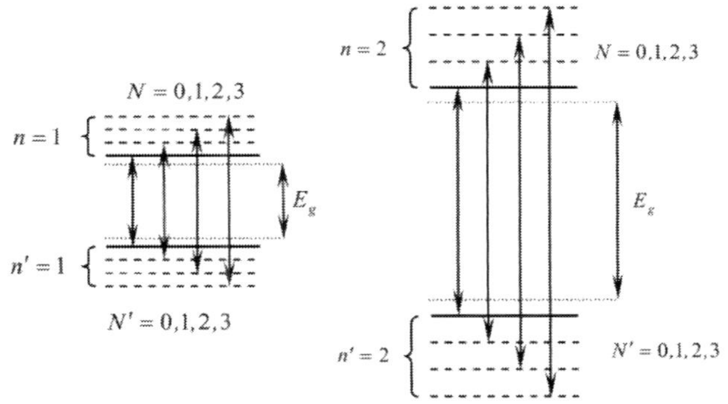

Figure 7. Schematic plot of corresponding interband transitions.

5. Conclusion

The energy levels of electrons in SOTEQD are shown to be equidistant and levels of "slow" subsystem are arranged on the levels of "fast" subsystem. These results are true only for the low levels of the spectrum or for small values of QNs, when the adiabatic approximation is applicable. It is shown that with increasing semiaxes the particle energy decreases, and it is more "sensitive" to the variations of the minor semiaxes due to SQ contribution. Red shift of absorption threshold has been observed depending on the values of geometrical sizes of QD.

6. References

[1] Harrison P 2005 *Quantum Wells, Wires and Dots. Theoretical and Computational physics*, Wiley, New York

[2] Bastard D 1989 *Wave Mechanics Applied to Semiconductor Heterostructures*, Les editions de physique, Paris

[3] Kazaryan E M, Petrosyan S G 2005 *Physical Principles of Semiconductor Nanoelectronics*, RAU, Yerevan

[4] Dvoyan K G, Kazaryan E M 2001 *Physica Status Solidi (B)*, Vol. **228**, 695

[5] Hayrapetyan D B 2007 *Journal of Contemporary Physics*, Vol. **42 (6)**, 293.
[6] Hayrapetyan D B, Dvoyan K G, Kazaryan E M 2007 *Journal of Contemporary Physics*, Vol. **42 (4)**, 151
[7] Sze S M 2009 *Semiconductor devices: Physics and technology*, Wiley India Pvt. Ltd
[8] Ramaniah L M, Nair SV 1993 *Phys. Rev. B* **47**, 7132
[9] Efros Al L, Efros A L 1982 *Sov. Phys. Semicond.* Vol. **16**, 772
[10] Dvoyan K G, Hayrapetyan D B, Kazaryan E M, Tshantshapanyan A A, 2009 *Nanoscale Research Letters* Vol. **4 (2)**, 130
[11] Sarkisyan H A 2002 *Modern Physics Letters B*, Vol. **16 (22)**, 835
[12] Aghekyan N G, Kazaryan E M, Petrosyan L S, Sarkisyan H A 2010 *Journal of physics: Conference Series*, Vol. **248 (1)**, 012048
[13] Dvoyan K G, Hayrapetyan D B, Kazaryan E M 2009 *Nanoscale Research Letters* Vol. **4 (2)**, 106
[14] Hwang T M, Lin W W, Wang W Ch, Wang W, 2004 *Journal of Computational Physics*, Vol. **196 (1)**, 208
[15] Sivakami A, Mahendran M 2010 *Physica B: Physics of Condensed Matter*, Vol. **405 (5)**, 1403
[16] Galitsky V M, Karnakov B M, Kogan V I, 1981 *Practical Quantum Mechanics*, Nauka, Moscow
[17] Anselm A I 1978 *Introduction to semiconductors theory*, Nauka, Moscow
[18] Kumar A, Biswas D, 2007 *Journal of Applied Physics*, Vol. **102 (8)**, 084305

International Symposium on Optics and its Applications (OPTICS2011) IOP Publishing
Journal of Physics: Conference Series **350** (2012) 012019 doi:10.1088/1742-6596/350/1/012019

Effect of radial electrostatic field on optical absorption in semiconductor nanotube

V A Harutyunyan[1] **and V A Gasparyan**[2]

[1]State Engineering University of Armenia, Gyumri Branch, 2 Mher Mkrtchyan Street, Gyumri 3103, Republic of Armenia,

[2]Russian-Armenian (Slavonic) State University, 123 Hovseph.Emin Street,Yerevan 375051, Republic of Armenia

[1]E-mail: volharut@gmail.com

Abstract. In the effective-mass approximation the single-electron states in a wide-band semiconductor nanocylindrical heterolayer in the presence of strong lateral-radial electrical field are considered. The explicit forms of the energy spectrum and envelope wave functions of single-particle states for charge carriers are obtained using variation approach. Corresponding absorption characteristics of interband and inter-subband optical transitions in the layer in the presence of strong radial field are calculated. The radial field leads to the explicit dependence of absorption intensity on the values of effective masses of charge carriers. It is shown that absorption's threshold frequencies depend explicitly on geometrical sizes of sample and intensity of external field.

1. Introduction

During the last decade along with various low-dimensional structures, semiconductor nanotubes (SNTs) are one of urgent objects of investigation and had been studied extensively both experimentally and theoretically (see reviews [1-3] and references therein). SNT is an emerging field and possesses potential to provide a wide range of functionalities. The precise controllability of structural and spatial positioning and versatile functionality make SNTs promising candidates for practical application in next generation nanoelectronic and nanophotonic devices. With the progress in synthesis and fabrication of different nanomaterials various kinds of tubular structures, such as CdX, HgX (X=S, Se, Te), compounds II_3-V_2 [4], InGaAs/GaAs, SiGe/Si [5,6], ZnO, GaN, AlN, AlGaN [7], InP/InAs/InP [8], BN, ZnS [9], SiGe/SiO/Ti, SiO_2,SiO_2/TiO_2 [10], SiGe/Si/SiN/Cr [11] and others are obtained and attract a lot of attention. It is necessary to note also that one of the distinctive features of SNTs is the compositional dependence of their electronic, structural and optical properties [12,13].

At the same time, it is now well recognized that the electronic, kinetic, optical and a number of other physical properties of low-dimensional structures (as well as those of bulk samples) are strongly affected by external static fields, in particularly, - by the external electrostatic fields. In the previous works of first author with co-authors the influence of external strong homogeneous electrostatic field and the influence of the weak (perturbating) radial electrostatic fields on the electronic and optical properties of SNTs are examined theoretically [14-18].

In this paper we investigate theoretically the influence of the strong radial electrostatic field on the energy spectrum of charge carriers and on the optical absorption spectrum in wide-band SNTs.

2. General approach and physical model

The system we consider is an infinite tube (along the *Z*-axis) with cylindrical symmetry. We will take into account the actual thickness of the layer and do not neglect the radial confinement of charge carriers.

Fig.1. shows schematically the composition under consideration. The thickness of layer equals $L = R_2 - R_1$, and the incident light wave $\vec{A}(\vec{r},t)$ is directed along the $Y-$ axis and is polarized linearly along the *X*- axis.

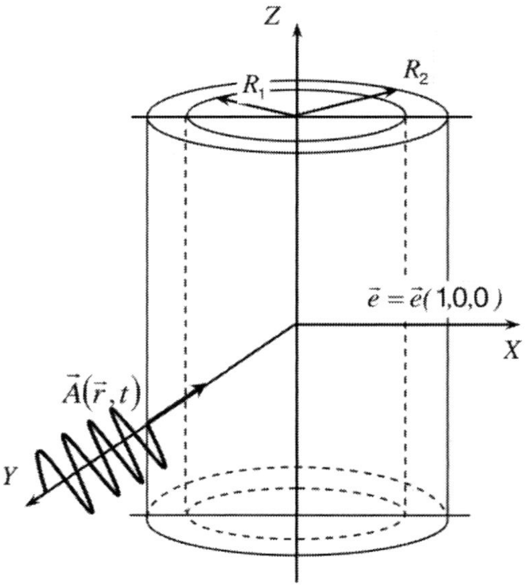

Fig.1 Geometry of the system

It has been assumed that the confinement potential of quantized heterolayer of the tube can be approximated, in radial direction, by the quantum well "rolled-up into a tube" [14-18]:

$$U_{conf}(r) = \begin{cases} 0, \; when \quad R_1 < r < R_2 \\ \infty, \; when \quad r \le R_1, r \ge R_2 \end{cases} \qquad (1)$$

It is assumed also that the layer is "thin enough" and lies fairly far from the symmetry axis (*Z*):

$$L << R_1, \; R_2; \qquad (2)$$

At the same time we will also restrict the analysis with the strong confinement (strong quantization) regime when the Coulomb interaction energy between the electron and hole in the layer can be neglected in comparison with the confinement energies of charge carriers. This means that the "strong quantization regime" takes place in the layer.

In the future, if the material of the quantum well layer has a relatively wide band gap ($E_g \ge 0,5\,\text{eV}$, for example in compositions CdS / HgS / CdS, CdSe / ZnSe / CdSe), we restrict ourselves to the approximation of isotropic effective mass(μ). In the absence of external fields to determine the envelope wave functions and energy spectrum of chage carriers in SNT, one can use the results of Ref[18] :

$$\psi_{n,m}^{(0)}(r,\varphi,z) = \phi_n^{(0)}(r) f_m(\varphi) w(p_z) = \sqrt{\frac{2}{L}} \frac{\sin \frac{\pi n}{L}(r - R_1)}{\sqrt{r}} \frac{e^{i|m|\varphi}}{\sqrt{2\pi}} \frac{\exp\left(\frac{i}{\hbar} p_z z\right)}{\sqrt{l}}, \qquad (3)$$

$$E_{n,m}^{(0)}(p_z) = \frac{\pi^2\hbar^2 n^2}{2\mu L^2} + \frac{\hbar^2 m^2}{2\mu R_n^2} + \frac{p_z^2}{2\mu} \equiv \left(E^{(0)}\right)_{rad} + \left(E^{(0)}\right)_{ang} + \left(E^{(0)}\right)_{tr}, \qquad (4)$$

Here $n = 1,2,3,...$; $m = 0,\pm1,\pm2,...$; $l-$is the normalizing size of the system along the axis of symmetry, and R_n is the effective radius of rotation [18].

3. Single-particle states in the presence of radial field

We will assume that the role of "source" of the radial field in the systems under consideration is played by the homogeneously charged core. Correspondingly, for the potential energy of charge-carriers in the layer we can write now:

$$U(r)=\begin{cases}\infty; & r<R_1;\ r>R_2;\\ \gamma\ln\dfrac{r}{R_1}; & R_1\le r\le R_2;\end{cases} \tag{5}$$

where γ is the "interaction constant" between the "source" of the field and charge carrier.

Taking into account the condition (2), and introducing a new variable $\rho=r-R_1$ for the electrostatic energy of the particle in the layer we can write:

$$U(\rho,\gamma)\cong\begin{cases}\infty; & \rho<0;\ \rho>L;\\ \gamma\dfrac{\rho}{R_1}; & 0\le\rho\le L;\end{cases} \tag{6}$$

Later on we will be interested in the case of strong radial field. In this case the field could be considered "strong" if the energy received by the particle from the field $\left(FL=\gamma L/R_1\right)$ is much more than the difference between the neighbouring levels of size quantization energy in the quantum well (1) ($\Delta\varepsilon_{n+1,n}=\varepsilon_{n+1}-\varepsilon_n;\ \varepsilon_n=\pi^2\hbar^2n^2/2\mu L^2;\ n=1,2,3,\ldots$). From the point of view of size quantization the quantum size effect is most clearly manifested at not much excited levels. Therefore, we present conditions that the external field is strong for the lowest states n = *1* and n = *2* since they are of real physical interest.

$$\frac{\gamma L}{3R_1\varepsilon_1}=\frac{FL}{3\varepsilon_1}>>1;\ \frac{\gamma L}{5R_1\varepsilon_1}=\frac{FL}{5\varepsilon_1}>>1 \tag{7}$$

We will use the variation approach to determine the analytical form of energy spectrum and wave functions of charge carriers in the heterolayer when the conditions (6)-(7) fulfil.

Let us consider now the single-particle states in the layer in the presence of a strong radial field. Using the results of Ref.[19] for the trial wave functions and energy of a particle can be written:

3.1.1. The ground state.

a) $\gamma<0$ (particle is attracted by the core)

$$\chi_1^{(-)}(\rho)=A\rho\exp\left(-\frac{\beta\rho^2}{2}\right) \tag{8}$$

b) $\gamma>0$ (particle is repulsed by the core)

$$\chi_1^{(+)}(\rho)=A(L-\rho)\exp\left[-\frac{\beta}{2}(L-\rho)^2\right] \tag{9}$$

3.1.2. First excited state

a) $\gamma<0$ (particle is attracted by the core)

$$\chi_2^{(-)}(\rho)=B\rho\left(1-\lambda\rho^2\right)\exp\left(-\alpha\frac{\beta\rho^2}{2}\right) \tag{10}$$

b) $\gamma>0$ (particle is repulsed by the core)

$$\chi_2^{(+)}(\rho)=B(L-\rho)\left[1-\lambda(L-\rho)^2\right]\exp\left[-\alpha\beta\frac{(L-\rho)^2}{2}\right] \tag{11}$$

In Exps.(8-11) A, B are the normalization constants, and α, β, λ are variation parameters:

$$A_{c,v}=2\left(\frac{\beta_{c,v}{}^3}{\pi}\right)^{\frac{1}{4}};\ \beta_{c,v}=\frac{\pi}{L^2}\left(\frac{2}{3}\frac{FL}{\varepsilon_{1c,v}}\right)^{\frac{1}{3}};\ B_{c,v}=A_{c,v}\left[\frac{12\alpha^{\frac{7}{2}}}{5(\alpha^2+1)-2\alpha}\right]^{\frac{1}{2}}\equiv A_{c,v}g;$$
$$\lambda_{c,v}=\frac{\beta_{c,v}(1+\alpha)}{3};\alpha\cong 0{,}814;\ \varepsilon_{1c,v}=\frac{\pi^2\hbar^2}{2\mu_{c,v}L^2};\ \chi(r)=\frac{\phi(r)}{\sqrt{r}};\ \rho=r-R_1; \tag{12}$$

From where for the energies of radial motion of ground state (E_{1rad}) and first excited state (E_{2rad}) in SNT we will obtain

$$E_{1rad}\cong\frac{3}{\pi}\left(\frac{3}{2}\varepsilon_1 F^2L^2\right)^{\frac{1}{3}};\quad E_{2rad}\cong\frac{4{,}109}{\pi^{2/3}}\left(\frac{3}{2}\varepsilon_1 F^2L^2\right)^{\frac{1}{3}}; \tag{13}$$

With regard to rotational motion, in the presence of strong radial field the adiabatic approximation holds with even greater accuracy than in the absence of the field.
For the effective radius of rotation, averaged over the states (10, 11) we obtain, respectively:

$$R_{eff}^{-2}=\begin{cases} R_1^{-2}\left[1-\dfrac{4}{\pi}\dfrac{L}{R_1}\left(\dfrac{3}{2}\dfrac{E_{1rad}}{FL}\right)^{1/3}\right] & when\,\gamma<0;\\ R_2^{-2}\left[1+\dfrac{4}{\pi}\dfrac{L}{R_1}\left(\dfrac{3}{2}\dfrac{E_{1rad}}{FL}\right)^{1/3}\right] & when\,\gamma>0\end{cases} \tag{14}$$

4. Discussion of results

The selection of trial wave functions in the form (8-11) is accurate enough and adequately describes the physical system under consideration. As an example of application of these results let us consider the interband and inter-subband transitions in SNT and examine the influence of strong radial electrostatic field on these transitions.
For the perturbation V, related to the light wave

$$\vec{A}(\vec{r},t)=\vec{e}A_0\exp\left[i(\omega t-\vec{q}\cdot\vec{r})\right]+\text{c.c.}$$

We will have, in dipole approximation [20]:

$$V=\frac{|q|A_0}{m_0c}(\vec{e}\cdot\vec{p}) \tag{15}$$

Here A_0 is the amplitude of vector-potential, $\vec{p}$ is a *3D* operator of momentum, m_o is the mass of free electron, c is the wave velocity in free space, and c.c. stands for the complex conjugation. As it was indicated in the beginning, the incident light wave with frequency ω is directed along the Y – axis: $\vec{q}=\vec{q}(0,q,0)$ and is polarized linearly along the X- axis: $\vec{e}=\vec{e}(1,0,0)$.

4.1. Interband dipole transitions

For the matrix element $M_{c,v}$ of interband $|v\rangle\to|c\rangle$ transitions in the general case we can write:

$$M_{c,v}=V_{c,v}\int\Psi_c^*(r,\varphi)\Psi_v(r,\varphi)rdrd\varphi\ , \tag{16}$$

where $V_{c,v}$ is the matrix element of operator (15), built on the Bloch amplitudes of v - and c- bands. Taking into account the form of angular and radial envelop wave functions from Exp.(3) and Exps.(8-11), respectively, and the corresponding energy values from Exp. (4) and Exp.(13), we will obtain the following expressions for the matrix elements $\left(M_{c,v}\right)$ and threshold frequencies $(\omega_{c,v})$:

4.1.a) Transitions $|1v,m_v\rangle \to |1c,m_c\rangle$.

$$M_{1,1} = \sqrt{8}V_{1,1}s^{\frac{1}{2}}\left(1+s^{\frac{2}{3}}\right)^{-\frac{3}{2}}\delta_{m_c,m_v} \tag{17}$$

$$\hbar\omega_{1,1} = E_g^L + \left(E_{1c}\right)_{rad} + \left(E_{1v}\right)_{rad} + \frac{\hbar^2\left(m^2-1/4\right)}{2R_1^2}\left(\frac{1}{\mu_v}+\frac{1}{\mu_c}\right); \tag{18}$$

4.1.b) Transitions $|1v,m_v\rangle \to |2c,m_c\rangle$.

$$M_{2,1} = \sqrt{8}gV_{2,1}s^{\frac{1}{2}}\left(s^{\frac{2}{3}}-1\right)\left(\alpha+s^{\frac{2}{3}}\right)^{-\frac{5}{2}}\delta_{m_c,m_v} \tag{19}$$

$$\hbar\omega_{2,1} = E_g^L + \left(E_{2c}\right)_{rad} + \left(E_{1v}\right)_{rad} + \frac{\hbar^2\left(m^2-1/4\right)}{2R_1^2}\left(\frac{1}{\mu_v}+\frac{1}{\mu_c}\right); \tag{20}$$

4.1. c) Transitions $|2v,m_v\rangle \to |1c,m_c\rangle$

$$M_{1,2} = \sqrt{8}gV_{1,2}s^{\frac{1}{2}}\left(s^{\frac{2}{3}}-1\right)\left(1+\alpha s^{\frac{2}{3}}\right)^{-\frac{5}{2}}\delta_{m_c,m_v} \tag{21}$$

$$\hbar\omega_{1,2} = E_g^L + \left(E_{1c}\right)_{rad} + \left(E_{2v}\right)_{rad} + \frac{\hbar^2\left(m^2-1/4\right)}{2R_1^2}\left(\frac{1}{\mu_v}+\frac{1}{\mu_c}\right); \tag{22}$$

4.1.d) Transitions $|2v,m_v\rangle \to |2c,m_c\rangle$

$$M_{2,2} = \sqrt{8}g^2V_{2,2}s^{\frac{1}{2}}\left[\alpha\left(1+s^{\frac{2}{3}}\right)\right]^{-\frac{3}{2}}\left\{\frac{5}{3}s^{\frac{2}{3}}(1+\alpha)^2\left[\alpha\left(1+s^{\frac{2}{3}}\right)\right]^{-2}-\frac{1}{\alpha}\right\}\delta_{m_c,m_v} \tag{23}$$

$$\hbar\omega_{2,2} = E_g^L + \left(E_{2c}\right)_{rad} + \left(E_{2v}\right)_{rad} + \frac{\hbar^2\left(m^2-1/4\right)}{2R_1^2}\left(\frac{1}{\mu_v}+\frac{1}{\mu_c}\right) \tag{24}$$

In Exps.(17-24) the selection rule $m_c = m_v \equiv m$ takes place, $s = \mu_v/\mu_c$, E_g^L is the band gap of bulk sample from the material of the layer.
As it follows from Exps. (17), (19), (21), (23) the corresponding matrix elements (intensities) of interband transitions do not depend explicitly on the intensity of the external field, but the external field leads to the explicit dependence of transitions' intensity on the values of charge carriers' effective masses. The Fig.2. shows the plots of functions $u_{c,v}\left(s\right)=\left|\frac{M_{c,v}}{\sqrt{8}V_{c,v}}\right|^2 \left(c,v=1,2\right)$.

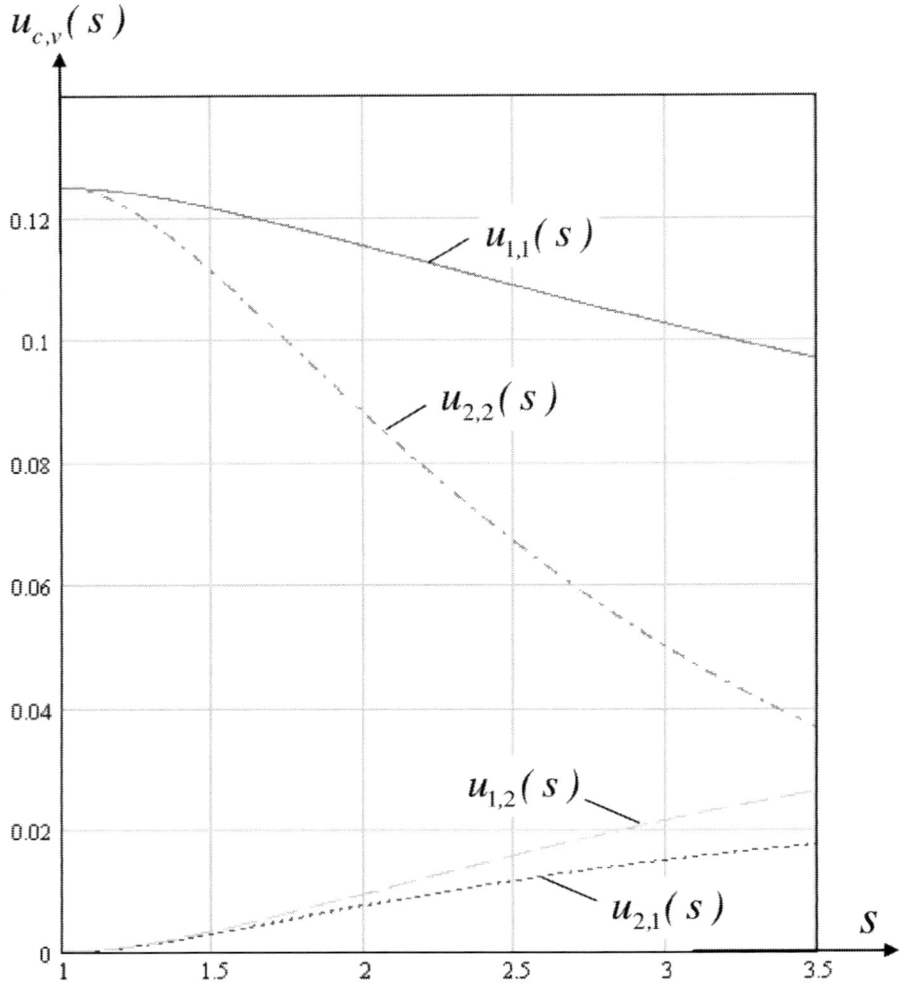

Fig.2. Plots of functions $u_{c,v}(s)=\left|\frac{M_{c,v}}{\sqrt{8}V_{c,v}}\right|^2$; $s=\frac{\mu_v}{\mu_c}$.

4.2. Inter-subband dipole transitions

In the general case we have for the matrix element $D_{f,i}$ of inter-subband $|i\rangle\rightarrow|f\rangle$ transitions

$$D_{f,i}=\int\Psi_f^*(r,\varphi)V\Psi_i(r,\varphi)rdrd\varphi \tag{25}$$

After calculations we obtain the following expressions for the corresponding matrix elements and threshold frequencies:

4.2.a) Transitions $|1c(v),m_i\rangle\rightarrow|2c(v),m_f\rangle$

$$D_{2,1}^{c(v)}=-i\hbar g\frac{|e|A_0}{3m_0L}\frac{3+\alpha}{(1+\alpha)^2}\left(\frac{2}{3}\frac{FL}{\varepsilon_{1c(v)}}\right)^{\frac{1}{3}}\delta_{m_f,m_i\pm1};\quad \frac{D_{2,1}^{v}}{D_{2,1}^{c}}=\left(\frac{\mu_v}{\mu_c}\right)^{\frac{1}{3}}=s^{\frac{1}{3}}; \tag{26}$$

$$\hbar\omega_{2,1}^{c(v)} = \left(E_{2c(v)}\right)_{rad} - \left(E_{1c(v)}\right)_{rad} + \frac{\hbar^2\left(2|m|+1\right)}{2\mu_{c(v)}R_1^2}; \quad \left(m \equiv m_i\right) \tag{27}$$

As one can see, in this case the selection rule $m_f = m_i \pm 1$ takes place. The absorption intensity of direct inter-subband transitions in the layer depends explicitly on the intensity of external field. The absorption increases with the increasing of external field. At the same time inter-subband absorption depends explicitly also on the values of effective masses of charge carriers.

5. Conclusions

On the results obtained in this study we can conclude the following:

- the suggested theoretical model is physically adequate to the system under consideration and allows to get results in analytical form;
- the external strong radial electrostatic field leads to a significant spatial separation of opposite charge carriers, which reduces the probability of radiative recombination of electron and hole;
- it is possible to modify the optical-energy parameters of the system by variation of geometrical sizes of the sample and the intensity of external field in a controlled manner;
- at the same time the dependence of absorption parameters upon effective masses of carriers opens a possibility to define its values experimentally in the presence of external radial electric field.

As a consequence, we can conclude that if the material and geometrical sizes of SNT are known, the results of above theoretical calculations could be used to carry out appropriate experiments.

In the application plan there is a real opportunity to use the theoretical results obtained for the geometrical and electrical tuning of optical and other parameters of SNT.

References

[1] Law M, Goldberger J, and Yang P 2004 *Annu. Rev.Mater. Res.* **34**, 83
[2] Shen G, Bando Y, and Golberg D 2007 *Intern. Journ. Nanotech.* **4**, 730
[3] Xiuling Li 2008 *Journ. of Physics D: Applied Physics* **41,** 193001
[4] .Niu H and M.Gao M 2006 *Angew.Chem. Int.Ed.* **45**, 6462 ; .Mahapatra A 2009 *J.Nanopart. Res.* **11**,467
[5] Prinz A and Prinz V,2003 *Surf. Sci.*, **532-535**, 911; .Prinz V 2004 *PhysicaE* **23,** 260
[6] Jin-Phillipp N, Thomas J, and Kelsch M 2006 *Appl.Phys. Lett.* **88**, 033113; Mohan P, Motohisha J, and Fukui T 2006 *Appl. Phys. Lett.* **88**, 013110
[7] Goldberger J *et al.*2003 *Nature* **422**, 599 ; Goldberger J et al., 2006 *Acc.Chem.* Res. **39**, 239 ; Zhang G *et al.*, 2007*Jpn. J. Appl. Phys.* **46** L730; de Almeida J 2010 *PhysicsLetters* **A 374** 877
[8] Mohan P, Motohisha J, Fukui T,. 2006 *Appl. Phys. Lett.* **88**, 133105
[9] Pan H and Feng Y 2008 *ACS Nano* **2** 2410
[10] Schmidt O *et al.*, 2001 *Adv. Mater.* **13**, 756; Aresipathi C, Feldhoff A,and Wark M 2010 *Journ. Mater. Science* **45**,1179
[11] S.V.Golod S V, Prinz V.Ya , *et al.*, 2004 *Appl. Phys. Lett.* **84**, 3391
[12] Musin R, and. Wang X.-Q 2006 *Phys. Rev.* **B 74** 165308
[13] Malkova N and Ning C 2007 *Phys. Rev.* **B 75** 155407
[14] Harutyunyan V *et al* 2004 *Physica* **E 21** 111
[15] Harutyunyan V *et al* 2005 *Semiconductors* **39** 805
[16] Harutyunyan V *et al* 2009 *Appl. Surf. Sci.* **256**, 455
[17] Harutyunyan V 2011 *J. Appl. Phys.* 109, 014325
[18] Harutyunyan V 2010 *Physics of Solid States* **52**, 1744 (2010)
[19] Harutyunyan V *et al* 2010 *Physica* **E 43** 614
[20] Harrison P 2000 *Quantum Wells, Wires, and Dots* (Wiley, New York); G.Bastard 1998 *Wave Mechanics Applied to Semiconductor Heterostructures* (Les Editionsde Physique, Les Ulis,1998).

Impurity states in ZnS/InP/ZnSe core/shell/shell spherical quantum dot

E M Kazaryan, A A Kostanyan and R G Poghosyan

Department of Physics and Technology, Russian-Armenian University, Yerevan, Armenia

E-mail: artavazd.kostanyan@gmail.com

Abstract. The impurity ground state wave function and energy, as well as electron binding energy are obtained in ZnS/InP/ZnSe core/shell/shell spherical quantum dot using the variational method. Considering the band structure of the system it is assumed that electron is localized in InP shell. Two parametrical potential is chosen as a confinement potential for the shell. It is assumed that the impurity is located in the center of quantum dot core (ZnS).

1. Introduction

In the terms of quantum effects the most interesting objects are quantum dots (QDs), in which low-dimensional effects are rather significant due to size quantization in all three directions. Up to now, QDs of various geometric shapes and sizes are implemented and theoretically studied (see e.g. [1-7]).

Recently core/shell quantum dots (CSQD) with spherical symmetry have been realized [9-14]. Most of them like CdSe/ZnS and InP/ZnS have type-I structure [9, 10]. In these structures the band gap of a shell material is greater than the band gap of a core. The given circumstance promotes substantial growth (improvement) of a quantum yield.

Nevertheless, there are just some examples of so-called *reverse* structures of type-I, such as CdS/HgS [12], CdS/CdSe and ZnSe/CdSe [13] at which, unlike the previous structures, the band gap of a shell material is less rather than the band gap of a core.

However, as cadmium possesses the considerable toxicity, it restricts their application, for example, in biomedicine. In this connection, as an alternative CSQD from InP/ZnS and $CuInS_2$/ZnS [14] have been offered. But their properties (including optical) are not to the full satisfactory for their wide application.

Recently reverse type cadmium free ZnSe/InP/ZnS core/shell/shell QDs (CSSQD) of type-I have been realized [15]. Optical properties (absorption and photoluminescence) of the QD are also investigated experimentally.

In the diagram form mentioned QD is presented on Fig. 1, where R_1 and R_2 are internal and external radiuses of InP shell. Energy diagram of the QD is presented on Fig. 2.

As it can be seen from figure, in contrast to the previous core/shell QDs, where carriers are localized in the core, in the reverse type QD localized states of carriers are in the shell of InP. Thereby, there is a need to consider the electronic states and energy spectrum of the shell for further discussions (e.g. to determine absorption coefficient).

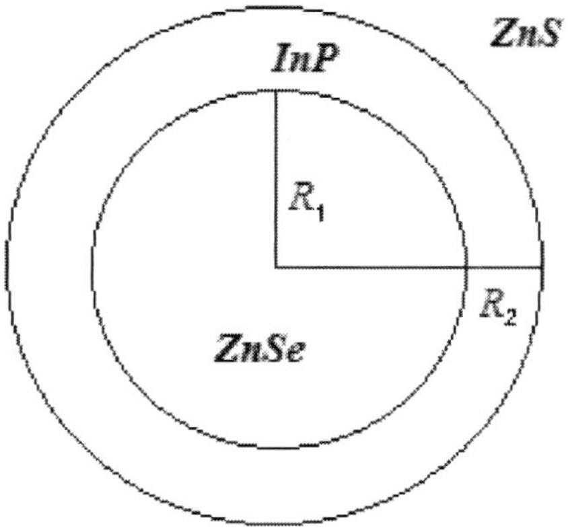

Fig. 1. ZnSe/InP/ZnS core/shell/shell quantum dot

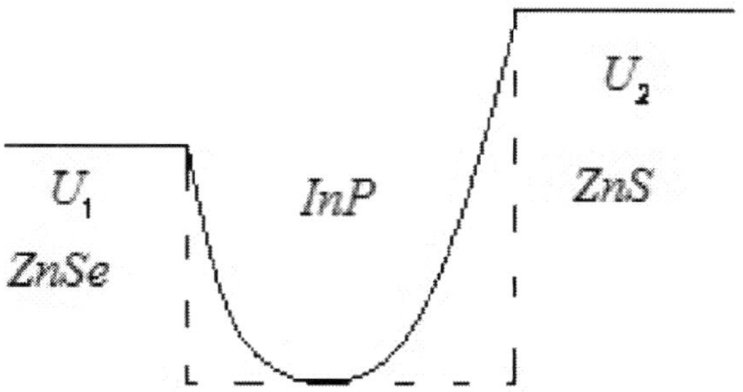

Fig. 2. Potential energy diagram of ZnSe/InP/ZnS core/shell material

In the simplest approximation confining potential of such a system can be proposed as rectangular (see Fig. 2, dotted line). However, based on the methods of producing these structures the following should be considered:

1. mixing of the components between the core and shell, shell and outer shell
2. the presence of impurities.

In the presented report impurity states are obtained in ZnS/InP/ZnSe core/shell/shell spherical quantum dot (CSSQD). According to Fig.2 electron is localized in the InP shell. Taking into account the mixing of components of the core and the shell it leads to a more realistic choice of the confining potential. For this particular problem the most suitable is a two parametrical potential of Smorodinsky-Winternitz [8]. Donor impurity ground state wave function and energy, as well as electron binding energy are obtained using the variational method. It is assumed that donor impurity is located in the center of QD core (ZnS). Dielectric constant mismatch between ZnS core and InP shell is neglected.

2. Electronic states in the shell

Two parametrical confining potential is chosen in a form

$$V(r)=\alpha r^2+\frac{\beta}{r^2}+2\sqrt{\alpha\beta}\,. \quad (1)$$

The Schrodinger equation for the radial wave function has the form:

$$\frac{d^2R}{dr^2}+\frac{2}{r}\frac{dR}{dr}+\frac{2m^*}{\hbar^2}\left(E-\frac{\hbar^2}{2m^*}\frac{l(l+1)}{r^2}-\alpha r^2-\frac{\beta}{r^2}+2\sqrt{\alpha\beta}\right)R=0, \quad (2)$$

Introducing the notations

$$\xi=\frac{\sqrt{2m^*\alpha}}{\hbar}r^2,\quad l(l+1)+\frac{2m^*\beta}{\hbar^2}=2s(2s+1),\quad \frac{1}{\hbar}\sqrt{\frac{2m^*}{\alpha}}\left(E-2\sqrt{\alpha\beta}\right)=4(n+s)+3,$$
$$s=\frac{1}{4}\left[-1+\sqrt{(2l+1)^2+\frac{8m^*\beta}{\hbar^2}}\right], \quad (3)$$

and taking into account the fact, that when $\xi\to\infty$ the desired solution behavior is alike $e^{-\xi/2}$, and when $\xi\to 0$ is alike ξ^s the solution must be taken in the form [16]

$$R(\xi)=e^{-\xi/2}\xi^s w(\xi)\,. \quad (4)$$

Function $w(\xi)$ is the confluent hypergeometric function

$$w = F\left(-n, 2s + \frac{3}{2}, \xi\right). \tag{5}$$

n must be a nonnegative integer. For the energy levels an infinite set of equidistant values is obtained

$$E_n = \hbar\sqrt{\frac{\alpha}{2m^*}}\left[4n + 2 + \sqrt{(2l+1)^2 + \frac{8m^*\beta}{\hbar^2}}\right] - 2\sqrt{\alpha\beta}, \tag{6}$$

where $n=0,1,2,\ldots$.

3. Impurity states

Let us consider spherical CSSQD containing hydrogen like impurity. Moreover, it is assumed that the impurity is located in the center of the core. As for previous case, confining potential is chosen in the form (1).

Taking into account the Coulomb field the Hamiltonian of the system has a form

$$\hat{H} = -\frac{\hbar^2}{2m^*}\Delta_r + V(r) - \frac{e^2}{\varepsilon r}, \tag{7}$$

where ε is permittivity, m^* – the effective electron mass.

Then the radial Schrodinger equation has the form

$$\frac{d^2R}{dr^2} + \frac{1}{r}\frac{dR}{dr} + \frac{2m}{\hbar^2}\left(E - V(r) - \frac{\hbar^2}{2m}\frac{l(l+1)}{r^2} + \frac{e^2}{\varepsilon r}\right)R(r) = 0 \tag{8}$$

The solution of equation (8) is obtained using the variational method, with the trial wave function of the ground impurity state in the form (see for example [17])

$$R_0(r) = N_0 e^{-\frac{1}{2}tr^2} t^s r^{2s} e^{-\lambda r} \tag{9}$$

where λ – a variational parameter , $t = \sqrt{2m^*\alpha}/\hbar$, s – positive value, which is given by (3), and N_0 – the normalization constant, which is determined from the normalization condition

$$N_0 = \frac{1}{\sqrt{4\pi I_0}}, \tag{10}$$

where

$$I_0 = \int_0^\infty t^{2s} r^{2s+2} \exp\left\{-tr^2 - 2\lambda r\right\} dr. \tag{11}$$

Ground state energy of the system can be obtained from the condition of minimizing the integral

$$\min_\lambda \int R_0(r)\hat{H}R_0(r)dV \tag{12}$$

After the minimization and taking into account (7) and (9),(10),(11), the following expression for the energy is obtained (dimensionless, in units of effective Rydberg energy)

$$\varepsilon(\lambda_0) = -A_1(\lambda_0) + A_2(\lambda_0) + A_3(\lambda_0) + A_4(\lambda_0) - 2a_0 b_0^{1/2}, \tag{13}$$

where A_1, A_2, A_3, A_4 – some functions depending on λ_0 , in which, to facilitate further calculations, the dimensionless quantities are introduced

$$\lambda_0 = \lambda a_b^*; \quad a_0 = a_b^*\sqrt{\alpha/R_y^*}; \quad b_0 = 2m^*\beta/\hbar^2 \tag{14}$$

The ground state energy is calculated after the minimization of the Eq. (12):

$$E_0^{imp} = \min_{\lambda_0} \varepsilon(\lambda_0) \tag{15}$$

After the binding energy as the difference

$$E_{bind} = E_0 - E_0^{imp} \tag{16}$$

where E_0 - the ground state energy of CSSQD without the impurity, which is given by (6). For the parameters (14) the following values for InP: the electron effective mass - $m^* = 0.073m_0$, the effective Bohr radius - $a_b^* = \frac{\varepsilon\hbar^2}{m^* e^2} = \frac{12.4}{0.073} a_b \approx 90\mathring{A}$, the effective Rydberg energy - $R_y^* = \frac{m^* e^4}{2\varepsilon^2\hbar^2} = \frac{0.073}{12.4^2} R_y \approx 6.46$ meV.

4. Energy levels and binding energy

Fig. 3 shows electron energy (in arbitrary units) dependence on outer radius of the shell. In this case, inner radius is fixed at the value of $R_1 = 1$ (in a_b^* units). The energy decreases monotonically with increasing of a width. Also energy increases with increasing values of *l*.

On Fig. 4 the dependence of binding energy on outer radius is presented. According to this figure, the value of binding energy monotonically decreases due to increase R_2 as a consequence of reducing the size quantization.

Fig.5 shows the dependence of electron energy on inner radius of the shell, but for this case outer radius value is fixed at $R_2 = 4a_b^*$. One may see that the energy increases with decreasing width, and the growth of the energy is quite slow, as in this case, the size quantization is rather weak. It is the result

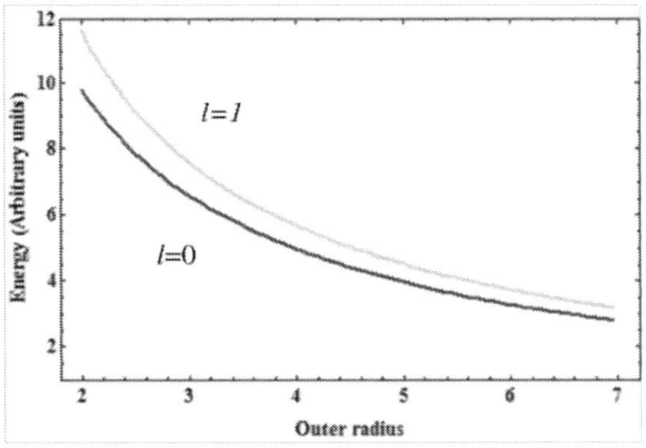

Fig. 3. The electron energy dependence on the shell outer radius

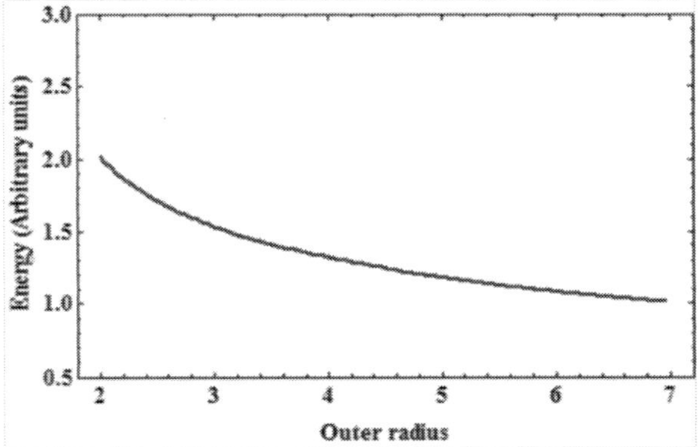

Fig. 4. The binding energy dependence on the shell outer radius

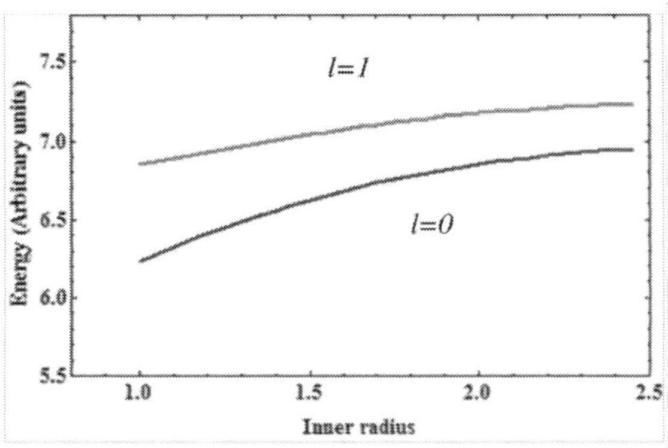

Fig. 5. The electron energy dependence on the shell inner radius

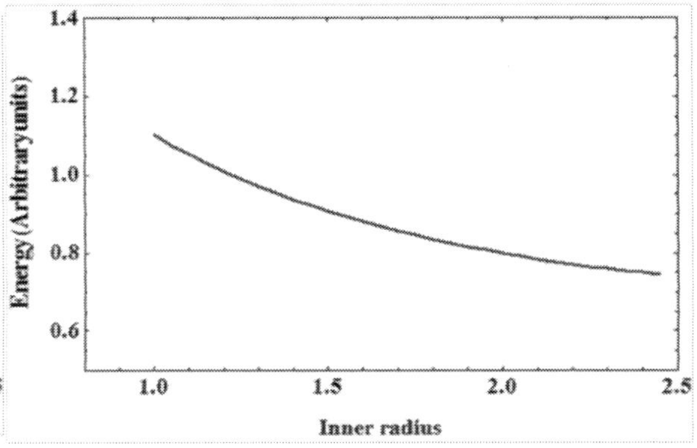

Fig. 6. The binding energy dependence on the shell inner radius

of the fact, that, although the thickness is reduced, but the volume of the sample rather large. As in previous case, the energy increases with increasing values of l.

Donor binding energy dependence on inner radius is shown in Fig.6. As it can be seen, when R_1 increases binding energy decreases, since an electron is moved away from the Coulomb center.

For the both cases it is assumed, that donor impurity is located in the center of CSSQD.

References

[1] Efros Al L and Efros A L 1982 *Sov. Phys. Semicond.* **16** 772
[2] Lorke A R, Luyken J, Govorov A O, Kotthaus J P, Garcia J M and Petroff P M 2000 *Phys. Rev. Lett.* **84** 2223
[3] Chakraborty T and Pietilainen P 1994 *Phys. Rev.* **B 50** 8460
[4] Sarkisyan H A 2004 *Modern Phys. Lett.* **B 16** 22 835
[5] Juharyan L A, Kazaryan E M and Petrosyan L S 2006 *Solid State Commun.* **139** 537
[6] Dvoyan K G, Kazaryan E M, Tshantshapanyan A A, Wang Zh M and Salamo G J 2011 *Applied Physics Letters* **98** 203109
[7] Kazaryan E M, Kostanyan A A and Sarkisyan H A 2007 *Journ. Contemp. Phys (Armenian Academy of Sciences)* **42** 4 218-226
[8] Kazaryan E M, Kostanyan A A, Petrosyan L S and Sarkisyan H A 2006 *Int. Conference – New Technologies for Development of Heterosemiconductors for Device Applications (Yerevan, Armenia, September 28-29)*
[9] Chen F Q, Gerion D 2004 *Nano Lett.* **4** 1827
[10] Peng X, Battaglia D 2002 *Nano Lett.* **2** 1027
[11] Xu S, Kumar S and Nann T 2006 *J. Am. Chem. Soc.* **128** 1054
[12] Braus M, Burda C and El-Sayed M A 2001 *J. Phys. Chem* **A 105** 5548
[13] Zhong X, Xie R, Basche Y, Zhang T and Knoll W 2005 *Chem. Mater.* **17** 4038
[14] Li L, Daou T J, Texier I, Chi T T K, Liem N Q and Reiss P 2009 *Chem Mater.* **21** 2422
[15] Kim S, Park J, Kim T, Jang E, Jun S, Jang H, Kim B and Kim S W 2011 *Small* **7** 11 70-73
[16] Landau L D, Lifshitz E M 1977 *Quantum Mechanics: Non-Relativistic Theory* vol 3 (3rd ed.) (Pergamon Press)
[17] Xiao Zh, Zhu J and He F 1996 *J. Appl. Phys.* **79** 12

The influence of polar optical phonon confinement on the binding energy of a hydrogenic impurity in quantum wires in the perpendicular electric and magnetic fields

A L Vartanian[1] , A Shahbandari[1] , M A Yeranosyan[1] and A A Kirakosyan[1]
[1]Department Of Solid State Physics,Yerevan State University, A.Manoogian 1, 375025 Yerevan, Armenia

E-mail: vardan@ysu.am

Abstract. The hydrogenic impurity binding energy in cylindrical quantum well wire with a finite confining potential including both barriers of finite height and an applied electric and magnetic fields are studied. The polaron effect on the ground-state binding energy are investigated by means of Landau-Pekar variation technique. The results for the binding energy as well as polaronic correction with taking into account polar optical phonon confinement effect are obtained as a function of the applied fields for different position of the impurity. Our calculations are compared with previous results in quantum wires of comparable dimensions.

1. Introduction

The coupling between electronic and vibrational excitations plays a central role in the physics of nanoscale semiconductor structures. In these systems the presence of an interface necessarily alters the phonon modes and their interaction with electronic excitations is also modified [1-2]. In particular, it is well known that the impurity bound states and excitons are strongly affected by the electron coupling with the polar optical phonons. These effects become more pronounced in the presence of external electric and magnetic fields [3-8].

In this paper we study the both effects of electric and magnetic fields on the bound polaron ground state energy in a polar cylindrical quantum wire (QW) with a finite height barrier for various positions of the impurity by taking into account polar optical (PO) phonon confinement effect. The coupling of phonons with the impurity-ion as well as the electron is considered in our calculations by means of Landau-Pekar variation technique [9].

2. Theory

We consider impurity bound polaron problem for the semiconductor QW embedded in a dielectric medium. In the presence of an electric field applied perpendicular to the axis of the wire (z-axis) and a magnetic field applied parallel with the z-axis the basic Hamiltonian for the single

[1] Corresponding author. A.L. Vartanian, Dept of Physics,YSU, A.Manoogian 1, 0025 Yerevan, Armenia Tel.: +37410 351879; Fax: +37410 554641.

conduction-band electron interacting with both the Coulomb impurity and the longitudinal optical phonon field can be written within the effective-mass approximation as

$$\begin{aligned} H = & \frac{1}{2m^*}(\mathbf{p}+\frac{e}{c}\mathbf{A})^2 + V(\rho) + eF\rho\cos\varphi - \frac{e^2}{\varepsilon_\infty r} + \\ & + \sum_{npq_z} \hbar\omega_{CO} a^+_{np}(q_z) a_{np}(q_z) + \sum_{sq_z} \hbar\omega_{IO} a^+_s(q_z) a_s(q_z) - \\ & - \sum_{npq_z} \left[\Gamma^{np}_{CO}(q_z)\phi_{np}(\rho) e^{im\varphi} e^{-iq_z z} a^+_{np}(q_z) + H.c. \right] \\ & - \sum_{sq_z} \left[\Gamma^{s}_{IO}(q_z) Q_s(\rho) e^{is\varphi} e^{-iq_z z} a^+_s(q_z) + H.c. \right], \end{aligned} \quad (1)$$

where

$$\phi_{np}(\rho) = \begin{cases} J_n\left(\frac{\kappa_{np}\rho}{R}\right), & \rho \le \mathrm{R} \\ 0, & \rho > \mathrm{R} \end{cases}, \quad (2)$$

$$Q_s(\rho) = \begin{cases} K_s(q_z R) I_s(q_z\rho), & \rho \le \mathrm{R} \\ I_s(q_z R) K_s(q_z\rho), & \rho > \mathrm{R} \end{cases}, \quad (3)$$

The first line of equation (1) is the Hamiltonian of a system consisting of an electron bound to a Coulomb impurity, inside a cylindrical QW, in the presence of external fields, the second line describes the non-interacting PO phonon system, and the third and the fourth lines represent the electron–confined optical (CO) phonon, electron–interface optical (IO) phonon interactions, respectively [10]. Here, m^* is the electron's effective mass, $\mathbf{p}$ is the momentum operator, $\mathbf{A}$ is the vector potential of the magnetic field, which is chosen as $\mathbf{A} = (0, Bx, 0)$ giving $\mathbf{B} = (0,0,B)$, the confining potential $V(\rho) = 0$ when $\rho < \mathrm{R}$, and $V(\rho) = V_0$, when $\rho \ge \mathrm{R}$, F is the value of the electric field, and $r = \sqrt{\rho^2 + \rho_i^2 - 2\rho\rho_i\cos(\varphi - \varphi_i) + z^2}$ gives distance of the electron from the impurity, $\mathbf{r}_i = (\rho_i, \varphi_i, 0)$ is the impurity position, $a^+_{np}(q_z)$, $a_{np}(q_z)$ and $a^+_s(q_z)\, a_s(q_z)$ are respectively the creation and annihilation operators of CO and IO phonons with frequency ω_{CO} and ω_{IO}, $\varepsilon_\infty(\varepsilon_0)$ is the optical (static) dielectric constant,

$$\omega_{\mathrm{CO}} = \omega_{\mathrm{LO}}, \qquad \omega_{IO} = \sqrt{1 + \frac{\varepsilon_0 - \varepsilon_\infty}{\varepsilon_\infty - \varepsilon}}\,\omega_{TO}, \quad (4)$$

$$\varepsilon(\omega) = -\frac{I_s(q_z R)\left[K_{s-1}(q_z R) + K_{s+1}(q_z R)\right]}{K_s(q_z R)\left[I_{s-1}(q_z R) + I_{s+1}(q_z R)\right]}\varepsilon_d. \quad (5)$$

$$\left|\Gamma^{np}_{CO}(q_z)\right|^2 = \frac{2e^2\hbar\omega_{CO}}{LJ^2_{n+1}(\kappa_{np})(\kappa^2_{np} + R^2 q_z^2)}\left(\frac{1}{\varepsilon_\infty} - \frac{1}{\varepsilon_0}\right), \quad (6)$$

$$\left|\Gamma^{s}_{IO}(q_z)\right|^2 = \frac{2e^2\hbar\omega_{IO}}{LK_s^2(q_z R) I_s(q_z R) q_z R\left(I_{s-1}(q_z R) + I_{s+1}(q_z R)\right)}\left(\frac{1}{\varepsilon - \varepsilon_0} - \frac{1}{\varepsilon - \varepsilon_\infty}\right), \quad (7)$$

$K_s(x)$ and $I_s(x)$ are the first and second kind modified Bessel functions, respectively, $J_n(x)$ is the Bessel function of the nth order, κ_{np} is the pth zero of $J_n(x)$, ε_d is the dielectric constant of the embedding medium, and R is the QW radius[11,12].

In the presence of a hydrogenic impurity and applied fields, the electron ground-state variational wave function can be written as

$$\Psi_0(\mathbf{r},\beta,\lambda)=\begin{cases} N_{01}F_1(-a_{01},1;\frac{1}{2}\gamma_c\rho^2)e^{-\frac{1}{4}\gamma_c\rho^2}e^{-\beta\rho\cos\varphi}e^{-\lambda r}, & 0\le\rho<R \\ N_0\dfrac{{}_1F_1(-a_{01},1;\frac{1}{2}\gamma_c R^2)}{U(-a'_{01},1;\frac{1}{2}\gamma_c R^2)}U(-a'_{01},1;\frac{1}{2}\gamma_c\rho^2)e^{-\frac{1}{4}\gamma_c\rho^2}e^{-\beta\rho\cos\varphi}e^{-\lambda r}, & \rho\ge R\end{cases} \quad (8)$$

where ${}_1F_1$ and U are the general forms of confluent hypergeometric functions and a_{01} and a'_{01} are the eigenvalues of the ground state for inside and outside of the wire, respectively, β and λ are the variational parameters. Notice that the variational parameters β and λ taken into account simultaneously the anisotropy of the hydrogen-like wave function in the presence of an electric field.

To obtain the polaron's ground state binding energy (BE), we used the Landau–Pekar theory [9]. As a result, the bound polaron's ground-state energy is obtained as

$$E^{el-ph}(\beta,\lambda)=-\sum_{npq_z}\frac{1}{\hbar\omega_{CO}}\left|\Gamma_{CO}^{np}(q_z)\right|^2\left|\langle\Psi_0(\mathbf{r},\beta,\lambda)|\phi_{np}(\rho)e^{-in\varphi}e^{iq_zz}|\Psi_0(\mathbf{r},\beta,\lambda)\rangle\right|^2 -\sum_{sq_z}\frac{1}{\hbar\omega_{IO}}\left|\Gamma_{IO}^{s}(q_z)\right|^2\left|\langle\Psi_0(\mathbf{r},\beta,\lambda)|Q_s(\rho)e^{-is\varphi}e^{iq_zz}|\Psi_0(\mathbf{r},\beta,\lambda)\rangle\right|^2+E(\beta,\lambda) \quad (9)$$

The energy of the bound polaron can be found by minimizing $E^{el-ph}(\beta,\lambda)$ with respect to β and λ. The BE of the bound polaron is determined as the difference between the ground-state energy of the electron confined in the QW in the absence of the impurity and the total energy of the polaron.

3. Numerical results and discussion

In the numerical calculations the parameters of GaAs and $Al_xGa_{1-x}As$ have been used. The value of the potential height is determined from the Al concentration x, through the expression for the energy-band

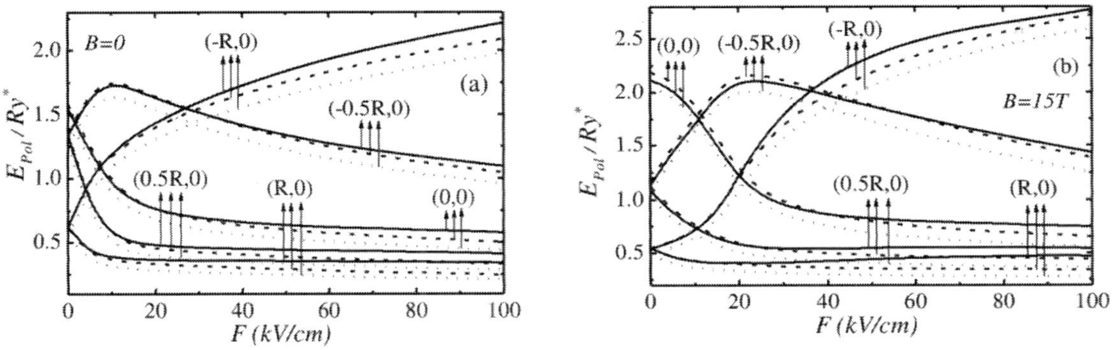

Figure 1. Reduced BE E_{Pol}/Ry^* for the impurity ground state with (solid and dashed lines) and without electron-PO phonon interaction (dotted lines) as a function of the electric field for different impurity positions and for various values of the magnetic field ($B=0$, $B=15T$). The solid lines represent the sum of the contributions due to confined and interface PO phonon modes. Dashed lines represent contributions bulk PO phonon modes.

gap discontinuity ε_d. The finite potential is taken as $V_0=Q(1.36x+0.22x^2)(eV)$ where $Q=0.6$. The Al concentration is taken to be $x=0.3$, $\alpha_{e-ph}=0.0681$, $\varepsilon_0=12.7$, $\varepsilon_\infty=10.9$ and $m^*=0.067m_0$ [13,14].

Figure 1(a),(b) shows the BE of a hydrogenic impurity in a QW as a function of applied electric field for different positions of the impurity, corresponding to magnetic fields $B=0$, and $B=15T$. The dashed and dotted lines in Figure 1(a),(b) represent the BE with and without electron-bulk PO phonon interaction respectively. Solid lines represent the sum of the contributions due to CO and IO phonon modes. It is clearly seen from the Figure 1(a),(b) that the BE for impurity positions (0,0), (0.5R,0) and (R,0) decreases with the increase of electric field, since electric field shifts the maximum of the electronic probability density far from mentioned positions of impurity. On the contrary, for impurity positions (-0.5R,0) and (-R,0) electric field pushes the maximum of the electronic probability density toward impurity positions, consequently, BE increases with increase of electric field, and when the locations of maximum of electronic charge density coincides with the positions of impurity, curves reach to the peaks values. It should be noted that in the presence of magnetic field the maximums of the curves occur for larger values of electric field.

The appropriate polaronic shift of BE due to confined (Figure 2 (a),(b)) and interface (Figure 2 (c),(d)) phonons for magnetic fields $B=0$, and $B=15T$ is presented as a function of applied electric field. The solid (dashed) lines correspond to the finite (infinite [15]) confinement potential case. In the finite confinement potential model, the penetration of the bound electron into the barriers is appreciable, which gives a reduction of the electron confinement and impurity BE. Also one may observe that the electric field dependence of the CO phonon caused poalronic shift in case of finite confinement potential has similar qualitative features as BE. In the case of infinite confinement potential the BE poalronic shift decreases for all positions of impurity (Figure 2 (a),(b)). In the model of finite confinement potential the polaronic shift of BE caused by electron-IO phonon coupling increases with increasing electric field whereas in the infinite potential model it decreases for impurity positions (0,0), (0.5R,0), (R,0), and increases for impurity positions (-0.5R,0) and (-R,0).

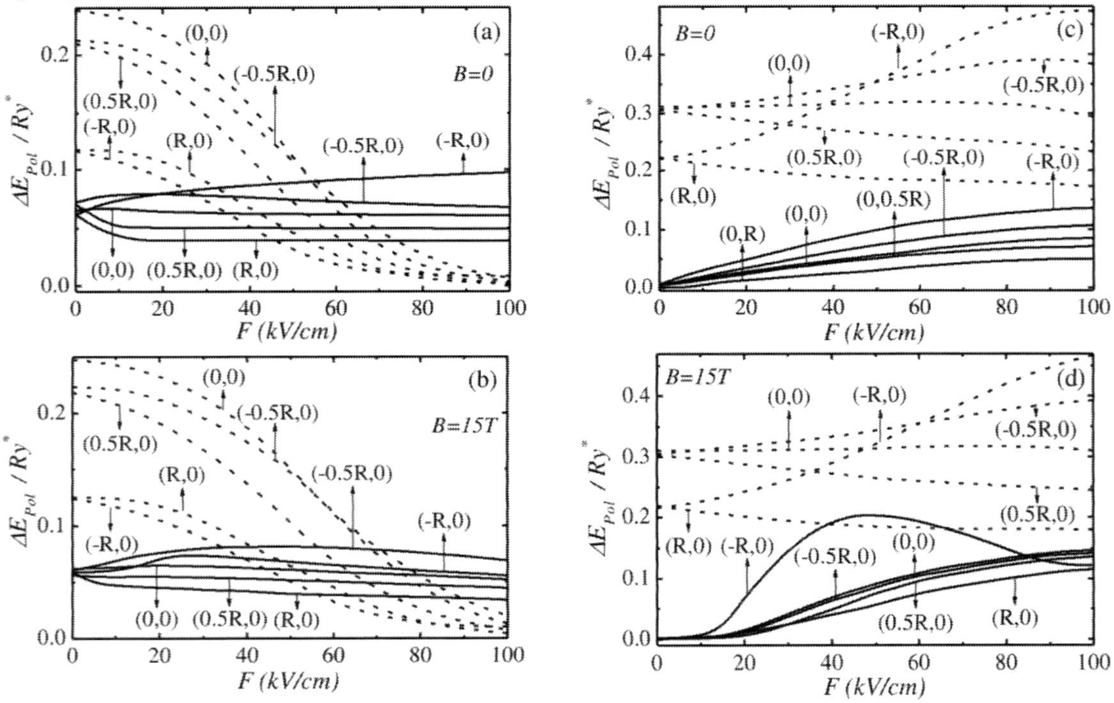

Figure 2. Reduced polaronic shift of the BE as a function of the electric field for different impurity positions and for various values of the magnetic field ($B=0$, $B=15T$). The solid and dashed lines represent the finite and infinite confinement potential cases, respectively.

The contribution of the electron-LO bulk phonon interaction in the impurity BE in magnetic field $B=15T$ in the model of finite (infinite) confining potential increases from 4.3% (6.06%) to 9% (15.6%,) and from 14.28% (10%) to 7.2% (12%) for impurity positions (0,0) and (-R,0) respectively, when the electric field increases from $F=0$ to $F=100kV/cm$. In the same conditions, the total contribution due to CO and IO phonon modes increases from 8% (16%) to 14.24% (22.4%,) and from 19.1% (21.7%) to 7.3% (26.5%) for impurity positions (0,0) and (-R,0), respectively.

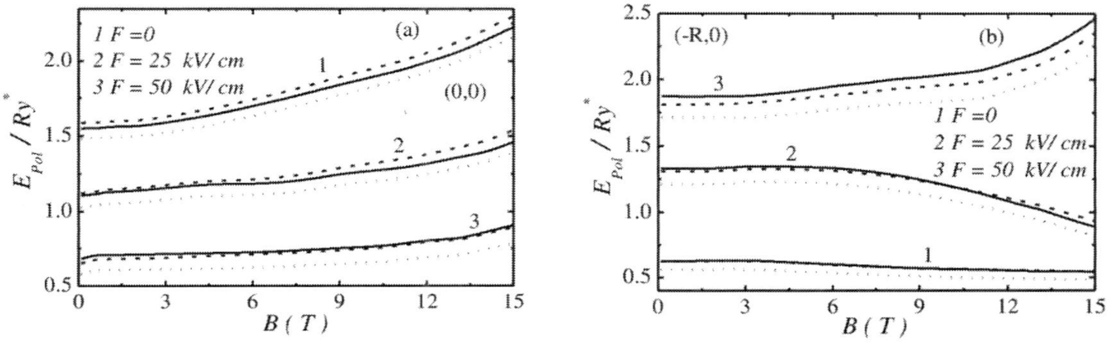

Figure 3. Reduced BE for the impurity ground state with (soled and dashed lines) and without electron PO phonon interaction (dotted lines) a function of the magnetic field for different impurity positions and for various values of the electric field ($F=0$, F=25kV/cm, F=50kV/cm). The solid lines represent the sum of the contributions due to confined and interface PO phonon modes. Dashed lines represent contributions bulk PO phonon modes.

The dependencies of the impurity BE on the magnetic field are shown in Figure 3 for two impurity positions ((0,0) and (-R,0)) at electric fields $F=0$, $F=25kV/cm$, and $F=50kV/cm$. It is clearly

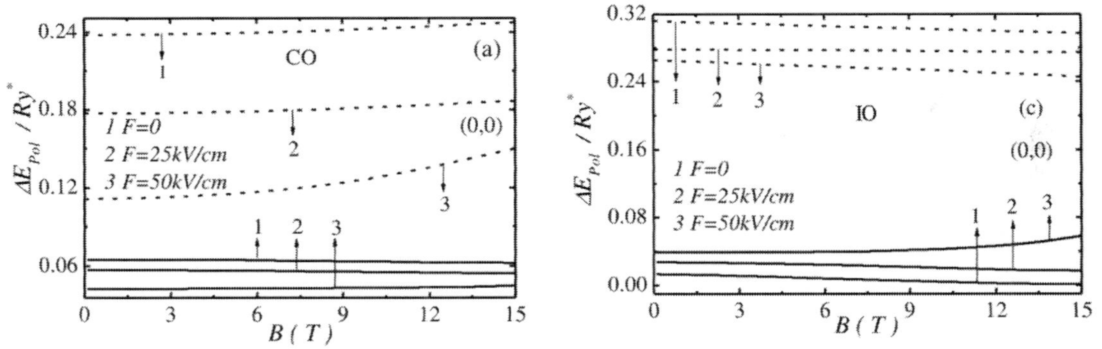

seen

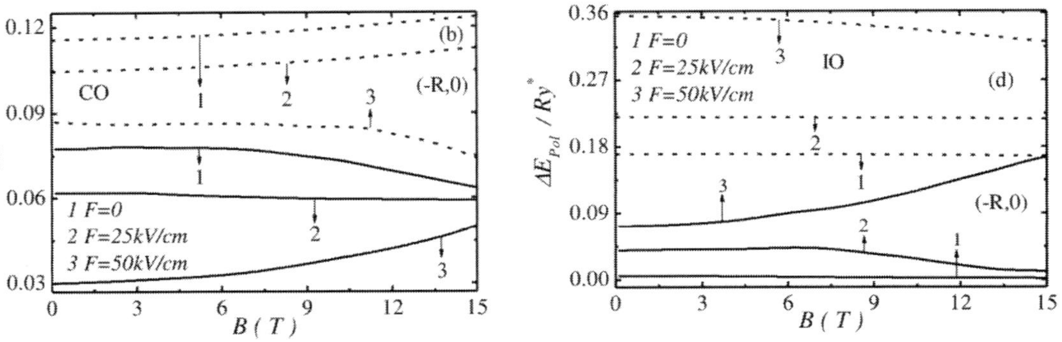

Figure 4 Reduced polaronic shift of the BE as a function of the magnetic field for different impurity positions and for various values of the electric field ($F = 0$, F = 25kV/cm, F = 50kV/cm).The solid and dashed lines represent the finite and infinite confinement potential cases, respectively.

from the Figure 3(a) that, the BE of the on-center impurity increases with the increase of magnetic field and decreases with increasing of electric field. It is a direct consequence of condensation and therefore of strengthened localization of electronic charge in the ground state around the axis of the cylinder on which the impurity is located. For electric field $F = 0$ the sum of the contributions due to CO and IO phonon modes is less than bulk phonon case and with increasing of electric it increases and become more than bulk phonon contribution at $F = 50kV/cm$. The curves in Figure 3(b) show that for impurity position (-R,0) when electric field is $F = 0$ and $F = 25kV/cm$, the binding energy decreases with increasing of magnetic field and increases at $F = 50kV/cm$. In Figure 4 we present the appropriate polaronic shift of BE due to confined (Figure 4 (a),(b)) and interface (Figure 4 (c),(d)) phonons for the impurity positions (0,0) and (-R,0). The solid (dashed) lines correspond to the finite (infinite [15]) confinement potential case. As we can see, the magnetic field dependence of the CO phonon caused polaronic shift in case of finite confinement potential has similar qualitative features as BE and for infinite confinement potential it decreases for all impurity positions (Figure 2 (a),(b)). In the model of finite confinement potential the polaronic shift of BE caused by electron-IO phonon coupling increases with increasing magnetic field whereas in the infinite potential model it decreases for impurity positions (0,0), (0.5R,0), (R,0), and increases for impurity positions (-0.5R,0) and (-R,0).

4. Conclusion

We have presented study of the ground state BE of a hydrogenic impurity in cylindrical quantum wire with a finite confining potential subjected to external electric and magnetic fields and electron PO phonon interaction with tacking into account phonon confinement effect. We have employed the Landau-Pekar variational method to obtain the polaronic corrections to the ground state binding energy. We have shown that the polar optical phonon confinement leads to a considerable enhancement of the polaron effect. The results of the polaronic effects on the binding energy are obtained as a function of the applied electric and magnetic fields for different positions of the impurity.

References

[1] Ipatova I P,Moslov A Yu and Proshina O V 2002 Surf. Sci 507-510 598
[2] Watt M, Torres C M S, Arnot, H E G and Beaumont S.P 1990 *Semicond.Sci.Technol* **5** 285
[3] Li Y C and Gu S W 1992 *Phys. Rev. B* **45** 12102
[4] Au-Yeung T C, Jong C C, Gu S W, Wong E M 1995 *Phys. Lett.* A **204** 155
[5] Vartanian A L, Yeranosyan M A, Kirakosyan A A 2005 *Physica B* **390** 256
[6] Vartanian A L, Vardanyan L A, Kazaryan E M 2007 *Phys. Let. A B* **360** 649

[7] Zhao Z.R, Liang X X 2009 *J. Appl. Phys.* **105** 083704
[8] Brains S V, Li G, Bajaj K K 1993 *Phys. Rev. B* **47** 1316
[9] Landau L D, Pekar S I and Eksp Zh 1946 *Fiz Teor* **16** 341
[10] Xie H J 2004 *Physica* E **22** 906
[11] Xie H J, Chen C Y, Ma B K 2000 *Phys. Rev*. B **61** 4827
[12] Vartanian A L, Vardanyan L A and Kazaryan E M 2008 *Phys. Stat. Sol.* B **245** 123
[13] Adachi S 1985 *J. Appl. Phys*. **58 R1**
[14] Bosio C, Staehli J L, Guzzi M, Burri G and Logan R A 1988 *Phys.Rev*.B **38** 3263
[15] Vartanian A L, Shahbandari A,Yeranosyan M A, Kirakosyan A A 2011 JLTP DOI: 10.1007/s10909-011-0401-x

Phonon confinement effect on the polaron basic parameters in nanowires in the presence of external fields

A L Vartanian, A L Asatryan, L A Vardanyan and V Y Arzumanyan
Department of Solid State Physics, Yerevan State University,
1, A.Manoogian, 0025 Yerevan, Armenia

E-mail: vardan@ysu.am; annaa@ysu.am

Abstract. The effect of electric and magnetic fields on the basic parameters of confined and interface polarons in cylindrical nanowires embedded in a non-polar matrix are studied theoretically for the first time. The analytical expressions for the quasi-one-dimensional Fröhlich polaron self-energy and effective mass are obtained as functions of the wire radius and the strength of the external fields; this can be used in the interpretation of optical phenomena related to polaron motion in cylindrical nanowire when the effect of applied fields competes with the spatial quantum confinement.

1. Introduction

Recent precise absolute far-infra-red magneto-transmission experiments performed in ultra-high magnetic fields on a low dimensional systems reveal new singular features related to expected polaronic effects [1-4]. Since the materials commonly used in the fabrication of the low dimensional systems are ionic semiconductors, the polaronic effects may strongly affect their physical properties. Peeters and co-workers [5] have studied the polarons in n-dimensional crystals and have found that the reduction of dimensionality leads to the enhancement of polaronic effects. Since then, a lot of attention has been paid to the various aspects of the polaronic process in quasi-one dimensional quantum wire systems (see e.g. [6-10]). Polarons in nanowires (NW's) are markedly different from those in bulk materials, due to the presence of wire potential, which confines the motion of the carriers in the plane transverse to the wire axis and due to the occurrence of confined and either surface or interface optical phonon modes when the confined structure is free standing or embedded in another material, respectively [11]. Polaron effects become more interesting in the presence of external electric and magnetic fields.

By using the Lee-Low-Pines (LLP) variational method [12], Lu et al. have studied the polaron energy shift due to the electron-phonon interaction [13] and polaron effective mass [14] in a rectangular GaAs quantum wire in the presence of a uniform strong magnetic field. In the calculations they have ignored the influence of the interface phonons. A complete theory of the quasiparticle properties, i.e., energy-momentum relation, damping, renormalization factor, and effective mass, for interacting electrons in lower-dimensional nanostructures using many-particle Green's function technique has been developed in [15]. Applying this theory to the Q1D magnetopolaron problem, an expression for the electron matrix self-energy in the subband space in the framework of a modified Hartree-Fock approximation has been derived. The free moving Fröhlich polaron properties in a rectangular quantum wire in the presence of an external electric field are studied in [16,17], where the analytical expressions for the polaron self-energy, the effective mass, and the average number of virtual phonons are obtained without including the effect of phonon confinement. In [18] the influence of phonon confinement as well as of an electric field applied perpendicular to the cylindrical NW axis on the polaron basic parameters has been investigated.

Till now, the problem of the effects of external electric and magnetic fields on the basic parameters (self-energy and effective mass) of confined and surface polarons in a cylindrical NW's remains unsolved to the best of our knowledge. In the present paper the influence of phonon confinement as well as that of magnetic and electric fields on the properties of Frohlich polaron confined in a polar CdSe quantum wire (embedded in a non-polar matrix) is studied by using the LLP variational approach. The dependencies of the polaron self-energy and effective mass on wire radius, electric field strength and magnetic field induction are obtained. We expect that the obtained results will be helpful in the interpretation of optical phenomena related to polaron motion in cylindrical NW when the effect of applied fields competes with the structural confinement.

2. Theory

Let us consider an electron moving in a polar cylindrical NW of radius R with infinite confining barriers. A uniform magnetic field applied parallel to the wire axis and an electric field applied perpendicular to the wire axis (z axis). The wire is surrounded by a non-polar matrix with dielectric constant ε_d. To obtain the polaron basic parameters, we perform the standard LLP variational method [12,16,17]. As a result, the following expressions for polaron self-energy and effective mass are obtained

$$E_{self}^{0}(F,B,R)=-\sum_{pq_z}\frac{\left|\Gamma_{CO}^{0p}(q_z)\right|^2\left|L_1(0,p,F,B,R)\right|^2}{\hbar\omega_{CO}+\dfrac{\hbar^2q_z^2}{2m^*}}-\sum_{q_z}\frac{\left|\Gamma_{IO}^{0}(q_z)\right|^2\left|L_2(0,q_z,F,B,R)\right|^2}{\hbar\omega_{IO}+\dfrac{\hbar^2q_z^2}{2m^*}},\tag{1}$$

$$m_{pol}=m^*\left(1+\sum_{pq_z}\frac{2\hbar^2q_z^2}{m^*}\frac{\left|\Gamma_{CO}^{0p}(q_z)\right|^2\left|L_1(0,p,F,B,R)\right|^2}{\left(\hbar\omega_{CO}+\dfrac{\hbar^2q_z^2}{2m^*}\right)^3}+\sum_{q_z}\frac{2\hbar^2q_z^2}{m^*}\frac{\left|\Gamma_{IO}^{0}(q_z)\right|^2\left|L_2(0,q_z,F,B,R)\right|^2}{\left(\hbar\omega_{IO}+\dfrac{\hbar^2q_z^2}{2m^*}\right)^3}\right),\tag{2}$$

where

$$\left|\Gamma_{CO}^{np}(q_z)\right|^2=\frac{4e^2\hbar\omega_{CO}}{LJ_{n+1}^2(\kappa_{np})(\kappa_{np}^2+R^2q_z^2)}\left(\frac{1}{\varepsilon_\infty}-\frac{1}{\varepsilon_0}\right),\tag{3}$$

$$\left|\Gamma_{IO}^{s}(q_z)\right|^2=\frac{4e^2\hbar\omega_{IO}}{LK_s^2(q_zR)I_s(q_zR)q_zR\left(I_{s-1}(q_zR)+I_{s+1}(q_zR)\right)}\left(\frac{1}{\varepsilon-\varepsilon_0}-\frac{1}{\varepsilon-\varepsilon_\infty}\right),\tag{4}$$

$$L_1(n,p,F,B,R)=\langle\Phi|J_n\left(\frac{\kappa_{np}\rho}{R}\right)|\Phi\rangle,\quad L_2(s,q_z,F,B,R)=\langle\Phi|K_s(q_zR)I_s(q_zr)|\Phi\rangle,\tag{5}$$

m^* is the electron's effective mass, L is the length of the wire, q_z is the phonon wave vector along wire direction, B is the magnetic field induction, F is the electric field strength, $\varepsilon_\infty(\varepsilon_0)$ is the optical (static) dielectric constant, $K_s(x)$ and $I_s(x)$ are the s th order first and second kind modified Bessel functions, respectively, $J_n(x)$ is the Bessel function of the n th order, κ_{np} is the p th zero of $J_n(x)$. We have used dispersionless LO phonons in the description of confined (CO) phonon modes for simplicity, but retained the wave-vector dependence in the case of interface (IO) modes

$$\omega_{CO}=\omega_{LO},\qquad \omega_{IO}=\sqrt{1+\frac{\varepsilon_0-\varepsilon_\infty}{\varepsilon_\infty-\varepsilon}}\,\omega_{TO},\tag{6}$$

$$\varepsilon(\omega) = -\frac{I_s(q_zR)\left[K_{s-1}(q_zR)+K_{s+1}(q_zR)\right]}{K_s(q_zR)\left[I_{s-1}(q_zR)+I_{s+1}(q_zR)\right]}\varepsilon_d. \quad (7)$$

For the trial wave function of an electron moving along the wire axis with momentum $\hbar k_z$ we have chosen a form $\Phi(\rho,\varphi,z) = N\,{}_1F_1\left(-a,1;\rho^2/2\alpha_c^2\right)\exp\left[-\rho^2/4\alpha_c^2\right]\exp\left[-\beta\rho\cos\varphi\right]\exp(ik_z z)$ [19], where N is the normalization constant, ${}_1F_1\left(-a,1;\rho^2/2\alpha_c^2\right)$ is the confluent hypergeometric function, β is the variational parameter which takes into account the application of an external electric field, $\omega_c = \hbar/m^*\alpha_c^2$ is the cyclotron frequency, $\alpha_c = (\hbar c/eB)^{1/2}$ is the cyclotron radius. The value of a is determined by the boundary condition that the wave function vanishes at the surface ($\rho = R$) of the NW. CO and IO phonon subsystems become independent and can be considered separately, therefore (1) and (2) allow us to investigate electron-CO phonon and electron-IO phonon interactions and calculate the total contributions in polaron self-energy and effective mass by simply adding up the two parts.

3. Numerical results and discussion

To provide a clear picture of the polaron effect in the cylindrical NW system in the presence of external fields, a numerical calculation is carried on a CdSe with a relatively high electron-phonon coupling constant ($\alpha = 0.46$). The medium outside the NW is vacuum, the material parameters are: $\varepsilon_0 = 9.56$, $\varepsilon_\infty = 6.23$, $\hbar\omega_{LO} = 26.46\,\text{meV}$, $m^* = 0.13m_e$, $\varepsilon_d = 1$ [20].

Figures 1a and 1b are respectively the plots of the polaron self-energy and effective mass versus wire radius in a magnetic field $B = 1T$ and at various values of an electric field ($F = 10kV/cm$, $F = 50kV/cm$, $F = 100kV/cm$). It can be clearly seen that the polaron self-energy monotonically decreases with the increase of R due to electron coupling with the IO and CO phonon modes. As a result, the total polaron self-energy, which incorporates the contributions of electron interaction both IO and CO phonons, also decreases. The same behavior is observed for the quasi-1D polaron effective mass (figure 1b).

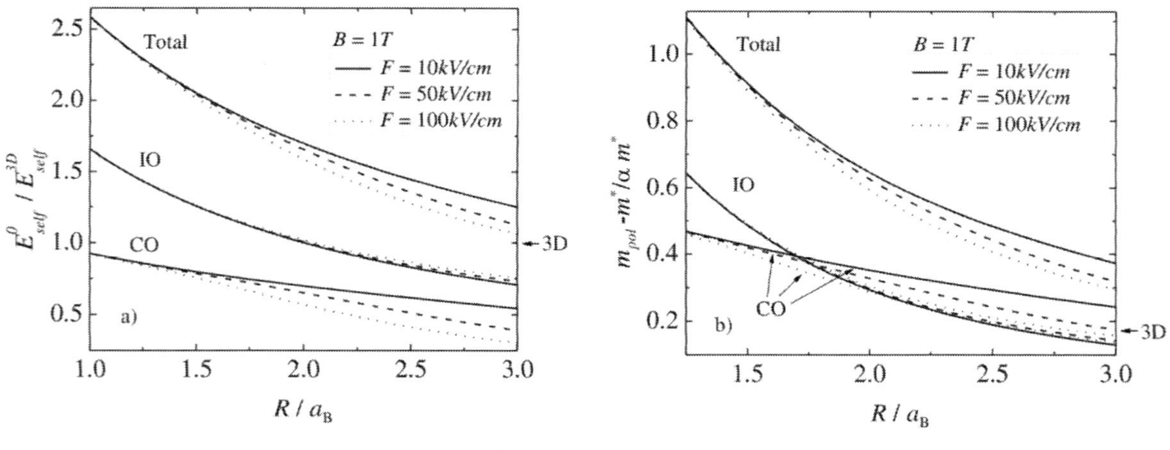

Figure 1. The polaron self-energy (a) and the effective mass (b) as a function of CdSe wire radius (expressed in terms of the Bohr radius: $a_B = \hbar^2\varepsilon_0/m^*e^2 \approx 39$ Å) for several values of the electric field and with a fixed magnetic field value $B = 1T$.

It should be noted that for the shown region of the wire radius the IO phonon contribution in the polaron self-energy dominates over the contribution of CO phonons and strongly depends on the NW radius. A similar result in the absence of external fields has been reported for narrow NW in [21]. In the polaron effective mass, starting with some value of R, CO phonon contribution becomes predominant. The IO phonon contributions (in percentage terms) at above mentioned values of external fields are given in table 1. As can be seen from the table with the increase of the electric field the contribution of IO phonons increases.

Table 1. IO phonon contribution to the total polaron self-energy and effective mass at different values of electric field and at magnetic field $B = 1T$.

	$F = 10kV/cm$	$F = 50kV/cm$	$F = 100kV/cm$
Self-energy	67,8% $(R = 0{,}5a_B)$ 58,9% $(R = 2a_B)$	67,8% $(R = 0{,}5a_B)$ 60,7% $(R = 2a_B)$	67,8% $(R = 0{,}5a_B)$ 64,1% $(R = 2a_B)$
Effective mass	73,7% $(R = 0{,}5a_B)$ 45,4% $(R = 2a_B)$	73,7% $(R = 0{,}5a_B)$ 47,6% $(R = 2a_B)$	73,7% $(R = 0{,}5a_B)$ 51,5% $(R = 2a_B)$

In figures 1a and 1b, the values of self-energy and effective mass of three-dimensional polaron are indicated by arrows. In the regime of weak electron-LO phonon interaction, these values are: $E_S^{3D} = -\alpha\hbar\omega_{LO}$, $m_p^{3D} = m^*(1+\alpha/6)$ [12]. We notice that for small wire radius the total polaron self-energy as well as the effective mass may be several times greater than the 3D values of considered quantities and the presented curves converge to the 3D limits, when R increases.

Figure 2 shows the polaron self-energy and the effective mass as a function of the electric field applied perpendicular to the NW axis in case of $R = 2a_B$ and $B = 1T$. In the considered region of the electric field the polaron self-energy (effective mass), due to electron interaction with the IO phonons, increases slowly with the increase of the electric field, while the CO part of the self-energy (effective mass) is reduces significantly. Therefore the total self-energy and the effective mass of the polaron

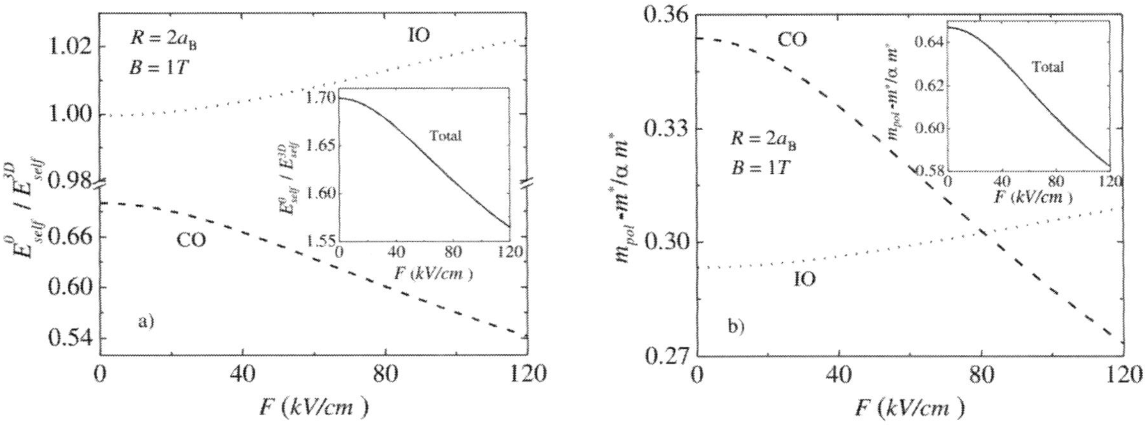

Figure 2. The polaron self-energy (a) and the effective mass (b) as a function of electric field for CdSe wire with radius $R = 2a_B$ and magnetic field $B = 1T$. Dashed (dotted) lines show the CO (IO) phonon contribution. Inset shows the total (CO and IO) contribution.

decreases, which can be seen more clearly in the insets to figures 2a and 2b. The electric field dependences of the IO as well as CO phonon caused polaron self-energy and effective mass can be explained by the fact that the maximum of the electronic charge distribution with increasing electric field displaces from the wire axis towards the barrier region. As a result, the role of IO phonons increases, while the role of CO phonons decreases. The figure 2 shows that the change in polaron self-energy and effective mass is about 8% and 10%, respectively, when the electric field increases up to $120kV/cm$.

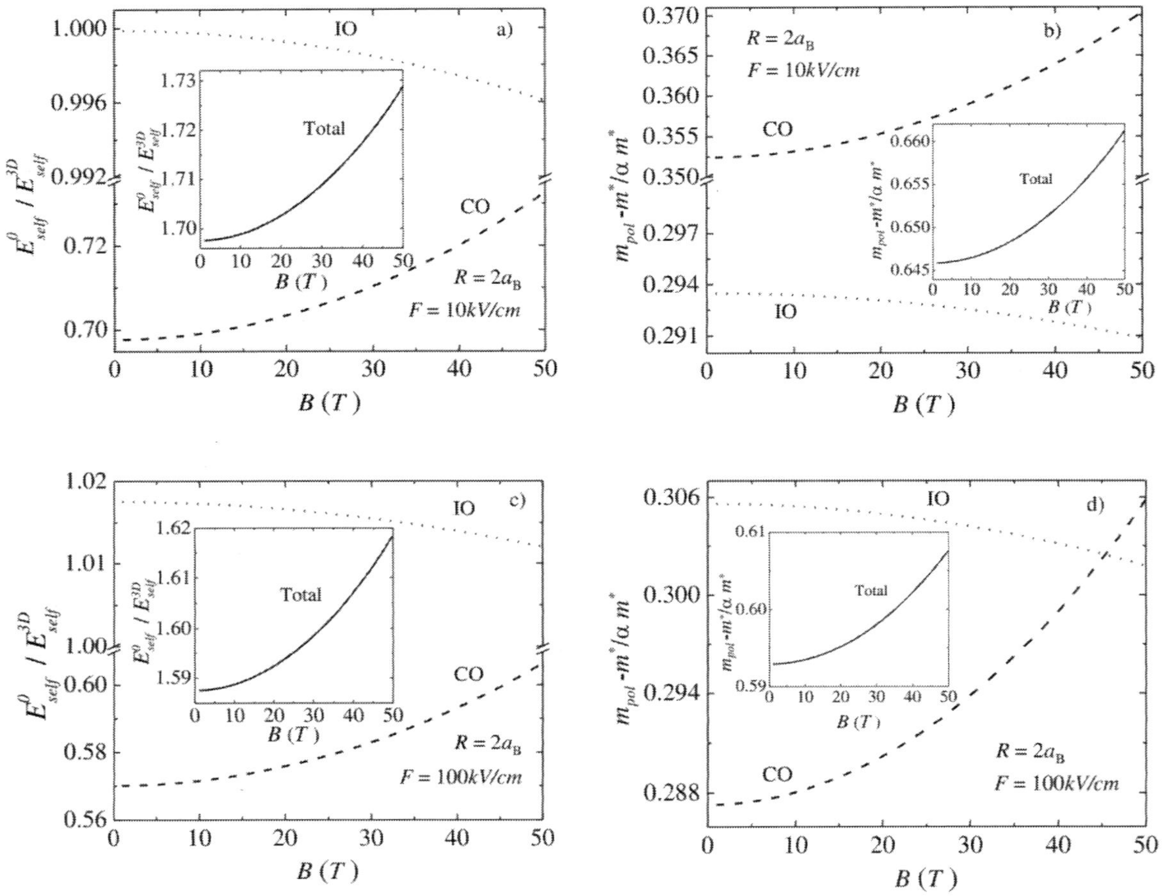

Figure 3. The polaron self-energy and the effective mass as a function of magnetic field for CdSe wire with radius $R = 2a_B$ and electric field values: $F = 10kV/cm$ (a,b); $F = 100kV/cm$ (c,d). Dashed (dotted) lines show the CO (IO) phonon contribution. Inset shows the total (CO and IO) contribution.

In figure 3 the magnetic field dependences of the polaron basic parameters are presented when the electric field strength is $F = 10kV/cm$ (figures 3a, 3b) and $F = 100kV/cm$ (figures 3c, 3d). The results presented in figure 3 clearly show that the IO phonon contributions in polaron basic parameters are decrease with the increase of magnetic field while the opposite behavior occurs for the CO phonons. This is caused by the electron localization close to the wire axis when the magnetic field is applied parallel to that axis. It can be seen from figures 3b and 3d that the relative arrangement of the curves, which represent CO and IO parts of polaron effective mass, changes when the electric field increases from $10kV/cm$ to $100kV/cm$. This indicates that the influence of electric field on the CO and IO parts of polaron effective mass appears in a variety of ways, which could be seen also in figure 2. It should be noted that for the relatively high values of the magnetic field and small values of the electric field the decreasing CO phonon caused part of the polaron self energy increases when the wire radius becomes larger. Particularly, at $B=30T$ and $F=10kV/cm$ the change of the behavior occurs at $R\approx 4,2a_B$. This could be explained by the competition between three factors: spatial confinement, Stark effect and magnetic confinement effect.

4. Conclusion

In conclusion, we have presented a systematic study of the Fröhlich polaron basic parameters in a cylindrical NW in the presence of external fields taking into account the confinement effect on the polar optical phonons. The numerical results on the CdSe material show that the polaron basic parameters strongly depend on the wire radius and the IO phonon modes play a major role in the phonon contributions, especially when the radius is relatively small. Two opposite behaviors of polaron self-energy (effective mass) due to electron interaction with different phonon modes are obtained depending on whether a magnetic or an electric field is applied. As far as we know, there are no experimental data available to compare with our theoretical results; we believe these results will be of importance in the understanding of future experiments in this subject.

Acknowledgement

This work has been supported by the Armenian National Science and Education Fund (Grant: condmatth-2553).

References

[1] Hameau S, Guldner Y, Verzelen O, Ferreira R, Bastard G, Zeman J, Lemaître A and Gérard J M 1999 *Phys. Rev. Lett.* **83** 4152
[2] Hameau S, Isaia J N, Guldner Y, Deleporte E, Verzelen O, Ferreira R, Bastard G, Zeman J and Gérard J M 2002 *Phys. Rev.* B. **65** 085316
[3] Carpenter B A, Zibik E A, Sadowski M L, Wilson L R, Whittaker D M, Cockburn J W, Skolnick M S, Potemski M, Steer M J and Hopkinson M 2006 *Phys. Rev.* B. **74** 161302(R)
[4] Yoon J, Girgis A M, Shalish I, Ram-Mohan L R and Narayanamurti V 2009 *Appl. Phys. Lett.* **94** 142102
[5] Peeters F M, Wu X -G and Devreese J T 1986 *Phys. Rev.* B **33** 3926
[6 Chen Q H, Ren Y H, Jiao Z K and Wang K L 1998 *Phys. Rev.* B **58** 16340
[7] Chen Q H, Ren Y H, Li T Z, Yu Y B and Jiao Z K 1999 *J. Phys.: Condens. Matter* **11** 4189
[8] Xie H -J, Chen C -Y and Ma B -K 2000 *Phys. Rev.* B **61** 4827
[9] Buonocore F, Iadonisi G, Ninno D and Ventriglia F 2002 *Phys. Rev.* B **65** 205415
[10] Xie H -J 2004 *Physica* E **22** 906
[11] Moriarty P 2001 *Rep. Prog. Phys.* **64** 297
[12] Lee T D, Low F E and Pines D 1953 *Phys. Rev.* **90** 297
[13] Lu T -Q and Gu S -W 1992 *Phys. Stat. Sol.* (b) **174** 427
[14] Lu T, Zheng Y, Zhang C and Wu X 1996 *Phys. Stat. Sol.* (b) **197** 399
[15] Wendler L 1998 *Phys. Rev.* B **57** 9214
[16] Vartanian A L and Vardanyan L A 2009 *Physica* E **41** 1417
[17] Vartanian A L 2009 *Phys. Stat. Sol.* B **246** 2279.
[18] Vartanian A L 2011 *Proc. of YSU: Phys. and Math. Sci.* **2(225)** 55
[19] Branis S V, Li G and Bajaj K K 1993 *Phys. Rev.* B **47** 1316
[20] Feddi E, Haouari M E, Assaid E, Stébé B, Khamkhami J E and Dujardin F 2003 *Phys. Rev.* B **63** 235313
[21] Maslov A Yu and Proshina O 2010 *Nanoscale Res. Lett.* **5** 1744

Investigation of Resonance Frequency Dependence on Cavity Geometry in Orthogonal Splitter with Photonic Crystal Structure

A. Abbasi[2], M. Noshad[1,2], R. Ranjbar[2] and R. Kheradmand[2*]
[1] Faculty of Electrical and Computer Engineering, University of Tabriz, Tabriz, Iran
[2] Research Institute for Applied Physics and Astronomy, University of Tabriz, Tabriz, Iran
*Corresponding Email: R_kheradmand@tabrizu.ac.ir

Abstract. We have investigated the dependence of resonance frequency of micro-cavity on its geometry for T-shaped beam splitter using finite difference time domain (FDTD) method. We derived resonance frequencies and output powers for elliptical cavity with various radii in both vertical and horizontal directions (X and Y directions). The results showed that resonance frequency is highly related to the cavity radii and broad range of frequencies is accessible. We also discussed T-beam splitter performance for a rotated cavity by different degrees. We observed that rotation of the elliptical cavity make changes only in power transmission and interestingly resonance frequencies become fixed. By setting the cavity angle at 45 degree we get the equivalent division of power with the transmission efficiency of 40% per each output port. Results showed that rotation of the elliptical cavity is an appropriate and easy way to modify the division of input power into two output waveguides.

1. Introduction

The photonic crystal (PhC) concept was proposed in 1987 [1, 2], and the first 3D experimental photonic crystal with full band gap was manufactured in 1991. The existence of band gap in PhC structures led to many prominent applications in integrated optics. Photonic crystals are nowadays used for different applications such as filters [3], modulators [4–6], switches [7], beam splitters [8-10], and super prisms [11–13] for multiplexing and demultiplexing for example.

Optical beam splitters have two common topologies; T-shaped, and Y-shaped. Usually, T-shaped photonic crystal splitters are made on the square lattice and Y-shaped photonic crystal splitters are based on the triangular lattice [14–17].

In this paper, we design a T-shaped photonic crystal splitter with a cavity in the junction. In new splitter geometry with an asymmetric cavity we control the energy division into output ports. Elliptical cavity has different modes and resonance frequencies. The cavity geometry determines the resonance frequency and transmission efficiency of the splitter. By modifying the two radii of the cavity, and rotating it, the resonance frequency and transmission efficiency change. In order to have the efficient transmission in any of the output waveguides, the input beam frequency should be adjusted to the resonance frequency of the cavity. Results show that rotation of the elliptical cavity is an appropriate and easy way to modify the division of input power into two output ports.

2. Design and Simulation

In this work a 2D photonic crystal lattice with cylindrical rods is used. The structure is based on a 23 × 23 square lattice with the lattice constant $a = 575\ nm$. The refractive index of the rods is 3.5 and the background refractive index is equal to 1. The radius of the rods is set to $0.2a$. As shown in figure 1, three waveguides are created by elimination of the rows of rods. These are connected to each other using an asymmetric cavity. Since the radii of the elliptical cavity are different, so, we will have two separate modes in the vertical and horizontal directions. To analyze the wave behavior, group velocity, bandgaps and defect modes, we have derived the dispersion diagram by the plane wave expansion method (PWE) shown in Figure 1. The input beam frequency is taken about 1.55 μm which is inside the band gap of the structure for TE mode.

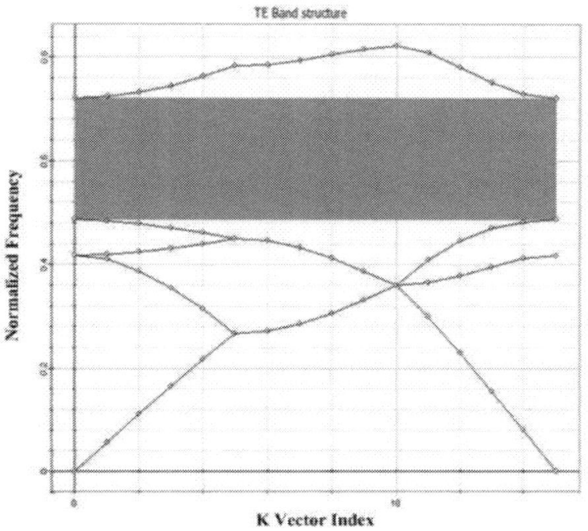

Figure 1. Band structure of the photonic crystal lattice

We have also applied numerical FDTD method to simulate the light propagation in the designed structure. Figure 2 shows the light propagation in the structure when the cavity has the radii of 0.32*a* and 0.27*a* with no rotation. In this case the transmission in *x*-direction (output port I) on the resonance wavelength of 1550 *nm* is 89%. Note that resonance wavelength in y-direction is 1649 *nm*, but we have no energy coupling between the input and output waveguides in y-direction via the cavity. In this case, power in output port II is about 0 for all frequencies.

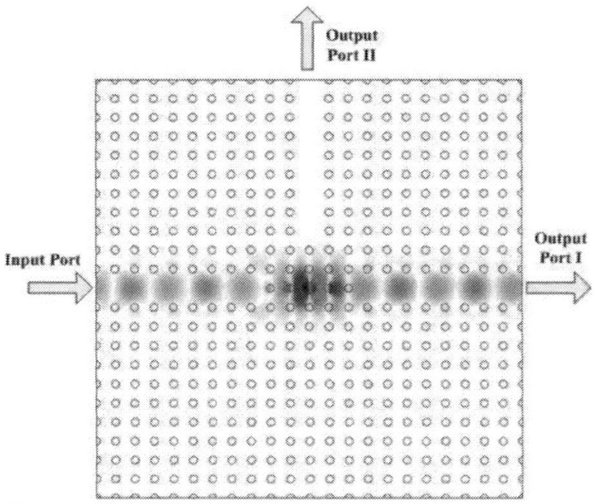

Figure 2. Light propagation in the structure when the cavity has the radii of 0.32a and 0.27

We demonstrate the resonance frequency dependence on cavity radii in figure 3. As shown in this figure, modification of one of the radii leads to linear changes in both X, and Y-direction resonance frequencies.

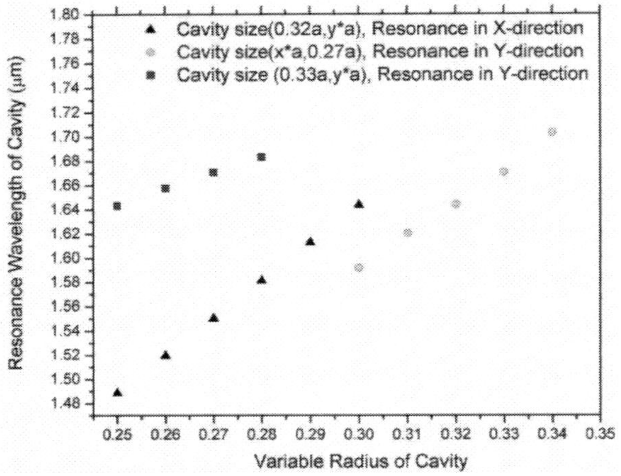

Figure 3. Resonance wavelength of cavity in both directions versus Variable Radius of Cavity in Terms of Coefficient of lattice constant a.

By rotation of the elliptical cavity we can split the input power into the output waveguides. Figure 4 represent the transmission efficiency in both output ports when we have various rotation degrees of cavity. It is remarkable to mention that rotation of the cavity doesn't alter the resonance frequencies of the cavity.
In Figure 4 (a) we see that by increasing the rotation degree, output power I decreases, and conversely, Output power II rises. This means that power division between two waveguides is controllable by rotating the cavity. In both Figures 4 (a) and (b) there is a maximum value for the power in Y-direction which is equal to 47%. The maximum transmission in this direction holds for the degrees of 60 and 40 in the resonance wavelengths of 1550 nm and 1649 nm respectively. In addition, for both resonance wavelengths, the rotation degree of 45 results in the equivalent division of power with the transmission efficiency of 40% per each output ports.

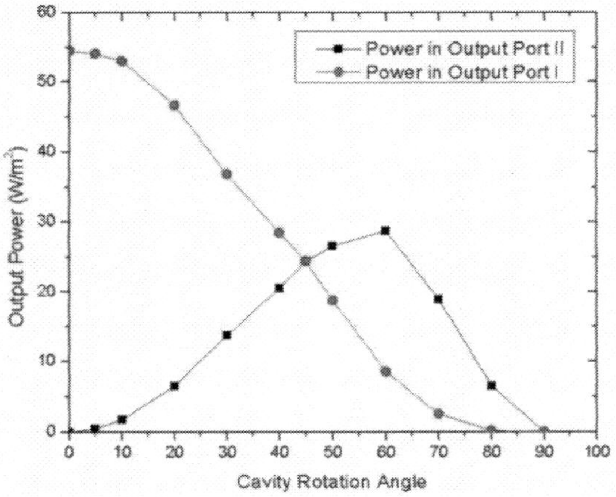

(a)

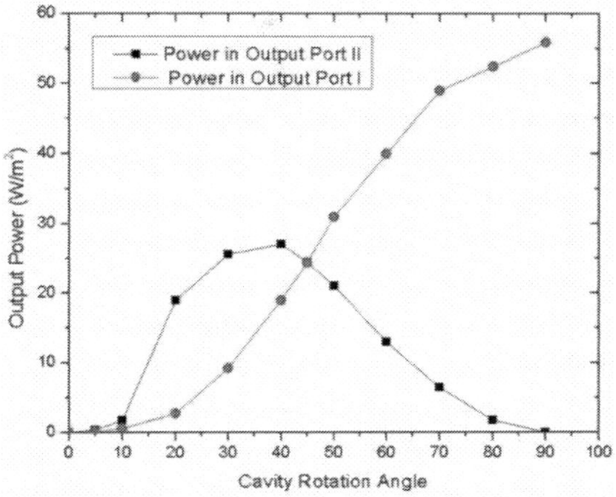

(b)

Figure 4 demonstration of the output power versus the cavity rotation angle for both of the output ports I and II in the (a) X-direction, and (b) Y-direction resonance frequencies.

3. **Conclusion**

In this work we designed an orthogonal splitter with a cavity in the junction for controlling of the power division in the output ports. The dependence of resonance frequency and transmission efficiency of this splitter on the cavity geometry was studied using a numerical method named FDTD. We discussed the cavity dimensions effect in resonance frequency of the cavity in both directions. Secondly, we took the angle of the elliptical cavity as a variable, and rotate the cavity by various degrees to observe the power division changes in the output waveguides. The results represented a maximum power transmission of 40% in output port II for the rotation angle of 60 degrees for X-direction resonance, and 40 degrees for Y-direction resonance. In the rotation angle of 45 degrees, the input power is divided equivalently in the output ports.

3. References

[1] Yablonovitch E 1987 "Inhibited spontaneous emission in solid-state physics and electronics" *Phys. Rev. Lett.* **58**, 2059–2062

[2] John S 1987 "Strong localization of photons in certain disordered dielectric superlattices" *Phys. Rev. Lett.* **58**, 2486–2489

[3] Qiu M, Mulot M, Swillo M, Anand S, Jaskorzynak B, Karlsson A, Kamp M, and Forchel A 2003 "Photonic crystal optical filter based on contra-directional waveguide coupling" *Appl. Phys. Lett.* **83**, 5121– 5123

[4] Fekete A, Kadlec F, Kuzel P, and Nemec H 2007 "Ultrafast optoterahertz photonic crystal modulator" *Opt. Lett.* **32**, 680–682

[5] Gu L, Jiang W, Chen X, Wang L, and Chen R T 2007 "High speed silicon photonic crystal waveguide modulator for low voltage operation" *Appl. Phys. Lett.* **90**, 071105

[6] Jiang Y, Jiang W, Gu L, Chen X, and Chen R T 2005 "80-micron interaction length silicon photonic crystal waveguide modulator" *Appl. Phys. Lett.* **87**, 221105

[7] Beggs D, White T, Cairns L, O'Faolain L, and Krauss T 2009 "Demonstration of an integrated optical switch in a silicon photonic crystal directional coupler" *Physica E* **41**, 1111–1114

[8] Lin S Y, Chow E, Bur J, Johnson S G, and Joannopoulos J D 2002 "Low-loss, wide-angle Y splitter at 1.6-_m wavelengths built with a two-dimensional photonic crystal," *Opt. Lett.*, vol. **27**, no. 16, pp. 1400–1402
[9] Takeda H and Yoshino K 2003 "Tunable light propagation in Y-shaped waveguides in two-dimensional photonic crystals utilizing liquid crystals as linear defects" *Phys. Rev. B*, vol. **67**, p. 073106
[10] Inoue K, Sugimoto K, Ikeda N, Tanaka Y, Asakawa K, Sasaki H, and Ishida K 2004, "Ultra-small photonic-crystal-waveguide-based Y-splitters useful in the near-infrared wavelength region," *Jpn. J. Appl. Phys.*, vol. **43**, no. 4A, pp. L 446–L 448
[11] Kosaka H, Kawashima T, Tomita A, Notomi M, Tamamura T, Sato T, and Kawakami S 1998 "Superprism phenomena in photonic crystals" *Phys. Rev. B* **58**, R10096
[12] Amet J, Baida F, Burr G, and Bernal M P 2008 "The superprism effect in lithium niobate photonic crystals for ultra-fast, ultra-compact electro-optical switching" *Photonic. Nanostruct.* **6**, 47–59
[13] Diziain S, Amet J, Baida F I, and Bernal M P 2008 "Optical far-field and near-field observations of the strong angular dispersion in a lithium niobate photonic crystal superprism designed for double (passive and active) demultiplexer applications" *Appl. Phys. Lett.* **93**, 261103
[14] Fan S, Johnson S G, Joannopoulos, Manolatou C, and Haus H A 2001 "Waveguide branches in photonic crystals," *J. Opt. Soc. Am. B*, Vol. **18**, 162–165
[15] Monifi F, Djavid M, Ghaffari A, and Abrishamian M S 2002 "Design of efficient photonic crystal bend and power splitter using super defects," *J. Opt. Soc. Am. B*, Vol. **25**, 1805–1810
[16] Boscolo S, Midrio M, and Krauss T F 2002 "Y junctions in photonic crystal channel waveguides: Hightransmission and impedance matching," *Opt. Lett.*, Vol. **27**, 1001–1003
[17] Park I, Lee H S, Kim H J, Moon K M, Lee S G, Park S G, and Lee E H 2004 "Photonic crystal power-splitter based on directional coupling," *Opt. Express*, Vol. **12**, 3599– 3604

Energy transfer processes in YAG:Er^{3+}-Ce^{3+} crystal

V G Babajanyan[1], R B Kostanyan, P H Muzhikyan

Institute for Physical Research, NAS of RA, Ashtarak 0203, Armenia

Email: bvahan@gmail.com

Abstract. In the work results of research of nonradiative energy transfer processes taking place in the impurity subsystem of the YAG: Er^{3+}-Ce^{3+} crystal under optical excitations of Er^{3+} and Ce^{3+} ions are presented at room temperatures. By means of time-resolved fluorescence spectroscopy, some kinetic characteristics of fluorescence responses of the crystal under optical excitations at 405 and 808 nm wavelengths are investigated. Comparison of the measured characteristics of fluorescence signals at 860, 980 and 1555 nm wavelengths with the theoretical values numerically calculated from appropriate sets of rate equations allowed to estimate the positive contributions of energy transfer processes into population dynamics of energy levels responsible for generation near 1.5 μm of Er^{3+} active ions in the YAG matrix.

1. Introduction

Laser radiation in eye-save spectral region around 1.5 μm has very high importance for applications in multi-kilometers optical communication systems, remote sensing, laser range finding, etc [1-4]. Thanks to the combination of good optical, thermal and mechanical properties, the yttrium aluminum garnet $Y_3Al_5O_{12}$ (YAG) crystals doped with Er^{3+} active ions are currently one of the most perspective activated medium for lasing in the mentioned spectral region.

Intensive spectroscopic studies have shown that, for improvement of optical excitation energy efficiency usage, it is very fruitful codoping of this crystal with sensitizer ions of trivalent cerium (Ce^{3+}), which have very intensive absorption bands in the blue and ultraviolet wavelengths regions, as well as have channels for effective nonradiative transfer of absorbed excitation energy (NTAEE) to active Er^{3+} ions in YAG matrix [5-19]. Our previous investigations on the influence of Ce^{3+} ions on the spectral characteristics and intensity of the luminescent bands of the YAG:Er^{3+}-Ce^{3+} crystal revealed significant increase in the efficiency of the fluorescence excitation near 1.5 μm on the ${}^4I_{13.2} \rightarrow {}^4I_{15.2}$ transition of the Er^{3+} ions under pumping of this material at 457 nm wavelength (about an order of magnitude compared with the YAG:Er^{3+} crystal when pumped near 980 nm) [18, 19].

Such kind of improvement is due to the fact that in this crystal excitation at 457 nm wavelength corresponds to the very strong absorption peak of the Ce^{3+} ion via interconfigurational 4f → 5d spin-allowed transition, as well as the fact that there is good spectral overlap between broad and intensive fluorescence band of the Ce^{3+} donor ions with the absorption bands of the acceptor Er^{3+} ions, corresponding to 4f→4f transitions from the ground level to excited.

For quantitative estimations of probabilities of NTAEE processes it is also necessary information about kinetic characteristics of fluorescent transitions in impurity subsystem of the material under study, which is absent in the literature.

The present paper is devoted to the kinetic investigations of fluorescence response of the YAG:Er^{3+}-Ce^{3+} crystal using time-resolved luminescence spectroscopy technique with optical excitations at 405, 457 and 808 nm wavelengths, allowing realization of selective excitation of either sensitizer (λ_{pump}=457 nm) or activator ions (λ_{pump}=808 nm) and simultaneous excitation of both impurity ions (λ_{pump}=405 nm).

To determine the degree of influence of the Ce^{3+} ion on the kinetic characteristics of the luminescent signal of Er^{3+} ions, similar investigations carried out on garnet crystals doped separately with Er^{3+} and Ce^{3+} impurity ions.

Comparison of measured kinetic characteristics with the theoretical values obtained by numerical solution of rate equations allowed us to estimate the contributions of certain NTAEE processes in the $^4I_{13.2}$ upper laser level population formation of the Er^{3+} ions in the matrix of YAG.

2. Samples and experimental technique

Studies of optical properties were performed at room temperatures on polished plates made from YAG crystal, doped with Er^{3+} (5% atomic), Ce^{3+} (0.15 atomic %) ions and codoped with Er^{3+} (5% atomic) + Ce^{3+} (0.15 atomic %) ions. More detailed information about the growth of these crystals has been published previously [19, 20].

Kinetic measurements were carried out with time-resolved luminescence spectroscopy method using short laser pulses (a few microseconds) at 457 (second harmonic of a pulsed YVO4:Nd^{3+} laser crystal, working at 914 nm wavelength), 405 and 808 nm (pulsed laser diode) wavelengths.

Selection of a fluorescence wavelength of crystals under study was carried out by a diffraction monochromator, at the exit slit of which depending on a spectral region corresponding high-speed radiation detector (photomultiplier or photodiode) was placed. The time constant of the registration system was about 1-2 microseconds. Electrical signals from the detectors delivered to the high-speed dual channel digital storage oscilloscope from which digital data transmitted to the on line personal computer for subsequent mathematical processing.

In the present work among kinetic characteristics of fluorescence responses the delay time between excitation pulse and the peak of the signal (Δt) and the decay time of the responses (τ_d) were chosen for detailed study.

3. Results

It is known that energy levels' population formation of the Er^{3+} ions in various matrixes is strongly dependent on dynamics of the processes associated with the $^4S_{3/2}$, $^4I_{11/2}$ and $^4I_{13/2}$ levels [6-19, 21-24], and, therefore, study of kinetic characteristics of their fluorescence was in the center of our attention. However, despite of numerous works, devoted to the study of kinetics of green fluorescence (near 550 nm, correspondent to the $^4S_{3/2} \rightarrow {}^4I_{15/2}$ transition) [10-16, 20-23], in the present work the fluorescence at 860 nm wavelength ($^4S_{3/2} \rightarrow {}^4I_{13/2}$ transition) was chosen as a test subject for our investigations. Such alternative choice seems convenient and useful for YAG:Er^{3+}-Ce^{3+} crystal study, since kinetic characteristics of fluorescence at both wavelengths are the same (because of initial level is the same) and it is not overlapped by the intense and wideband fluorescence of the Ce^{3+} ions (peaked near the 590 nm in the YAG host), which could result in considerable distortion of the green fluorescence dynamics. The 980 and 1555 nm wavelengths are correspondent to the well-known $^4I_{13/2} \rightarrow {}^4I_{15/2}$ and $^4I_{13/2} \rightarrow {}^4I_{15/2}$ transitions of Er^{3+} ions in the YAG crystal.

Results of room temperature measurements of above mentioned kinetic characteristics from fluorescence responses, registered at 860, 980, 1555 nm wavelengths of the YAG:Er^{3+} and YAG:Er^{3+}-Ce^{3+} crystals under short optical excitations, are presented in the table 1. It will be noted that studies of kinetic parameters of YAG:Er^{3+} and YAG:Er^{3+}-Ce^{3+} crystals fluorescence under 457 nm wavelength pumping did not give additional useful information and they are not included in the table, since the duration of the laser's optical excitation we used (about 10 μs) was much longer than the excitation energy transfer "duration" from Ce^{3+} donor to the Er^{3+} acceptor ions.

Journal of Physics: Conference Series **350** (2012) 012024 doi:10.1088/1742-6596/350/1/012024

Table 1. Kinetic characteristics of fluorescence responses of the investigated crystals depending on optical excitations at 405 and 808 nm wavelengths.

YAG:Er^{3+} pump at 405 nm			
Lumin., nm	**860**	**980**	**1555**
$\tau_d/\Delta t$, µs	14/4	97/22	6060/350
YAG: Er^{3+} pump at 808nm			
$\tau_d/\Delta t$, µs	-	94/4	7460/700
YAG:Er^{3+}-Ce^{3+} pump at 405 nm			
$\tau_d/\Delta t$, µs	14/4	92/6	6000/280
YAG:Er^{3+}-Ce^{3+} pump at 808 nm			
$\tau_d/\Delta t$, µs	-	78/2	6081/600

Figure 1 shows a simplified energy level diagram of Er^{3+} and Ce^{3+} ions in the matrix of YAG, which was used for compiling of corresponding systems of rate equations for the populations of the energy levels of impurity ions under different excitation conditions.

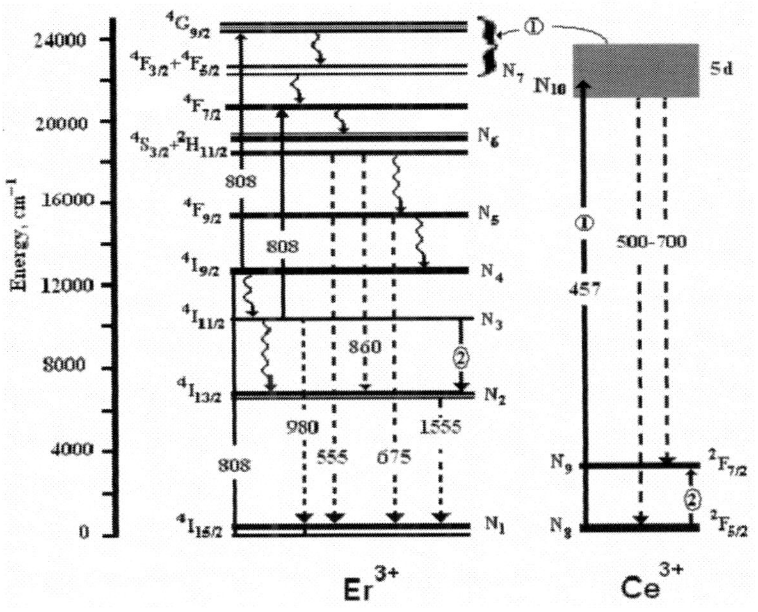

Figure 1. Energy levels of Er^{3+} and Ce^{3+} ions in YAG.

Here the N_i symbols denote the populations of the levels, solid and dashed lines correspond to the optical excitations and observed fluorescent transitions, respectively, and curly curves point the non-radiative transitions. The well-known process of excitation transfer from Ce^{3+} donor to Er^{3+} acceptor ions (5d→4f [Ce^{3+}]:$^4I_{15/2}$$^4F_{3/2}$+$^4F_{5/2}$ [Er^{3+}]) is labeled by symbol 1 in the circle and possible cross-relaxation back-transfer process ($^4I_{11/2}$→$^4I_{13/2}$ [Er^{3+}]:$^2F_{5/2}$→$^2F_{7/2}$ [Ce^{3+}]) is marked by symbol 2 in the circle.

Dynamics of the processes taking place in the impurity subsystem of the YAG:Er^{3+}-Ce^{3+} crystal under optical excitations can be described by the following set of rate equations:

$$\begin{aligned}
\dot{N}_1 &= A_{21}N_2 + A_{31}N_3 + A_{41}N_4 + A_{51}N_5 + A_{61}N_6 + A_{71}N_7 - C_1^{cr}N_{10}N_1 \\
\dot{N}_2 &= A_{32}N_3 + A_{42}N_4 + A_{52}N_5 + A_{62}N_6 + A_{72}N_7 + w_{32}N_3 - A_{21}N_2 + C_2^{cr}N_3N_8 \\
\dot{N}_3 &= A_{43}N_4 + A_{53}N_5 + A_{63}N_6 + A_{73}N_7 + w_{43}N_4 - A_{31}N_3 - A_{32}N_3 - w_{32}N_3 - C_2^{cr}N_3N_8 \\
\dot{N}_4 &= A_{54}N_5 + A_{64}N_6 + A_{74}N_7 + w_{54}N_5 - A_{41}N_4 - A_{42}N_4 - A_{43}N_4 - w_{43}N_4 \\
\dot{N}_5 &= A_{65}N_6 + A_{75}N_7 + w_{65}N_6 - A_{51}N_5 - A_{52}N_5 - A_{53}N_5 - A_{54}N_5 - w_{54}N_5 \\
\dot{N}_6 &= A_{76}N_7 + w_{76}N_7 - A_{61}N_6 - A_{62}N_6 - A_{63}N_6 - A_{64}N_6 - A_{65}N_6 - w_{65}N_6 \\
\dot{N}_7 &= -A_{71}N_7 - A_{72}N_7 - A_{73}N_7 - A_{74}N_7 - A_{75}N_7 - A_{76}N_7 - w_{76}N_7 + C_1^{cr}N_{10}N_1 \\
\dot{N}_8 &= A_{98}N_9 + w_{98}N_9 + A_{8,10}N_{10} - C_2^{cr}N_3N_8 + C_1^{cr}N_{10}N_1 \\
\dot{N}_9 &= -A_{98}N_9 - w_{98}N_9 + A_{10,9}N_{10} + C_2^{cr}N_3N_8 \\
\dot{N}_{10} &= -A_{10,8}N_{10} - A_{10,9}N_{10} - C_1^{cr}N_{10}N_1
\end{aligned} \tag{1}$$

Here A_{ij} and w_{ij} are probabilities of radiative and nonradiative transitions between i and j energy levels, correspondingly, C_i^{cr} are macroscopic transfer coefficients of appropriate cross-relaxation processes.

Taking into account that the nonradiative transitions dominate in the transitions between adjacent energy levels for multiplets above $^4I_{9/2}$, the system of rate equations (1) in case of excitation at 808 nm for YAG:Er^{3+}-Ce^{3+} crystal can be simplified to the following form

$$\begin{aligned}
\dot{N}_1 &= A_{21}N_2 + A_{31}N_3 \\
\dot{N}_2 &= A_{32}N_3 + w_{32}N_3 - A_{21}N_2 + C_2^{cr}N_3N_8 \\
\dot{N}_3 &= w_{43}N_4 - A_{31}N_3 - A_{32}N_3 - w_{32}N_3 - C_2^{cr}N_3N_8 \\
\dot{N}_4 &= -w_{43}N_4 \\
\dot{N}_8 &= w_{98}N_9 - C_2^{cr}N_3N_8 \\
\dot{N}_9 &= -w_{98}N_9 + C_2^{cr}N_3N_8 \\
&N_1(0) = N_{Er}; \quad N_2(0) = N_3(0) = 0; \quad N_4(0) = N_4^0; \\
&N_8(0) = N_{Ce}; \quad N_9(0) = 0 \\
&N_1 + N_2 + N_3 + N_4 = N_{Er}; \quad N_8 + N_9 = N_{Ce}.
\end{aligned} \tag{2}$$

In the Figure 2 the measured response (points) and the theoretical curve (solid) calculated with system of rate equations (2) by use of the Wolfram Mathematica 8 software are shown for the luminescence response at 980 nm of the YAG: $Er^{3+}Ce^{3+}$ crystal under optical excitation at 808 nm. In the numerical calculations the following data were used: A_{21}=156.2s^{-1}, A_{31}=131.7s^{-1}, A_{32}=29.4s^{-1}, w_{32}=9838.9s^{-1}, w_{43}=2·10^{7}s^{-1}, A_{51}=1449.5s^{-1}, A_{52}=56.6s^{-1}, A_{53}=73.2s^{-1}, A_{54}=5.2s^{-1}, w_{54}=8.3·10^{5}s^{-1}, A_{61}=814.3s^{-1}, A_{62}=548.4s^{-1}, A_{63}=39s^{-1}, A_{64}=73.8s^{-1}, A_{65}=0.8s^{-1}, w_{65}=5.9·10^{4}s^{-1}, $A_{41} = A_{42} = A_{43} = A_{71} = A_{72} = A_{73} = A_{74} = A_{75} = A_{76} = 0$. The initial conditions are N_{Er}=6.7·10^{20} cm^{-3}, N_{Ce}=1.667·10^{19} cm^{-3}, N_4^0=2·10^{18} cm^{-3}. By use of known technique of the root-mean-square best possible fit was achieved when the value of C_2^{cr} cross-relaxation transition rate was equal to 4·10^{-17}s^{-1} cm^3.

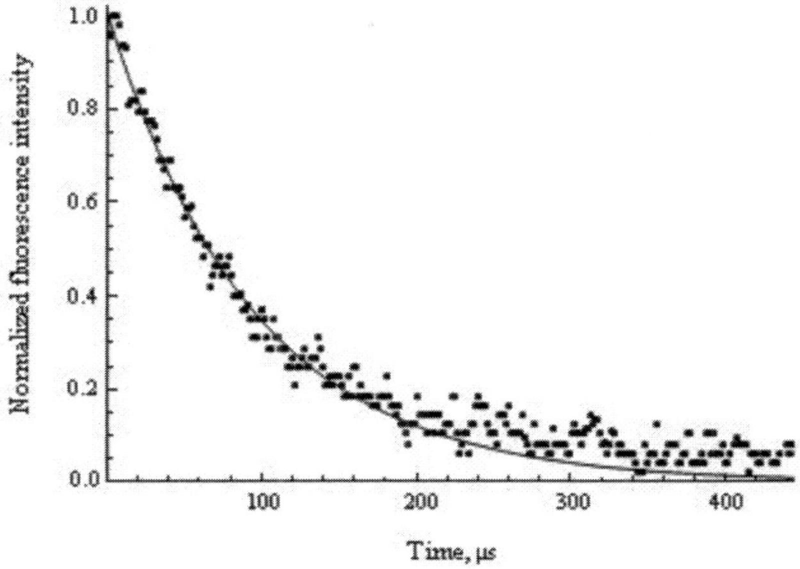

Figure 2. Comparison between measured (points) and calculated (solid curve) decay of fluorescence at 980 nm for the YAG:Er-Ce crystal under excitation at 808 nm.

4. Discussion

Analysis of measured kinetic characteristics gathered in Table 1 shows that the decay time of the fluorescence from $^4I_{11/2}$ energy level of Er^{3+} ions in the YAG crystal is reduced in the presence of Ce^{3+} ions under optical excitation at both 405 and 808 nm. Such behaviour can be explained by taking into consideration processes resulted in depletion of this level. It seems to us, that such de-excitation can be induced by the back-transfer cross-relaxation process taking place according to the following mechanism: $^4I_{11/2}\rightarrow{}^4I_{13/2}$ [Er^{3+}]:$^2F_{5/2}\rightarrow{}^2F_{7/2}$ [Ce^{3+}]. This effect is somewhat interesting and unexpected since concentrations of the impurity ions in our crystals are rather small and it was observed in other materials at considerably larger content of impurity ions [6, 11, 14]. Such reduction is probably due to better spectral overlap between the luminescence band of Er^{3+} donor ions at the $^4I_{11/2}\rightarrow{}^4I_{13/2}$ transition with the absorption band of Ce^{3+} acceptor ions at the $^2F_{5/2}\rightarrow{}^2F_{7/2}$ transition in the YAG host than in other matrixes. It will be noted that this process should have a positive influence to achieve lasing near 1.5 µs, since it suppresses the radiative transition $^4I_{11/2}\rightarrow{}^4I_{15/2}$ and enhances nonradiative transition of $^4I_{11/2}\rightarrow{}^4I_{13/2}$ and hence promotes more rapid establishment of population inversion between initial and final laser levels with smaller losses of excitation energy.

From the same Table 1 one can see that delay times of the peaks of fluorescence responses relative to the end of excitation pulses have strong dependence on the excitation wavelength. This behavior is apparently due to the involvement of different mechanisms of optical excitation energy redistribution in the impurity subsystem of the crystal under study and thus can serve as a sensitive indicator of presence of various cooperative processes. Unfortunately, we were not able to deduce an analytical expression describing the dependence of this parameter on probabilities of various cooperative processes.

5. Conclusion

Kinetic characteristics of the fluorescent responses from the $^4S_{3/2}$, $^4I_{11/2}$ and $^4I_{13/2}$ energy levels of Er^{3+} ions in YAG:Er^{3+} and YAG:Er^{3+}-Ce^{3+} crystals under optical excitation at wavelengths of 405, 457 and 808 nm are studied. For the first time a marked shortening of the $^4I_{11/2}$ level decay time due to the presence of cerium ions in the YAG matrix is revealed. The possible mechanism of back-transfer cross-relaxation process ($^4I_{11/2}\rightarrow^4I_{13/2}$ [Er^{3+}]:$^2F_{5/2}\rightarrow^2F_{7/2}$ [Ce^{3+}]) is suggested, which satisfactorily explains the observed reduction of the decay time of the fluorescence signal at 980 nm. Comparison of experimental data with theoretical model numerically calculated by use of rate equation formalism allows to estimate the probability of such process.

The high sensitivity of the delay times to the involvement of various cooperative mechanisms in impurity subsystem of doped crystals is pointed out.

Acknowledgments

This work was made possible in part by a research grant from the Armenian National Science and Education Fund (ANSEF) based in New York, USA.

References

[1] Kaminskii A A and Antipenko B M 1989 Multilevel functional schemes for crystal lasers (Moscow: Nauka)
[2] Di Bartolo B and Forte O 2006 Advances in spectroscopy for lasers and sensing (The Netherlands: Springer)
[3] Liu G and Jacquier B 2005 Spectroscopic properties of rare-earths in optical materials. (Berlin: Springer)
[4] Murina T M 1989 Laser on crystals of yttrium-erbium aluminium garnets. *Proc. of the IOFAN* **19** (Moscow: Nauka)
[5] Denisov A, Umyskov A and Sherbakov I 1992 *Sov.J. Quantum Electron.* **19** 167
[6] Meng Zh, Yoshimura T, Fukue K, Higashihata M, Nakata Y and Okada T 2000 *J. Appl. Phys.* **88** 2187
[7] Meng J X, Cheah K W, Shi Zh P and Li J Q 2007 *Appl. Phys. Lett.* **91** 151107
[8] Meng Z, Kamebayashi J, Higashihata M, Nakata Y, Okada T, Kubota Y, Nishimura N and Teshima T 2002 *IEEE Photon. Techn. Lett.* **14** 609
[9] Sani E, Toncelli A and Tonelli M 2005 *J. of Appl. Phys.* **97** 123531
[10] Tanner P, Fu L, Ning L, Cheng B-M and Brik M 2007 *J. Phys; Condens. Matter* **19** 216213
[11] ZhouY, Wang J, Dai Sh Xu T, Nie Q and Huang Sh 2009 *J. Lumin.* **129** 1
[12] Kuzmicheva G, Ageev A, Rybakov V, Panyutin V, Yue Y and Chizhikov V 2002 *J. Cryst. Growth* **237–239** 637
[13] Zhou Y, Gai N and Wang J 2009 *J. of Phys. Chem. Sol.* **70** 261
[14] Sani E, Toncelli A and Tonelli M 2005 *Optics Express* **13** 8980
[15] Dantelle G, Mortier M, Vivien D and Patriarche G 2006 *Opt. Mater.* **28** 638
[16] Yu Y, Zhang S, Tie S and Song M 1995 *J. Alloys and Comp.* **217** 148
[17] Doroshenko M, Ivanov M, Sigachev V and Timoshechkin M 1992 *Sov. J. Quantum Electron.* **22** 588
[18] Babajanyan V, Kostanyan R, Muzhikyan P and Petrosyan A 2011 *J Contemp. Phys. NAS of RA* **46** 85
[19] Babajanyan V, Kostanyan R, Muzhikyan P, Petrosyan A and Sargsyan R 2011 *Proc. of SPIE* **7998** 799806-1
[20] Ovanesyan K L, Shirinyan G O and Petrosyan A G 2008 *Proc. Conf. on Laser Physics -2007* (Ashtarak) (Yerevan, Gitutiun) p 20
[21] Lupei A, Lupei V, Georgescu S, Ursu I, Zhekov V, Murina T and Prokhorov A 1990 *Phys. Rev. B* **41** 10923
[22] Haigui Y, Zhenwen D and Tao S 2006 *J. Rare Earths* **24** 519
[23] Xu H L and Kroll S 2005 *J. Lumin.* **111** 191
[24] Bogdanov V K, Booth D J and Gibbs W E K 2003 *J. Non-Crystal. Sol.* **321** 20

Journal of Physics: Conference Series **350** (2012) 012025 doi:10.1088/1742-6596/350/1/012025

Peculiarities of photonic lattices recorded by Bessel beam technique in $LiNbO_3$:Fe crystal

A Badalyan, R Hovsepyan, V Mekhitarian, P Mantashyan and R Drampyan
Institute for Physical Research of National Academy of Sciences of Armenia
Ashtarak-2, 0203, Ashtarak, Armenia

E-mail: rdramp@ipr.sci.am

Abstract. The aim of this work is the investigation of the peculiarities of recording of the refractive photonic lattices by Bessel beam technique in photorefractive Fe doped $LiNbO_3$ crystal. Optical C-axis of the crystal was oriented along the crystal surfaces. The Gaussian laser beam at 532nm wavelength with 17.7mW is transformed into the Bessel beam by an axicon. The intensity pattern of non-diffracting Bessel beam is imparted into the photorefractive medium, being irradiated, via electro-optic effect, thus creating radially modulated refractive index photonic lattices. The annular refractive photonic lattice created inside the LN:Fe crystal has 10 μm period in radial direction. The recorded photonic lattice has been tested by probe beam. The direct observation of recorded lattice by phase microscope was also performed. The azimuthal dependence of created lattice was observed. The qualitative explanation is given.

1. Introduction

Materials with spatial periodic structures such as the photonic crystals currently find applications in many fields of physics and optical device engineering, including guiding and trapping systems, optical devices, telecommunications, information storage, etc. Among different methods for the fabrication of the artificial periodic structures in dielectric materials the holographic technique [1] is one of the promising methods for the fabrication of photonic lattices. Holographic technique is based on the creation of spatially periodical structures by intensity modulated light beams in photosensitive materials. There are two main elements which are important for holographic recording: the method of creation of intensity modulated light beams and materials suitable for recording of photonic lattices. Numerous investigations are devoted to the study of dynamic and permanent optical refractive gratings using classic two-beam interference arrangement in atomic vapors [2], crystals [3, 4] and liquid crystals (see, for example, Ref. [5] and references therein).

The doped photorefractive crystals are very convenient materials for holographic recording. The illumination of photorefractive medium by spatially modulated beam leads to the refractive index modulation via electro-optic effect, thus creating refractive lattices. The light excites the electrons from impurity ion state to conduction band. Electrons migrate in the conduction band and finally are trapped by ions. The redistribution of the charges builds up an internal electric field E and so changes the refractive index $\Delta n_i = r_{ij}E_j$, where r_{ij} is the component of electro-optic tensor. The charge transport is mainly due to photovoltaic effect and diffusion mechanism [6-9].

Bessel beams [10] are very convenient for the creation of artificial periodical structures in photorefractive materials. Recently we developed travelling and counter propagating Bessel beam technique [11] for the formation of 1D and 2D micrometric scale photonic lattices, respectively. The aim of this work is the investigations of the peculiarities of photonic lattices recording by Bessel beam technique in photorefractive Fe doped $LiNbO_3$ (LN:Fe) crystal. Taking into account the annular symmetry of the recording beam the formation of annular photonic lattice will depend on the orientation of the C-axis of the crystal, i.e. whether C-axis is oriented along (Y-orientation) or perpendicular (Z-orientation) to the surfaces of the crystal. For this purpose, annular photonic lattice recorded in Y-cut LN:Fe crystal was tested both by probe beam technique and by phase microscope.

2. Experiment

2.1. Formation of Bessel beam by axcion

Bessel beams or diffraction-free beams are new type of coherent beams [10]. Bessel beams have a feature of conserving their transverse intensity distribution, expressed by the zeroth-order Bessel function, while they propagate in free space. The simplest diffraction-free beams can be formed by superposition of plane waves whose wave vectors lie on the cone. One of the ways for the creation of Bessel beams is the use of an optical element- axicon [10].

The scheme of the Bessel beam formation from Gaussian beam at 633 nm wavelength by an axicon with aperture cone angle 175° is shown in figure 1. The convergence angle of the beams behind the axicon was adjusted by moving the output lens of the beam expander back or forth, thus varying the convergence angle within ~ 3-4°, which, in turn, changes the spacing between the concentric rings in the range of 10 - 25 μm. An ideal Bessel beam has no intensity gradient along the propagation axis and can be schematically represented as a set of co-axial hollow light cylinders surrounding the central light rod. The profile of Bessel beam is a set of concentric rings (figure 2). Bessel beam becomes divergent behind the overlapping zone and forms a ring pattern in the far field (figure 1).

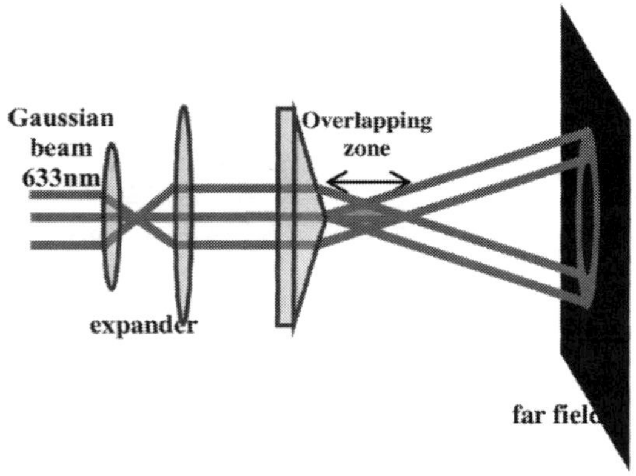

Figure 1. Experimental scheme illustrating the formation of Bessel beam by axicon.

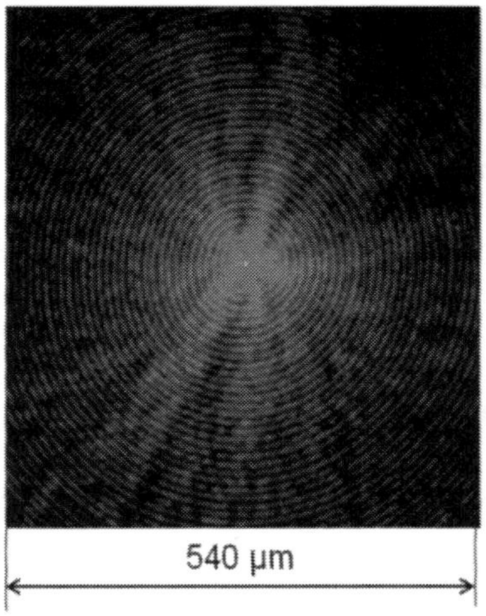

Figure 2. Fragment of radial intensity distribution of Bessel beam formed by axicon with aperture cone angle 175° in the overlapping zone of the beams.

The spacing between the concentric rings, measured by beam profiler, shows their equidistant disposition, except for few central rings. The period of annular structure shown in the figure equals

~10 μm for certain position of output lens of the beam expander. The number of rings reaches up to 1000. The annular ring pattern is two-dimensional, however, as an annular grating this structure is one-dimensional with the period determined by the spacing between rings.

2.2. Recording of annular photonic lattice by Bessel beam technique in LN:Fe crystal
The experimental scheme of recording of annular photonic lattice is shown in figure 3. The laser source was single-mode second harmonic of cw YAG:Nd laser at 532 nm wavelength with linear polarization and 100 mW power. Bessel beam obtained by scheme illustrated in figure 1 illuminated LN:Fe crystal which was placed in the overlapping zone of the axicon. Optical C-axis of the crystal was oriented along the crystal surfaces (Y-orientation). The laser beam polarization was directed along the C-axis. LN crystal doped with 0,05wt% Fe had 15mmx10mmx2mm dimensions.

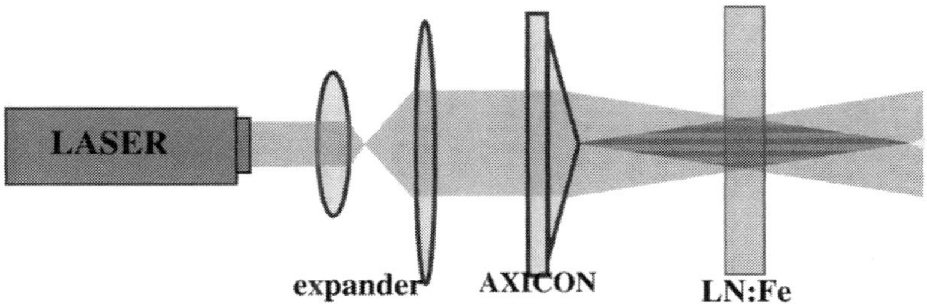

Figure 3. Schematic for the creation of annular photonic lattice by single axicon

2.3. Testing of recorded refractive photonic lattice
The recorded annular photonic lattice was tested using red laser beam by observing the diffraction pattern from the photonic lattices in the far field. Figure 4 shows the readout scheme for testing of lattice recorded inside the crystal. The testing was performed by red beam to avoid the erasure of the grating during readout [4, 6].

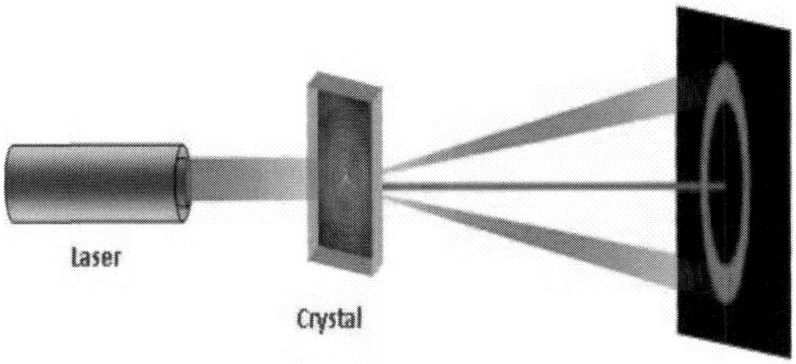

Figure 4. Readout scheme for testing by red laser of photonic lattices recorded by green laser. The screen situated in the far field shows the hypothetic diffraction pattern.

Figure 5 shows the result of testing by Gaussian beam of photonic lattice recorded in Y-cut LN:Fe. The far field transmitted diffraction pattern consists of two opposite disposed segments of a ring. Thus the diffraction pattern from photonic lattice has pronounced azimuthal dependence of intensity distribution with higher diffracted intensity along the C- axis of the crystal. To investigate the azimuthal dependence of intensity distribution of diffraction pattern the direct observation of photonic lattices by phase microscope was also performed.

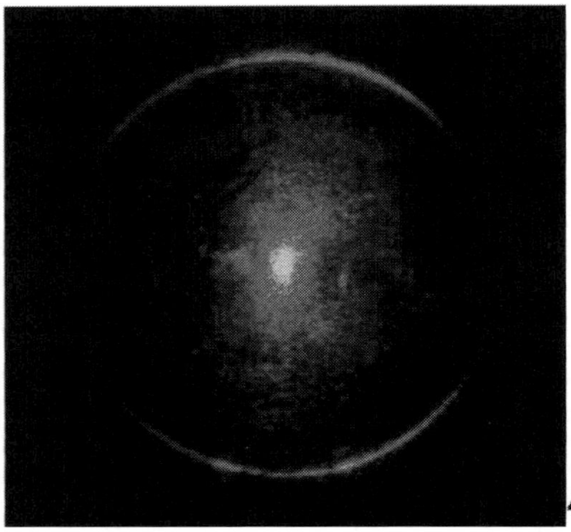

Figure 5. Far field transmitted diffraction pattern from photonic lattice recorded in Y-cut LN:Fe during 60 min, for nearly orthogonal incidence of the probe Gaussian beam at 633 nm to the crystal surface.

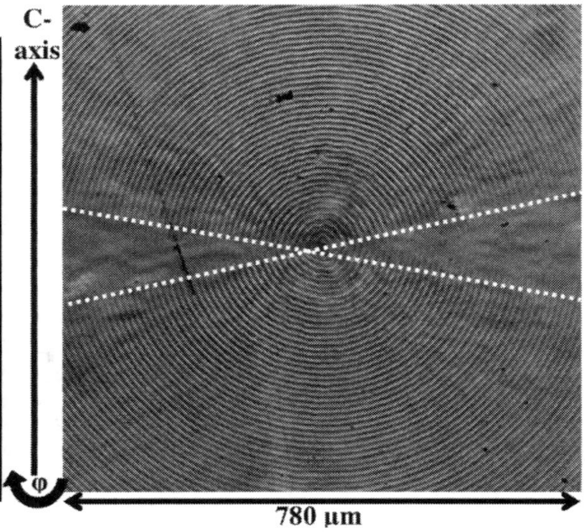

Figure 6. Phase microscope image of the annular photonic lattice inside the LN:Fe crystal. Vertical arrow shows the direction of optical C-axis of the crystal. Circular arrow shows the direction of azimuthal angle φ

Figure 6 shows phase microscope image of the photonic lattice inside the LN:Fe crystal. The white dashed lines mark the areas where the grating is recorded with high contrast (upper and lower sectors) and is not recorded at all (right and left sectors).

3. Discussion

The physical mechanism of the formation of holographic lattices in photorefractive materials is based on the electro-optic effect [6-9]. Fe ions occur in LN crystal in different valence states: Fe^{2+} and Fe^{3+}. The corresponding band diagram is shown in figure 7. The green light excites the electrons from Fe^{2+} to conduction band. Electrons migrate in the conduction band and finally are trapped by Fe^{3+}.

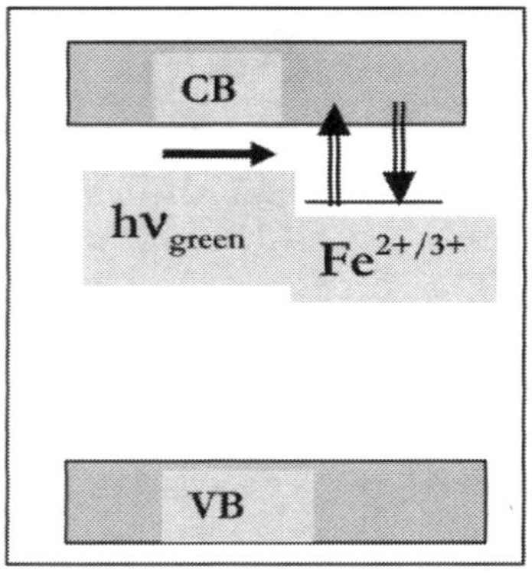

Figure 7. Band diagram of lithium niobate doped with iron. CB is the conduction band, VB is the valence band.

The redistribution of the charges builds up an internal electric field *E* and so changes the refractive index. Thus, the inhomogeneous illumination of photorefractive materials leads to the modulation of refractive index. Two main mechanisms – photovoltaic effect and diffusion of photo-induced carriers are responsible for formation of refractive lattices in photorefractive crystal [6-9.] The diffusion effect can be neglected for lattice spatial frequencies less than 10^5 lines/cm [8] which is in case of the present experiment. The electric field induced by photovoltaic effect is due to the charge separation taking place along the C-axis of the crystal [8].
In LN crystal the change of extraordinary index is larger than the change of ordinary index by a factor of four [7] and the induced refractive index change Δn is mainly due to the distortion of the extraordinary index of refraction.

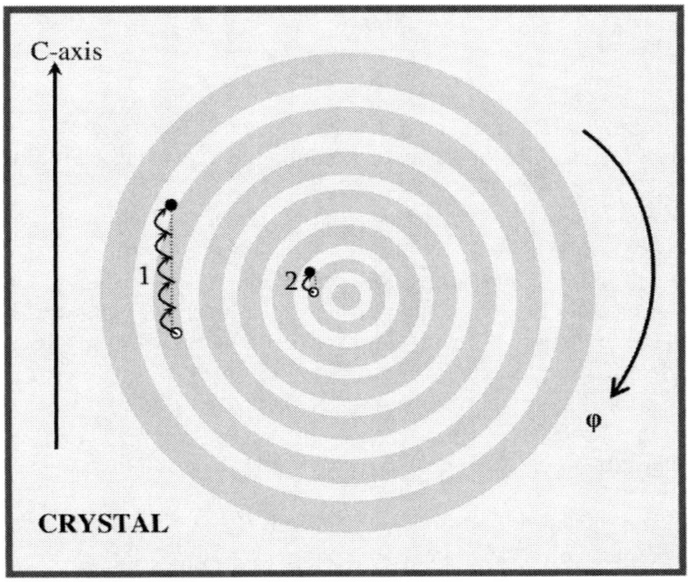

Figure 8. Schematic of space-charge field formation in photorefractive crystal during the illumination by Bessel beam. White and green circles are non-illuminated and illuminated regions of Bessel beam, respectively. Bent arrows schematically illustrate migration of electrons on periphery (1) and central part (2). Open and filled circles show schematically an electron and trap, respectively

Figure 8 shows schematically the migration and space charge formation in different regions of annular intensity distribution of Bessel beam inside the photorefractive Y-cut crystal. The appearance of azimuthal dependence of recorded lattice is due to the predominant migration of the electrons along the C-axis of the crystal. In $LiNbO_3$:Fe crystals the displacement of the electron is 0.5 Å per one absorbed photon for λ=0.53 μm wavelength [8]. The distance between bright and dark zones along C-axis is approximately 100 times larger on periphery compared with the central part of illuminated region. The diameter of Bessel beam on the crystal surface was measured 5mm. The period of lattice is 10 μm. Thus probability of the migration and final trapping of the electrons in the dark zone is higher in the central region compared with the periphery. As a consequence, left and right sectors of the lattice will have more contrast than upper and lower sectors which lead to the azimuthal dependence of the recorded lattice (see Fig.6). The suggested model requires detailed quantitative study taking into account the recording beam power, period of grating, geometrical size of recording lattice etc. Azimuthal dependence of recorded photonic lattices can be avoided by use of Z-cut photorefractive crystal as recording medium. The recording of the 2D lattices in Y-cut crystal by

Bessel standing wave [11] with half wavelength period in axial direction will also reduce the azimuthal dependence of recorded circular structure due to switching on the diffusion mechanism of recording which takes place in all directions but with less efficiency compared with photovoltaic effect [8]. These experiments are in progress and the results will be published elsewhere.

4. Conclusion

Annular refractive photonic lattice was created by Bessel beams technique in photorefractive Y-cut LN:Fe crystal. The testing of the lattice by probe beam showed azimuthal dependence of intensity distribution of far field diffraction pattern. Further observation by phase microscope showed that recorded lattice has pronounced azimuthal dependence. The appearance of azimuthal dependence of recorded lattice is the result of the predominant migration of the electrons due to the photovoltaic effect along the C-axis of the crystal. The qualitative explanation is given.

5. References

[1] Collier R J, Buckhard Ch B, Lin L H, *Optical holography* 1971 Academic press New York

[2] Korneev N, Benavides O, Mechanisms of holographic recording in rubidium vapour close to resonance, 2008 *JOSA B* **25** 1899 – 1906

[3] *Photorefractive materials, effects and devices. Control of light and matter*, 2009, *Applied Physics B*, special issue, **95** N.3

[4] Adibi A, Buse K, Psaltis D, Two-center holographic recording, *JOSA B*, 2001 **18**, 584-601

[5] Pagliusi P, Macdonald R, Bush S, Chipparrone G, Kreuzer M, Nonlocal dynamic gratings and energy transfer by optical two-beam coupling in a nematic liquid crystal owing to highly sensitive photoelectric reorientation, 2001 *JOSA B* **18** 1632-38

[6] Adibi A, Buse K, Psaltis D, The role of carrier mobility in holographic recording in $LiNbO_3$, 2001 *Appl. Phys. B* **72** 653–9

[7] Chen F S, Optically induced change of refractive indices in $LiNbO_3$ and $LiTaO_3$, 1969 *J Appl. Phys* **40** 3389-96

[8] Glass A M, D von der Linde, Negran T J, High-voltage bulk photovoltaic effect and the photorefractive process in $LiNbO_3$, 1974 *Appl Phys Lett* **25** 233-235

[9] Avanesyan G T, Vartanyan E S, Mikaelyan R S, Hovsepyan R K, Pogosyan A R, Mechanisms of photochromic and photorefractive effects in doubly doped lithium niobate crystal, 1991 *Phys Stat Sol (a)* **126** 245 – 252

[10] Durnin J, Mikely J J, Jr Eberly J H, Diffraction-free beams, 1987 *Phys Rev Lett* **58**, 1499-1501

[11] Badalyan A, Hovsepyan R, Mekhitaryan V, Mantashyan P, Drampyan R, New holographic method for formation of 2D gratings in photorefractive materials by Bessel standing wave, in "Fundamentals of Laser Assisted Micro- and Nanotechnologies 2010, edited by Vadim P.Veiko, Tigran A. Vartanyan, *Proceedings of SPIE*, **7996** (SPIE Bellingham, WA, 2011) 799611-1-9

[12] Mantashyan P, "Photochromic effect and holographic recording in doubly doped Li NbO_3 crystals", International Conference on laser Physics 2010, Edited by Aram V.papoyan, *Proceedings of SPIE*, 7998, (SPIE, Bellingham, 2011) OJ-1-9

Acknowledgments

The authors are grateful to Dr. E Kokanyan for providing the LN:Fe crystal. The work was supported by International Science and Technology Center Grant, Project A - 1517.

International Symposium on Optics and its Applications (OPTICS2011) IOP Publishing
Journal of Physics: Conference Series **350** (2012) 012026 doi:10.1088/1742-6596/350/1/012026

On the possible key experiments to show the *singlet*-to-*triplet* conversion of the Cooper pairs by the use of a SFCO-method

S G Gevorgyan[1-2], H G Shirinyan[2], V S Gevorgyan[1], S T Muradyan[1-2], G S Gevorgyan[1], B K Kurghinyan[1], A A Polyanskii[3]
[1] Solid State physics Chair, Yerevan State University, Yerevan, Armenia,
[2] Institute for Physical Research, National Academy of Sciences, Ashtarak-2, Armenia,
[3] Applied Supercond. Center, NHMFL, Florida State University, Tallahassee, USA

E-mail: gevs_sam@ysu.am

Abstract. The problem of electron pairing above Meissner expel arose when high-T_c superconductive (*HTS*) materials were discovered. And scientists are admitting know that there is need to consider 2 processes for *HTS*: electron pairing and long-range phase coherence among Cooper pairs (*pair condensation*) – separately and independently each other. Otherwise, it is admitted, that in *HTS* electrons become paired above Meissner expel (above T_0, starting from T_c) & start forming *SC*-condensate only at T_0, while in low-T_c *SC* materials (*LTS*) the pairing and pair condensation start simultaneously (T_0=T_c). But, since discovery of *HTS* new data appeared and importance of some earlier data was recognized, indicating analogy among the *HTS* and *LTS* in connection with these 2 processes. Here we show that there are no differences among these materials regarding the said processes: electron pairing and pair condensation are separate & independent even in *LTS*. Difference is in a temperature scale. For *LTS* these processes run in a narrow range (*~30mK*), while in *HTS* the scale is longer (for *YBaCuO*, more than *3K*). This is the reason why separation of T_c from T_0 in *LTS* is so hard so far. The problem is still open also due to lack of methods for "nonperturbative" study of the beginnings of *SC* transition in clean (*tiny*) objects with small signal. Analysis of data permits to conclude that becomes urgent sensitive study of the *Fulde-Ferrell-Larkin-Ovchinnikov* superconductivity in heavy-fermion materials that have layered electronic structure. And, the *SFCO* method may stand key tool for sensitive study of the pair formation in *FFLO* state. Such a study may show *singlet*-to-*triplet* conversion of Cooper pairs. It may also allow separating the *ideal conducting* ($R=0$) & *ideal diamagnetic* ($B=0$) states, which are important for true interpretation of a real nature of superconductivity.

1. Introduction

Majority of scientists is admitted know that in high-T_c superconductive (HTS) material electrons become paired above Meissner expel (above T_0, starting from T_c) and start forming the *SC* condensate only at the T_0, while in low- T_c *SC* material (LTS) it is assumed that the pairing process and the Cooper-pair condensation (*long-range phase coherence among pairs*) take place simultaneously (at a same temperature – T_0=T_c): due to much larger pair size it is assumed that their wavefunctions are overlapping in a *LTS* material [1]. Shortly after discovery of *HTS* materials [2], however, a "*paramagnetic*" (PM) precursor to superconductivity was detected in *LTS* tin (Sn) [3]. Weakly expressed this key

[1] Samvel Gevorgyan; gevs_sam@ysu.am

effect, and related to it "*caloric*" precursor to *SC* transition [4] (seen on *Corak*'s heat capacity vs. temperature curves, before the specific heat's "*jump*" – first, also detected in *LTS* tin) provide weighty arguments to have the reverse opinion regarding the *LTS* materials, discussed below. There are the reasons why these two fine effects have the same physical origin. To validate such a global correlation between the electromagnetic and thermal properties of the *SC* materials (regardless it is *HTS* or *LTS*) an advanced idea was putted forward by us in [5], admitting existence of two types of Cooper pairs, both in *HTS* and *LTS* materials – *singlet* and *triplet*. According to main conclusions we came to in [5], they show absolutely different non-traditional temperature behaviour upon cooling of the *SC* material. For the *YBaCuO* composition *HTS* material, for example, respective curves are shown in a figure 1.

In this connection, it was urgent searches and detection of the said "*paramagnetic*" and "*caloric*" precursors of superconducting state in *HTS* material. We tried to do that, and much more precision recent experiments by the use of a single-layer flat-coil-oscillator (SFCO) based highly sensitive our unique technique [6-7] enabled to detect in a *YBaCuO* composition *HTS* material the said "*paramagnetic*" effect [3] (see figure 2), at the beginnings of the *SC* phase transition, which precedes Meissner expel. Next, it enabled to detect in this material the said "*caloric*" effect, discussed below. But, before turn to it let's list *Meissner*-state precursor other phenomena, detected by various groups, which all are important for the true understanding of the real nature of the superconductive phenomenon.

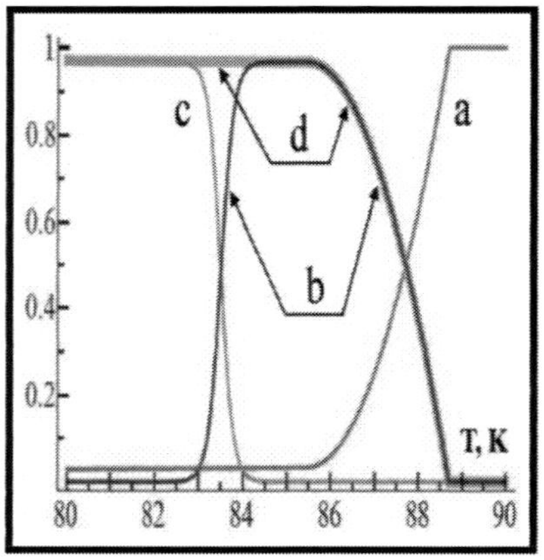

Figure 1. Temperature dependances of normal electrons (a), *triplet* (b) & *singlet* (c) Cooper pairs & all pairs (d) [5]

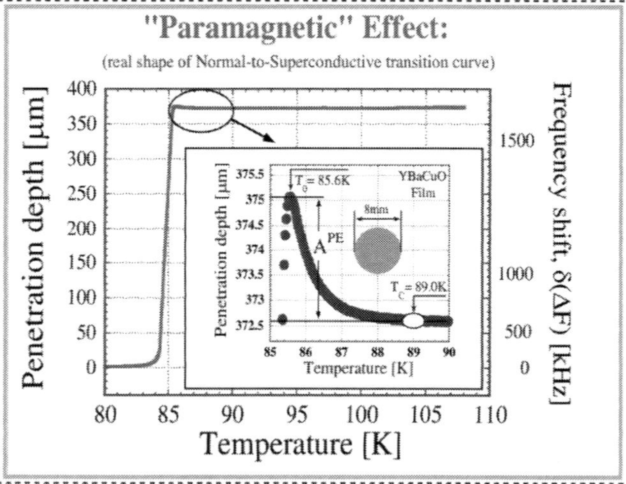

Figure 2. NS transition of the *YBaCuO* film [3]. Inset: enlarged view of the "*ParaMagnetic*" effect (PE) detected at the beginnings of *SC* phase transition, which precedes Meissner expel. The A^{PE} is the *PM* effect's height.

Superconductor is a double ideal material, since it becomes *ideal conductor* and gets properties of the *ideal diamagnetic* below some temperature. The latter behaves also as an *ideal conductor* – reverse is not true. But, must *SC* material obtain such properties at the same temperature? And why transitions of different nature (*1-st is assumed to be connected with electron-pairing, followed by zeroing of the pair momentum, the 2-nd, with pair condensation – due to collection of enough singlet pairs – arguments see in* [8]) should occur at the same temperature? And also, are there other *different*-nature effects happening at the same moment in a Nature? Such questions acquired meaning since the "*paramagnetic*" effect was detected in micron-size tin grains [3] (T_c ~*3.72K*), pointing out the real physical onset of the Meissner expel. It precedes the diamagnetic ejection and corrects the shape of the normal-to-superconductive (N/S) phase transition curve. The origin of the above questions relates also with a "*preceding*" effect, opened in a percolating *YBaCuO* (in ceramics [2] and in films [9] with a granular structure of the material). According to it, *resistive* transition ends before the start of Meissner expel.

The effect was seen also in a *BiCaSrCuO* crystal [10], but doesn't attract a proper attention of the author – perhaps due to lack of assurance in accordance among *temperature*-scales of conducted tests, performed in different set-ups. Questions were deepened when a "*diamagnetic activity*" was revealed in *LaSrCuO* film [11] – at temperatures much higher transition temperature of material established by the onset point of Meissner expel. A *super*-sensitive *scanning*-SQUID microscope was used for those tests. Such a flux activity is interpreted by the author as the effect *precursor* to the Meissner state.

Among other *Meissner*-precursor effects let's stop on the study conducted by *Tonica Valla* [12]. It shows that a "*pseudo-gap*" in the energy level of *HTS* materials' electronic spectrum is the result of electrons being bound into Cooper pairs above the transition temperature to *SC*-state, but unable to *superconduct*, because pairs move incoherently. As to *LTS* (*which act closer to Absolute Zero*), "superconductivity in these materials occur as soon as pairs are formed", *Valla* says. "In the case of *HTS* materials, however, electrons, though paired, "*do not 'see' each other* above some temperature," *Valla* thinks [12], "so, they can't establish the phase coherence, with all the pairs behaving as a '*collective*'".

Listed here data indicate, however, that such is the case for all types of *SC* materials. And also, NO indication on any microscopic physical mechanism for establishment of the phase coherence among *SC* pairs in *Valla*'s works, and in papers of other scientists. While, as a hint to possible mechanism for absence (*at higher temperatures*) and presence of the phase coherence among the pairs (*at further cooling*) might be discussed presence of 2 types of pairs in *SC* materials (regardless it is high- or low-T_c) – *singlet* and *triplet*, with different angular momentum [8]. They have different *temperature*-behavior upon cooling of a material (qualitatively illustrated by figure 1 [5]). Such an approach may result in the *ideal conductivity* (it starts with a pair formation from the T_c, and acts for both types of pairs [8]). That is because pairs are quasiparticles with a Zero momentum, so, with an infinitely large *de Broglie* wavelength, due to which they can move in a material without scattering, ignoring defects of a crystalline structure. At a later cooling of a material, starting from temperature T_0, such an advanced approach may also result in the *superconductivity* (*ideal diamagnetism*) – but, for only *singlet* pairs.

And finally, it is very important to note that recently existence of the superconductive Cooper pairs with parallel spin direction was detected experimentally for the first time, in works [13-14].

2. Results and discussion

So, even based on the listed above data (*among which, many ours*) one may believe, that there are no visible differences between the *HTS* and *LTS* materials regarding the said 2 processes. Apparently, the electron pairing and the Cooper-pair condensation (*phase coherence*) are separate and independent even in a *LTS* material. Difference is in a temperature scale. In a *LTS* material these processes are running in a very narrow temperature range (*~30mK* [3-4]), while in *HTS* the scale is much longer. For the *YBaCuO*, for example, the scale of the "*paramagnetic*" effect estimated by *Gantmakher* (*that indicates the scale of the event selection for the above 2 physical processes*) is about *1K* [15], while much more precise experiments by our group evidence that it is even broader (more than *3K* [3] – figure 2). Possibly, this is the reason why separation of the T_c from the T_0 in a *LTS* material was so much hard so far – that is done only qualitatively up to now [3-4]. The problem is still open due to lack of methods for "*non-perturbing*", sensitive study of the *SC* transition in thin *LTS* objects with small signals – especially at very beginnings of the phase transition, where even a sensitive *SQUID* technique can not "*notice*" such small changes in a normal-state "*skin*"-depth (i.e., start of the "*PM*" effect – figure 2).

And so, in connection with the problem of electron pairing above Meissner expel, detection of the said "*caloric*" precursor to superconductivity in *HTS* material was so urgent. That is why we now turn to present our data related with the searches and detection of the said fine thermal peculiarity in high-T_c superconductive film. It was not so easy, however, to measure the heat capacity (*moreover, its too much little changes at the beginnings of the SC phase transition*) in so small volume objects, such as *film*-structures. However, we tried to do that and could detect so much small effect in *YBaCuO* composition *HTS* film (figure 3). For such fine measurements we have further upgraded the cryogenic laser scanning microscope suggested by our joint group earlier [16-17]. Schematics of an improved imaging technique are shown in figure 4. Generally, that is combination of the well-focused laser scanning microscopy [18] and the *SFCO* method-based sensitive measurement technique [6-7]. One of advantages of the *SFCO*-method is its ability to "*notice*" details of the phase transition between the normal and *SC* phases in thin, plate-like objects – with an *angstrom*-scale absolute resolution at measurement of the penetration depth (*by frequency shift of the SFCO-based testing tunnel diode* (TD) *oscillator* [6-7]), and with about a *nano*-watt resolution at absorption measurements (*by amplitude changes of the same oscillator* [19]). Besides, it can serve as a sensitive temperature probe with better than $0.1mK$ resolution [20]. The imaging technique uses a focused laser beam, as a probing signal. It is capable of imaging properties of *HTS film*-structures with $\sim 1\mu m$ spatial resolution. It operates as follows: the *HTS film*-bridge is illuminated point by point with a focused laser beam ($\Phi_{beam} \sim 1\mu m$), which results in a slight local heating. In our case the substrate of the film is transparent, and so, the amount of laser power passed to the flat-coil based thermal sensor depends on peculiarities of the heat capacity (*or heat conductivity*) of the material under test in a lighted point of the film, positioned on the flat face of the detecting coil – leading to some changes in the amplitude (*and frequency*) of the testing *TD*-oscillator. Moving position of the "*X-Y*" stage along both coordinates by stepper motors (with $\sim 0.5\mu m$ step) and controlling these movements by the PC (in LabVIEW environment), we could make micrometer precision positioning of the probing laser beam and scan it over the surface of the film (figure 4). This enabled to get *2D*-images of the *N/S* phase transition of thin, plate-like *HTS* materials (e.g., *film*-structures) with $1\text{-}2\mu m$ spatial resolution. The curve in figure 3 is the *2D*-raster image of the laser beam power transmitted through the substrate, obtained by summarizing of the *amplitude change* signals all over the film. In other words, in figure 3 the *2D*-raster image of the heat distribution in the *HTS* film under investigation is shown. Since the created laser scanning microscope is an instrument of the bolometric nature (*the flat pick-up coil detects the residual power of the laser beam that has not been dissipated in the test film*), the signals detected by the flat pick-up coil may be indirectly related to anomalies in the heat capacity (*as in the first approximation, the heat conductivity may not have any visible anomalies in such a narrow range of the temperature in the vicinity of the SC phase transition*).

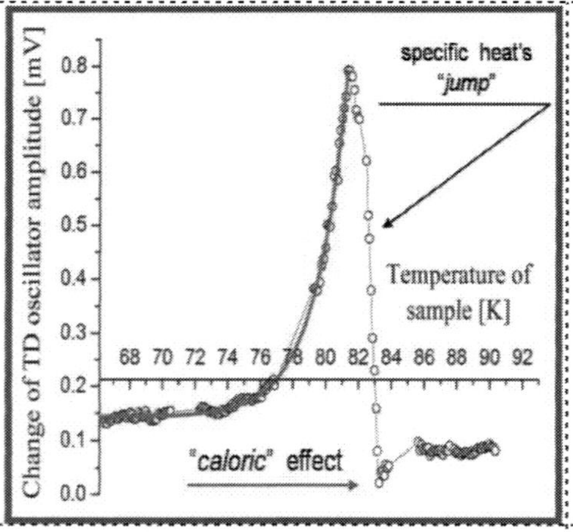

Figure 3. Detection of a "*caloric*" effect in *YBaCuO* film by an imaging technique created in this work based on the *SFCO* method [6-7]. Circles are measured data, while solid line – exponential fit of data.

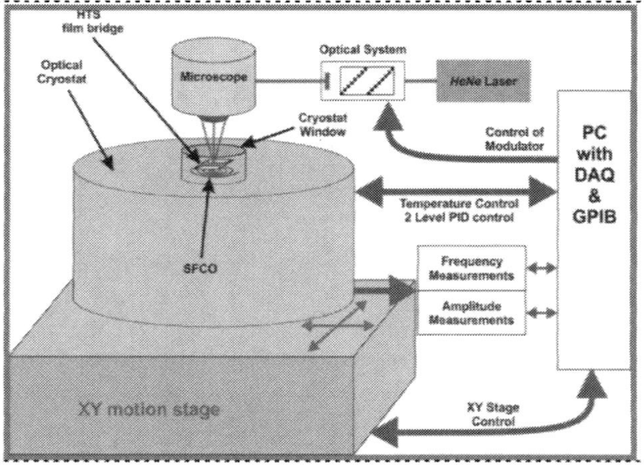

Figure 4. Schematics of the created and used cryogenic laser scanning microscope, based on the focused ($\Phi_{beam} \sim 1\mu m$) laser scanning microscopy [18] and the single-layer flat geometry pick-up coil-based sensitive tunnel diode oscillator (SFCO-technique) [6-7].

Superconductivity is a macroscopic quantum effect where electrons of opposite spin & momentum condense into pairs – according to original explanation by *Bardeen, Cooper & Schrieffer* (BCS) [21]. Large fields destroy superconductivity by coupling to the orbital motion of electrons. This critical field (*orbital limit*) separates the uniform superconducting state from the normal metallic state. Magnetic field can, however, couple predominantly to the spins of electrons. Superconductor is then in a *paramagnetic limit*. It was shown in 1964 by *Fulde & Ferrell* [22] & independently, *Larkin & Ovchinnikov* [23], that this superconductive state would be basically different from the *conventional BCS* case. In this new state, magnetic field tries to polarize opposite spins of *SC*-pairs. In response, the *SC* order parameter develops nodes in a real space, leading to the alternating regions of *SC*-layers and spin-polarized magnetic walls. This *FFLO* state manifests itself as a "*wedge*" in the field-temperature (*B-T*) phase diagram at very low temperatures, below the critical field. In figure 5 the constructed in 2003 *B-T* phase diagram is shown, obtained from the heat capacity measurements for the plane parallel orientation [24-25]. Blue squares denote the critical field, separating the normal-metallic state from the superconductive one. The "*wedge*" at low temperatures & high magnetic fields is the *FFLO* phase, with the red squares separating uniform from non-uniform superconductivity. The exact shape of *B-T* diagram and the *FFLO* "*wedge*" area as well depend on many microscopic parameters, such as the Fermi surface geometry, dimensionality of the host crystal, impurities, and ratio of orbital-to-paramagnetic effects [22-24]. The material in which first was observed the *FFLO* state in 2003 is the $CeCoIn_5$.

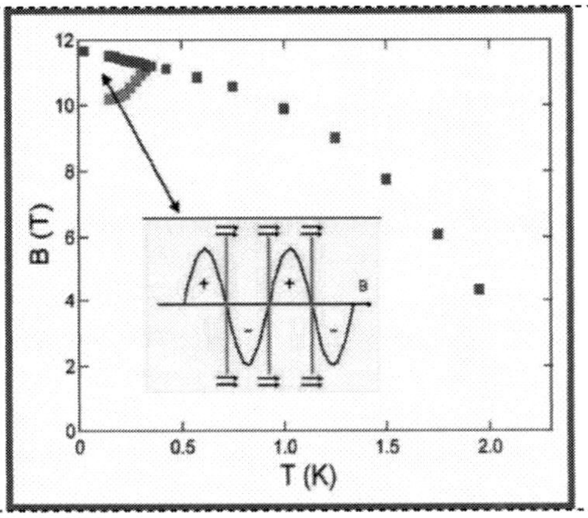

Figure 5. Critical field (blue) & *FFLO* transition line (red). Inset: schematics of *FFLO* order parameter with magnetic walls [24-25].

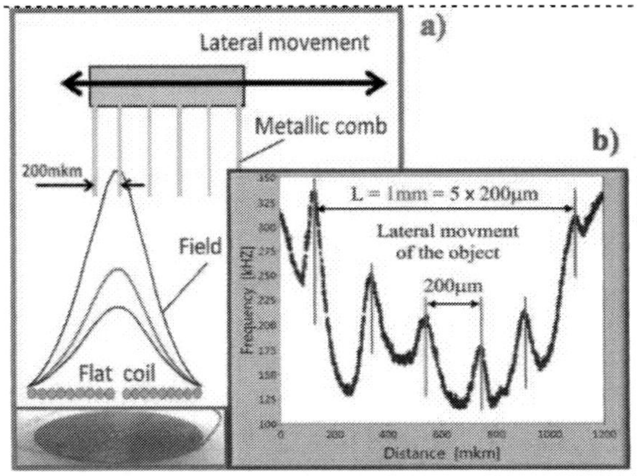

Figure 6. *TD*-oscillator frequency shift $\delta(\Delta F)$ (b) vs. the lateral position of a 1D grid-shaped metallic object (a) relative to flat coil face.

Taking into account above data, one may come to the conclusion that sensitive study of the *FFLO* state in heavy-fermion materials becomes urgent. That is important for understanding of microscopic mechanisms of electron pairing. In this regard, we suggest start to newly study it in a $CeCoIn_5$ material by involving methods enabling much higher resolution compared to earlier tests discussed in [24-25]. This material has a critical temperature of *2.3K*, and the layered electronic structure. Such anisotropy resists orbital motion of electrons for fields applied parallel to conducting *ab*-planes of this material. Besides, the large spin susceptibility favors paramagnetic limitation. So, this material fulfills delicate balance of properties needed to detect and study the *FFLO* superconductivity. The *SFCO*-method [6-7] (with its high capabilities enabling to reveal many fine effects in superconductivity during the last decade [3, 16, 19, 26]), could be serve as a unique instrument for precision study of the Cooper pair formation in *FFLO* state. That may permit demonstration of the *singlet*-to-*triplet* conversion of Cooper pairs, and also, separation of the *ideal conducting* ($R = 0$) and *superconducting* (*ideal diamagnetic* – $B = 0$) states. These are among matter of principle questions for true interpretation of the real nature of superconductivity. Such a study may also clarify microscopic mechanisms of the electron pairing (*physical mechanisms for the establishment of the long-range phase coherence among the SC pairs*).

Alongside with the heat capacity measurements, *TD*-oscillator tests were done in 2003 by *Radovan* et al., with a solenoid pick-up coil [25]. They provide evidence that the *FFLO* transition can be qualitatively seen also in penetration depth studies – even with a less sensitive solenoid pick-up coil. How-

Journal of Physics: Conference Series **350** (2012) 012026 doi:10.1088/1742-6596/350/1/012026

ever, while studying the *SC*-phenomenon in *YBaCuO* composition *HTS* material during the last 10-15 years we could demonstrate in [3, 6-8, 17, 19, 26], that for the flat geometry (plate-like) objects (*films, crystals*) a flat-coil based *TD*-oscillator (our *SFCO* technique) shows higher resolution by at least 3-4 orders of a value, compared to its solenoid-coil based analog used in above pioneering works [24-25]. This permits to believe, that the *SFCO* technique with about *1mm* in diameter pick-up coil may satisfy technical conditions needed for much more sensitive detection and investigation of the *FFLO* state.

In this connection, our research shows also that the flat-coil oscillator can be activated also with its internal capacitance [27] (*without an external capacitance in its resonant circuit*). That is due to relatively high value of the internal capacitance of single-layer flat coils (compared to their parasitic capacitances with respect to the surrounding radio-technical environment). This opens exotic areas for flat-coil oscillator application. Namely, a "*needle-like*" testing magnetic field of such a flat coil (see figure 6a), used as a pick-up in such a stable oscillator, enables a new method (*new approach*) for surface probing, based on replacement of short-range solid-state probes of acting microscopes (*such as needles or cantilevers of the tunneling* [28-29] *& atomic-force* [30] *microscopes*) by the long-range action non-solid-state ones. Such a unique probe shows strong dependence of a detected signal on the size of the spatial-gap between the probe & the surface of the object – crucial for the probe microscopy (PM) [31]. This opens a chance for creation "*magnetic-field*" probes with a *RF* power applied to the sample lying in the range of *1nW* to *5μW*. The gap between such "*probe-formative*" coil and the object can be larger than *100μm*, compared to *1nm* gap of the acting probe microscopes [31]. In our tests we reached a lateral resolution *~1μm* – even for the relatively large diameter ($2R_{coil}$~*14mm*) flat-coil technique.

Such a *SFCO*-probe could also "*notice*" & distinguish details of relief of a normal-metallic object, with *~10μm* spatial-resolution ((see figure 6). To demonstrate this we performed an experiment with one-dimensional (1D) metallic grid made of 6 copper wires (see figure 6a): each wire was *20-30μm* in dia., and was positioned with an average interval of *~200μm* between the wires. Wires distort coil *RF*-field configuration when they move (*or when coil moves relative to the grid*), leading to changes of the oscillator frequency or/and amplitude. The effect is maximal when each wire reaches to the coil center. Figure 6b illustrates detected dependence of the oscillator frequency shift, $\delta(\Delta F)$, vs. the lateral position of the metallic-comb relative to the coil face (*relative to "magnetic-field" probe*). Average distance between detected 6 vertical neighboring peaks on the curve in figure 6b is *~200μm* – just in agreement with the experimental setup in figure 6a. That is why we think, that our *SFCO*-probe may also distinguish and image (*both by amplitude and frequency of the TD-oscillator*) details of the relief of the nodal structure in *FFLO*-state, in a real space (*consisting of alternating regions of the SC-layers and the spin-polarized magnetic walls*). For that aim, we believe the best way is creation and use of a said *SFCO* method-based "*magnetic-field*" probe, with a lithographically made coil of about *1mm* in dia. [32] – as an effective "*needle-type*" probing instrument with better than *100nm* predicted lateral resolution. Such a probe may have much larger work-distances (more than *100μm*) between the probe and the surface of the testing object, which enables a "*visual*" control of the local probing area of the object, and, if needed, application of test perturbations (for example, exposition to laser radiation).

3. Conclusions

Resuming the said above, let's underline importance and feasibility of key experiments for sensitive study of the *Fulde-Ferrell-Larkin-Ovchinnikov* superconductivity (*FFLO*-state) [22-23] by the use of a *SFCO*-technique [6-7] – that stands urgent to demonstrate the *singlet*-to-*triplet* conversion of Cooper pairs. The material in which we suggest to study this phenomenon at temperatures below *350mK* is a heavy-fermion superconductor $CeCoIn_5$. That was investigated earlier [24-25], but with much less resolution compared to what we offer to do. The unique *SFCO*-method (*with its unprecedented capabilities we demonstrated while studying the SC-phenomenon in YBaCuO during the last 10-15 years* [3, 6-8, 17, 19, 26]) we suggest to use as the main research instrument for precision study of the pair formation in *FFLO* state. Such a study (to be carried out approx. by the scenarios implemented in original works [24-25]) may enable demonstration of the *singlet*-to-*triplet* conversion of Cooper pairs, and also separation of the *ideal conducting* ($R = 0$) and *superconducting* (*ideal diamagnetic* – $B = 0$) states, which are important not only for true interpretation of the nature of superconductivity (*in a whole*), but also for correct understanding of the microscopic mechanisms of the electron *pairing* (*physical mechanisms for the establishment of the long-range phase coherence among the superconductive pairs*).

At this, distortion of a testing *RF*-field configuration near the flat coil face (*taken by the frequency increase of the SFCO technique*) may be used for checking whether the *ideal diamagnetic* state (with $B = 0$) is established or not. While, amount of the absorption of a testing *TD*-oscillator power by the sample (*taken by the amplitude increase of the testing SFCO technique*) one may use to check whether the *ideal conducting* state (with $R = 0$) is established or not. Alongside with that, establishment of the *ideal conducting* state one may check also by the traditional *AC 4-probe* measurement technique.

Acknowledgments
The work is supported by the Armenian *National Foundation of Science & Advanced Technologies* & the U.S. *Civilian Research & Development Foundation* under *Grants ## ISIPA 01-04 & UCEP 07/07.* The work was supported also by state sources of Armenia in frames of Projects 301-0046 and 72-103.

References
[1] Mourachkine A 2004 Principles of Superconductivity (*Room Temperature Superconductivity*) (Cambridge: Internat. Science Publishing) chapter 4 – arXiv:***cond-mat/0409240***
[2] Bednorz J G and Muller K A 1986 *Z. Phyzik* B **64** 189–93
[3] Gevorgyan S G, Kiss T, et al. 2001 *Supercond. Sci. & Technol.* **14**(12) 1009-13
[4] Corak W S and Satterthwaite C B 1956 *Phys. Rev.* **102**(3) 662–6
[5] Kurghinyan B K, Gevorgyan S G and Ohanyan V R 2008 *Proc. of the Yerevan State University* (Natural Sciences) Physics **3** 82–90 – in Armenian (arXiv:1109.0462v1 [cond-mat. supr-con])
[6] Gevorgyan S G, Kiss T, Movsisyan A A, et al. 2000 *Rev. Sci. Instrum.* **71**(3) 1488–94
[7] Gevorgyan S, Kiss T, Ohyama T, Movsisyan A, Shirinyan H, et al. 2001 *Physica* C **366**(1) 6–12
[8] Gevorgyan S G 2003 *J. of Contemporary Physics* (National Acad. of Sci. of Armenia), Allerton Press Inc. **38**(1) 41–53 and **38**(2) 45–50 – ***cond-mat/***papers/0501/***0501416***.pdf
[9] Avagyan A A, et al. 1990 *Superconductivity:* Physics, Chemistry, Technique **3**(1) 89–92
[10] Morris P A, Bonner W A, Bagley B G, Hull G W, et al. 1988 *Appl. Phys. Lett.* **53**(3) 249–51
[11] Iguchi I, Yamaguchi T, Sugimoto A 2001*Nature* **412** 420–23
[12] Valla T. 2005 *Proc. of SPIE* **5932** 593203 DOI: 10.1117/12.621626 – cond-mat/0507653v1
[13] Sprungmann D et al 2010 *Science:* Physics (*PHYSorg.com*) Dec.1 – physorg.com/news/2010-12
[14] Sprungmann D, et al. 2010 *Phys. Rev.* B **82** 060505(R) – DOI:10.1103/PhysRevB.82.060505
[15] Gantmakher V F, Golovko N I, Naumenko I G, et al. 1990 *Physica* C **171** 223–30
[16] Gevorgyan S, Shirinyan H, et al. 2004 *Nucl. Instrum. & Meth. in Phys. Res.* **520**(1-3) 314–19
[17] Gevorgyan S G, et al. 2009 *Proc. of Yerevan State University*, Phys. & Math. Sci. **2**(219) 50-4
[18] Zhuravel A P, Sivakov A G, Turutanov O G, et al. 2006 *Low Temp. Physics* **32**(6) 592–607
[19] Gevorgyan S G, Kiss T, Shirinyan H G, Movsisyan A A, et al. 2001 *Physica* C **363**(2) 113–18
[20] Shirinyan H G 2002 *J. of Contemp. Physics* (NAS of Armenia), Allerton Press Inc. **37** 387–90
[21] Bardeen J, Cooper L N and Schrieffer J R 1957 *Phys. Rev.* **108** 1175–201
[22] Fulde P and Ferrell R A 1964 *Phys. Rev.* **135** A550
[23] Larkin A I and Ovchinnikov Y N 1964 *Zh. Eksp.Teor.Fiz.* **47** 1136 (1965 *Sov. JETP* **20** 762)
[24] Martin C, Agosta C C, et al. 2005 *Phys. Rev.* B **71** 020503 – *cond-mat/0309125*(v2) 20 Apr
[25] Radovan H A, et al. 2004 *Nature* **427** Issue 6977 802 (see also: 2003 *NHMFL Reports* **10**(3))
[26] Gevorgyan S G, Kiss T, Inoue M, Movsisyan A A, et al. 2002 *Physica* C **378-381**(P1) 531–36
[27] Muradyan S, et al. 2008 *J. of Contemp. Physics* (NAS of Armenia), Allerton Press Inc. **43**(2) 97
[28] Binning G, et al. 1980 "*STM microscope*," **U.S. Patent 4,343,993** Aug.10, 1982. Field: Sep.12
[29] Binning G, et al. 1982 *Appl. Phys. Lett.* **40** 178
[30] Binning R, Quate C, et al. 1986 *Phys. Rev. Lett.* **56** 930
[31] Bikov V 2000 *Topical rev. on Probe Microscopes* Sov. Doctoral Dissert. Moscow (in Russian)
[32] Spiral Chip Inductors (U.S. Microwaves production – www.usmicrowaves.com)

Journal of Physics: Conference Series **350** (2012) 012027 doi:10.1088/1742-6596/350/1/012027

Characterization of Cu_2O thin films prepared by evaporation of CuO powder

V A Gevorkyan, A E Reymers, M N Nersesyan and M A Arzakantsyan

Russian-Armenian (Slavonic) University, 375051, H.Emin str.123, Yerevan, Armenia

E-mail: vgev@rau.am

Abstract. Among the potential photovoltaic devices based on semiconductor oxides as active layer is cuprous oxide (Cu_2O). This oxide semiconductor shows many attractive characteristics useful for solar cells production such as low cost, nontoxicity, high mobility and diffusion length of minority carriers, high absorption coefficient and direct energy gap. In this work we report our results of optical and structural investigations of Cu_2O thin films fabricated by thermal vacuum evaporation of CuO powder. The effects of the deposition velocity on structural and optical properties of Cu_2O films were investigated. The X-ray investigations have shown that at low deposition velocity the films consist only of Cu_2O phase without any interstitial phase and have a nano-grain structure. The grains have an average dimensions about (25-30) nm and all these grains showed (200) preferential crystallographic orientation. Optical investigations have shown that the absorption edge of prepared films is due to a direct allowed transition. The value of determined optical band gap is 2.05 eV which corresponds to band gap of bulk Cu_2O.

1. Introduction

Thin films technologies allow developing next and more efficient generation of solar cells. Among the potential photovoltaic devices based on semiconductor oxide as active layer is cuprous oxide (Cu_2O). The theoretical energy conversion efficiency of Cu_2O solar cell is about 20% [1]. This oxide semiconductor shows many interesting characteristics useful for solar cells production such as low cost, nontoxicity, good mobilities, fairly high minority carrier diffusion length, high absorption coefficient and direct energy gap.

Cu_2O is spontaneously a p-type semiconductor since it contains negatively charged copper vacancies and probably interstitial oxygen [2]. A large number of elements were tested as doping impurities. Up to now n-type conductivity is not demonstrated conductivity and a possible explanation for this negative result is based on the self-compensation mechanism [3] or the low solubility of the tested doping impurities [4].

Although the theoretical limit of Cu_2O solar cell efficiency is about 20% the highest efficiency obtained up to now is 2% [5]. This is due to a limited amount of work devoted to this semiconductor and only during last a few years PV device composed of p-Cu_2O and n-ZnO has received broadens attention as a real candidate of next generation thin film solar cells [6-9]. Only recently, conversion efficiency 3.83% has been obtained in ZnO/Cu_2O heterojunction solar cells [10].

The optimization of Cu_2O solar cell is slowed down by the lack of clear understanding of the electrical and crystalline structure properties of this material and defects arising on heterojunction, as well as limited technological methods applied up to now.

Several methods of deposition have been used in attempts to obtain thin films of Cu_2O, such as an electro-deposition [11,12], a chemical methods [13,14], sputtering method [15,16] and method of Cu_2O powder thermal evaporation [17]. The thermal oxidations of Cu sheets and sputtering methods have attracted much attention because up to now these two methods provide high mobility and structure performance of Cu_2O.

The objective of the present paper is to fabricate Cu_2O thin films by thermal evaporation of small portions of CuO powder and investigate the dependence of structural and optical properties of these films on deposition conditions.

2. Experimental

Thin films of Cu_2O were fabricated by continuous thermal evaporation of CuO small particles. Schematic view of vacuum evaporation setup presented in figure 1. CuO particles have different size from 100 to 500 micrometers. These particles were prepared by crush of pallet which was obtained by pressing commercially available CuO powder (99,9% purity) and sintering at 700 ^{0}C in air 1 h. The vacuum chamber before evaporation was pumped to a base pressure of $1.5{\cdot}10^{-5}$ mm Hg. The thermal evaporation was carried out by dosed supply of CuO particles to the molybdenum boat heated to the temperature 1300^0 C. Process of CuO particles evaporation occurred as follows. After contact to boat the particles of CuO in process of their heating at first convert to Cu_2O by the reaction:

$$4CuO \rightarrow 2Cu_2O + O_2 \qquad (1)$$

and then convert to Cu by the reaction:

$$2Cu_2O \rightarrow 4Cu + O_2 \qquad (2)$$

Since the temperature of boat was above the temperature of Cu evaporation (1260 ^{0}C) after thermal decomposition of Cu_2O to Cu and O_2 a full evaporation of Cu was occurred. This method of evaporation provides in average the same flows of copper and oxygen atoms due to different temperature of evaporating particles.

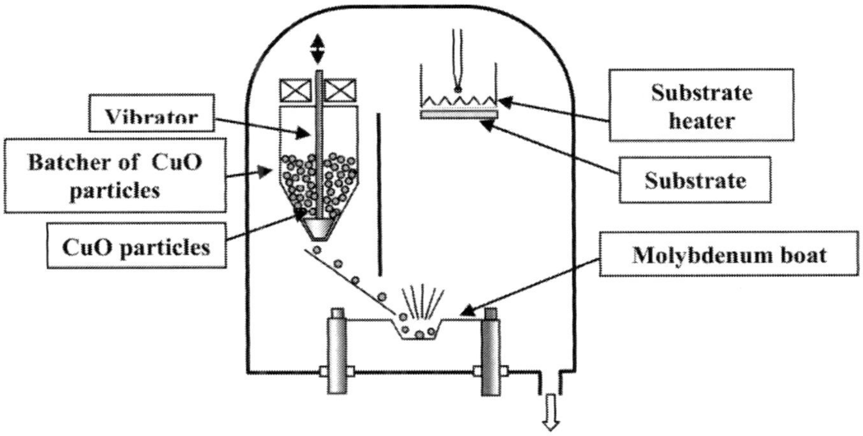

Figure 1. Schematic representation of the experimental setup.

The films were deposited on sapphire substrate. The temperature of substrates was 700 ^{0}C. During the evaporation process the base pressure in vacuum chamber was changed from base pressure $1.5{\cdot}10^{-5}$

mm Hg to the partial pressure of oxygen $1 \cdot 10\text{-}4$ mm Hg. After the film deposition the substrate was cooled to the room temperature during 1 hour with the chamber pressure $6 \cdot 10\text{-}6$ mm Hg. The film thickness and duration of deposition were typically 1-3 micrometer and 3 hours.

The structural properties of samples were studied by X-ray diffractometer URD-6 in the θ-2θ mode using Cu - Kα radiation as well as by optical Olympus microscope with magnitude 500^x. Transmittance and reflectance, over the wavelength from 300 to 1000 nm were measured using double beam Specord M-50 spectrophotometer.

3. Results and discussion

3.1. Phase formation

We have investigated the characteristics of films depend on the deposition velocity at the temperature of substrate 700^0 C. Figure 2 shows the photo of surface morphology of two films series A and B which differ from each other by the velocity of deposition. For low deposition velocity less than 5 nm/min the films of series A consist of smaller size grains than the films of series B (figures 2a and 2b).

Figure 2. The photo of surface morphology of (a) films series A, and (b) films series B (500^x).

3.2. Structural analysis (XRD)

Anglar XRD patterns of films series A and B are shown in figure 3. For films series A, there is one strong XRD peak at $2\theta = 42.4^0$, corresponding to the (200) crystal plane of Cu_2O (figure 3a). This XRD pattern indicates that films series A is single phase Cu_2O film with crystalline structure. These films consist of nano-size grains and all these grains have identical orientation. Full width of half maximum (FWHM) of the XRD peak was used to estimate the grain size of the films series A. The values of the grain size calculated using Scherrer's formula [18]:

$$D = \frac{0.9\lambda}{d \cdot \cos\theta} \tag{3}$$

where λ is the x-ray wavelength, θ is the Bragg diffraction angle, and d is full-width at the half-maximum (FWHM) of the peak corresponding to θ.

For the peak corresponding to the (200) plane of Cu_2O, the resultant grain size is 25 nm. The XRD pattern for films series B presented in figure 3b contains one Cu_2O - related diffraction (111) peak ($2\theta = 36.38^0$) and two Cu - related diffraction peak (111) ($2\theta = 42.9^0$) and (200) ($2\theta = 49.8^0$).

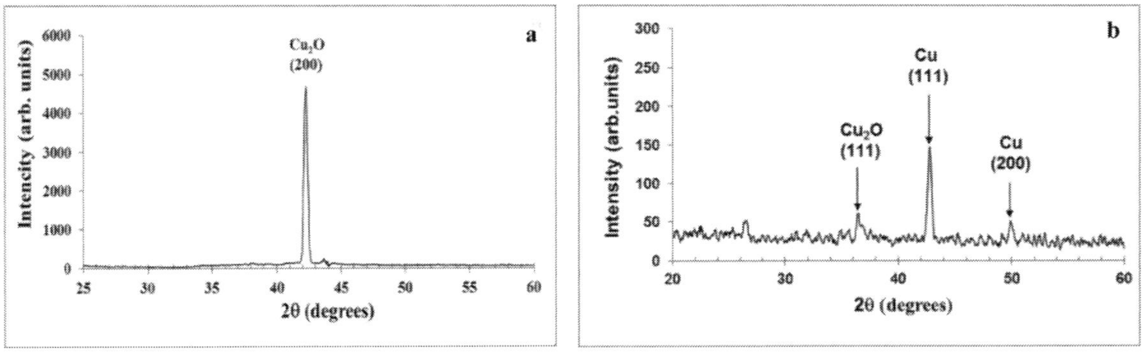

Figure 3. XRD patterns of (a) films series A, and (b) films series B.

The composition of deposited layer formed due to reaction between cupper and oxygen. Figure 4 shows the equilibrium phase diagram of the Cu-O system [19]. The partial pressure and temperature of substrate corresponding to the films deposition conditions are shown in figure 4 (point A). At low velocity of deposition Cu atoms have enough time in order to completely react to oxygen. This time determined by the time of kinetic reaction for formation Cu_2O phase. In this case the composition of deposited film determined by the equilibrium Cu-O phase diagram. It is confirmed by XRD pattern of films series A where observe only Cu_2O - related diffraction peaks. We believe that at high velocity of deposition Cu atoms have not enough time to completely convert in Cu_2O phase (figure 3b).

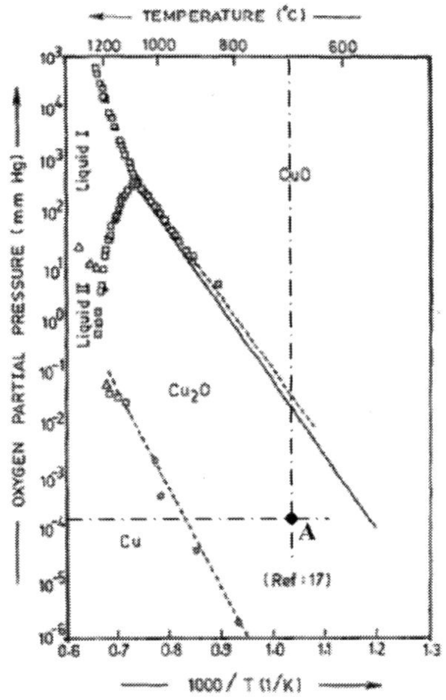

Figure 4. The cooper-oxygen equilibrium phase diagram.

3.3. *Optical properties*

The transmission and absorbance spectra were taken to obtain information on the optical properties of the copper oxide thin film. In the fundamental absorption region, the optical absorption coefficient (α) can be calculated from the transmittance and reflectance data by the equation,

$$\alpha = \frac{1}{d}\ln\frac{(1-R)}{T} \tag{4}$$

where d is the thickness of the film.

Above the fundamental absorption edge the dependence of the absorption coefficient on the incident photon energy is given by Taue's model [20]:

$$\alpha \cdot hv = B\left(hv - E_g\right)^n \tag{5}$$

where hv is the photon energy, E_g is the optical band gap and B is a constant and n is an exponent that depends on the type of optical transitions. As shown in figure 5 the best linear relationship is obtained by plotting $(\alpha \cdot hv)^2$ against photon energy (hv), indicating that the absorption edge in this film is due to a direct allowed transition. The linear portion of the curve is fitted using linear regression analysis. The value of optical band gap was determined from the value of intercept of the straight line at $\alpha = 0$. The value of determined optical band gap is 2.05 eV which corresponds to the band gap of bulk Cu_2O [21].

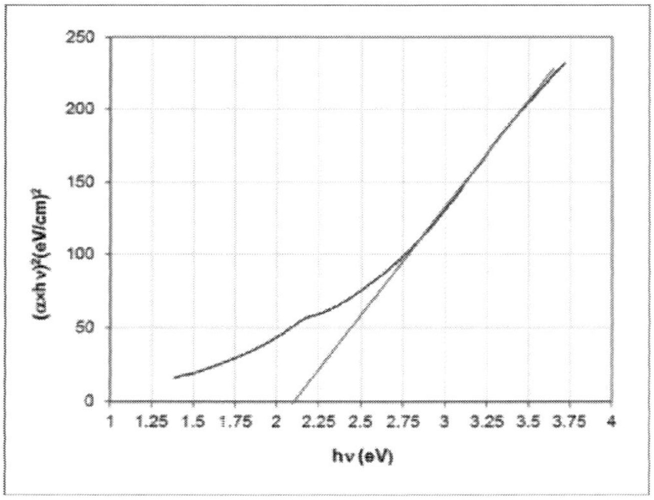

Figure 5. Variation of $(\alpha hv)^2$ as a function of photon energy.

4. Conclusion

The films of copper oxide were deposited on sapphire substrate by the continuous thermal evaporation of CuO small particles. The effects of deposition velocity on the structural and optical properties of Cu_2O films were investigated. It was shown that at deposition velocity less than ~ 5 nm/min the films deposited on the 700 ^{0}C substrate consist only of Cu_2O phase. These films have a nano-grain structure and all these nano-grains have (200) preferential crystallographic orientation with average dimensions about (25-30) nm.

It was found that the optical band gap of fabricated Cu_2O films is 2.05 eV which corresponds to the band gap of bulk Cu_2O.

Reference

[1] Olsen L C, Addis F W and Miller W 1982-1983 *Solar Cells* **7** 247
[2] Porat O and Riess I 1995 *Solid State Ionics* **81** 29
[3] Tsur Y and Riess I 1999 *Phys. Rev.* B **60** 8138
[4] Tsur Y and Riess I 1998 *Z. Phys. Chem.* **207** 181
[5] Mittiga A, Salza E, Sarto F, Tucci M and Vasanti R 2006 *Appl. Phys. Lett.* **88** 163502
[6] Hsueh T J, Hsu C L, Chang S J, Guo P W, Hsiehc J H and Chen I C 2007 *Scr. Mater.* **57** 53
[7] Paul G K, Ghosh R, Bera S K, Bandyopadhyay S, Sakurai T and Akimoto K 2008 *Chem. Phys. Lett.* **463** 117
[8] Wang R C and Lin H Y 2010 *Sens. Actuators* B **149** 94
[9] Lv R, Zheng W, Lin L, Peng F, Huanf Z and Lai F 2011 *Physica* B **406** 1253-1257
[10] Minami T, Nishi Yu, Miyata T and Nomoto J 2011 *Applied Physics Express* **4** 062301
[11] Hu F, Chan K C, Yue T M 2009 *Th. Sol. Films* **518** 120–125
[12] Wijesundera R P, Hidaka M, Koga K, Sakai M and Siripala W 2006 *Th. Sol. Films* **500** 241 – 246
[13] Necmi S, Tulay S, Seyda H and Yasemin C 2005 *Semicond. Sci. Technol.* **20** 398–401
[14] Nair M T S, Guerrero L, Arenas O L and Nair P K 1999 *Appl. Surf. Sci.* **150** 143–151
[15] Itoh T and Maki K 2007 *Vacuum* **81** 904 –910
[16] Sheng H, Hong-Ying C, Lien-Teng K, Cheng-Hsien T 2008 *Th. Sol. Films* **517** 1195–1199
[17] Al-Kuhaili M F 2008 *Vacuum* **82** 623–629
[18] Klug H P and Alexander L E 1974 *X-Ray diffraction Procedure for Polycrystalline and Amorphous Materials* (New York: Wiley)
[19] Weichman F L and Reyeas J M 1980 *Can.J.Phys.* **58** 325
[20] Tauc J 1970 *Optical properties and Solids* (Amsterdam: North-Holland)
[21] Nakamura N, Ishiwary S and Tanaka A 1998 *Appl.Phys.Lett.* **73** 23-25

Journal of Physics: Conference Series **350** (2012) 012028 doi:10.1088/1742-6596/350/1/012028

Thermoelectric single-photon detector

A A Kuzanyan, V A Petrosyan, A S Kuzanyan

Institute for Physical Research, National Academy of Sciences, Ashtarak 0203, Armenia

E-mail: kuzanian@yahoo.com

Abstract. The ability to detect a single photon is the ultimate level of sensitivity in the measurement of optical radiation. Sensors capable of detecting single photons and determining their energy have many scientific and technological applications. Kondo-enhanced Seebeck effect cryogenic detectors are based on thermoelectric heat-to-voltage conversion and voltage readout. We evaluate the prospects of CeB_6 and $(La,Ce)B_6$ hexaboride crystals for their application as a sensitive element in this type of detectors. We conclude that such detectors can register a single UV photon, have a fast count rate (up to 45 MHz) and a high spectral resolution of 0.1 eV. We calculate the electric potential generated along the thermoelectric sensor upon registering a UV single photon.

1. Introduction

In recent years the progress in telecommunications and information systems is frequently connected with the advent of quantum cryptography and information processing by quantum systems (quantum computers). In this context the interest towards single-photon detectors of optical radiation, allowing to register single quanta of light with high efficiency, has grown significantly. The employment of single-photon detectors can also favour the creation of a new generation of laser tomographs and lidars, X-ray analysers and defectoscopes. Without single-photon detectors it is hard to imagine the progress in such areas of science, as cosmic astronomy and particle physics.

Detectors based on photomultiplier tubes (PMT) and avalanche photodiodes (SPAD) are capable of operating in a single-photon regime. But, unfortunately, they do not possess wide spectral sensitivity, high energy resolution and count rate. At this point of time the superconducting single-photon detectors (SSPD) are considered the most promising among single-photon detectors [1–4]. They have exceptional technical characteristics but they too have deficiencies, such as the need of a rigid temperature regime and the complexity of construction. The low-temperature thermoelectric single-photon detectors can serve as an alternative to SSPDs. Detectors based on CeB_6 and $(La,Ce)B_6$ crystals can detect a single UV photon, have very fast count rate (up to 1.8 GHz in case of thin-film geometry) and a spectral resolution of 0.1 eV. These detectors do not require strict temperature conditions and are simpler in construction [5–7]. The comparison of features of currently available single-photon detectors and the thermoelectric detectors with a thin-film sensor is presented in [7]. In this work we will focus in more detail on the possibility of creating a thermoelectric detector with a monocrystalline sensor and will discuss its characteristics.

Journal of Physics: Conference Series **350** (2012) 012028 doi:10.1088/1742-6596/350/1/012028

2. The crystalline thermoelectric single-photon sensor

2.1. The structure of the crystalline thermoelectric single-photon sensor
In principle the sensitive element of the thermoelectric detector on the basis of single crystals can be composed of only two elements (Figure 1): a thermoelectric monocrystal and an absorber, as well as two contacts through which one can register the voltage generated upon absorbing a photon.

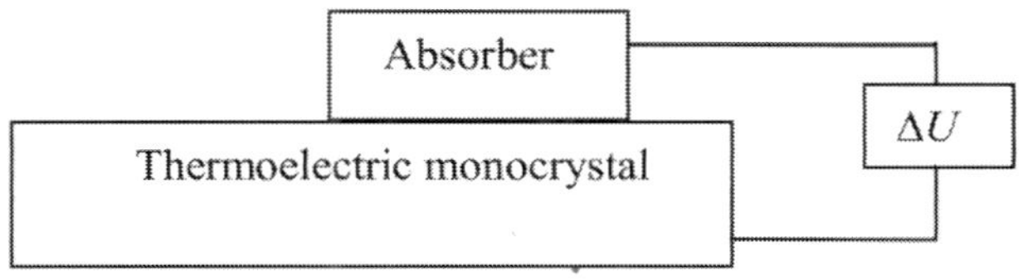

Figure 1. The schematic of a thermoelectric single-photon detector based on single crystals.

The method of preparing the monocrystalline sensitive element is protected by an Armenian patent [8] in which one of the two possible constructions of the monocrystalline sensor is described and a method of its fabrication is proposed.

2.2. The spectral resolution
The most important characteristics of a detector are the photon count rate and the spectral resolution. The physical processes occurring in a pixel of a thermoelectric detector after a single photon is absorbed are described in detail in [6]. The full-width at half-maximum expression of spectral resolution of a monocrystalline thermoelectric sensor based on the observable response to monoenergetic photons is:

$$\Delta E_{FWHM} = 2.35\{2k_B T^2 C_{abs}[1+(ZT)^{-1}]\}^{1/2}, \qquad (1)$$

where C_{abs} is the absorber heat capacity (an element in the construction of the detector which absorbs the photon), ZT is the thermoelectric figure of merit ($ZT=TS^2\sigma/k$, were S is the Seebeck coefficient, σ and k are, respectively, the electrical and thermal conductivity, T is the absolute temperature). For a 1 μm^3 tungsten absorber at 9 K the C_{abs} is equal to ~ 10^{-15} J/K. If we take into account the ZT_{max} ~ 1.19 (for CeB_6 crystals [7]), we can find at 9 K the spectral resolution to be ~ 12.7 eV. This means that a thermoelectric detector built on CeB_6 crystals may be able to distinguish UV range photons (which have an energy of 3–124 eV) of not very low energy, as well as X-ray photons. That, in turn, means that the detector will be able to detect a single X-ray photon and can be used in cosmic X-ray astronomy, in X-ray microanalysis, etc.

In case of $(La_{0.99},Ce_{0.01})B_6$ we can choose C_{abs} ~ 10^{-17} J/K (at 0.5 K temperature and 0.25 μm^3 volume) and, taking into account that the ZT for this hexaboride is 0.1, we can find the value of energy resolution to be ~ 0.1 eV [6]. The single-photon detector built on $(La,Ce)B_6$ will be capable of registering even low-energy single UV photons.

2.3. The count rate
The second important characteristic of the detector is its photon count rate. Before proceeding to the calculation of this parameter, we need to clarify the characteristics of the absorber, namely the absorber material and its geometric dimensions which are required for a reliable absorption of a photon of certain energy. In many detectors heavy metals are used as an absorber. We opted for the tungsten; now let us see what absorber thicknesses are required.

The absorption of incident radiation is determined by the Beer–Lambert law:

$$I = I_0 e^{-\mu l}, \qquad (2)$$

often written as $I = I_0\, e^{-(\mu/\rho)\rho l}$, where μ is the linear attenuation coefficient (it depends on photon energy), ρ is the absorber density (for tungsten ρ = 9.3 g·cm^{-3}), l is the absorption depth.

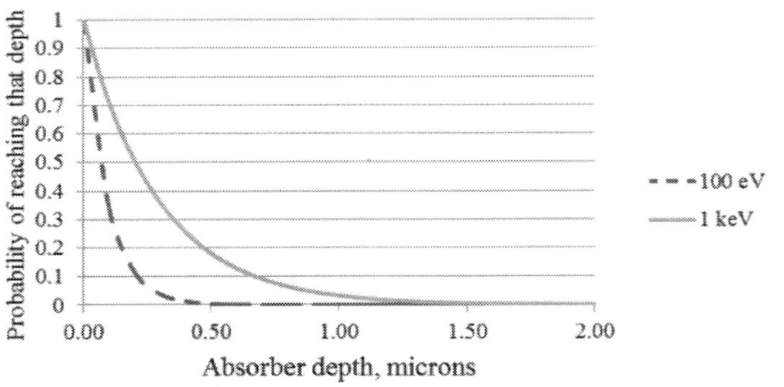

Figure 2. The probability of 100 eV and 1 keV energy photons reaching a certain depth in tungsten.

Figure 2 demonstrates the probability of reaching a certain depth in tungsten for photons of 100 eV and 1 keV energy. The graph shows that the maximum depth the photons with energies of 1 keV can reach is about 1.5 μm, whereas photons with an energy of 100 eV cannot overcome the 0.5 μm thickness.

The count rate of the thermoelectric single-photon detector with a crystalline sensor is determined by the speed of heat outflow from absorber to crystal. This time is determined by the following expression [6]:

$$\tau_{\text{heat flow}} = C_{\text{abs}}/G = C_{\text{abs}} l/kA, \qquad (3)$$

where C_{abs} is the absorber heat capacity, G is the heat conductance of the crystalline sensor, l is its length, k is its thermal conductivity and A is its cross-sectional area. So, the count rate is strongly dependent upon the detector's geometry. If we choose the crystal size as 0.3 mm × 1 mm × 1 mm and the absorber size as 1 μm × 1 mm × 1 mm, then taking into account that at 9 K $C_{\text{abs}} = 10^{-9}$ J/K and k = 10 mW/cm K [9], we obtain for the heat flow time $\tau_{\text{heat flow}}$ = 300 ns, which corresponds to a count rate of about 3 MHz.

Obviously, by adjusting the geometric dimensions of the crystal and the absorber one can achieve higher count rates. With today's capabilities of crystal processing one can have crystals sized only few tens of microns. In Table 1 we demonstrate the count rates corresponding to different absorber–sensor geometries. It is notable, that with 20 μm × 20 μm × 20 μm sized crystals a count rate of 45 MHz is achievable.

Table 1. The count rate of the thermoelectric detector with different geometries

Absorber size (length×width×height)	Sensor crystal size (length × width × height)	Absorber heat capacity	Count rate with a CeB_6 sensor at 9 K
1 μm × 1 mm × 1 mm	0.3 mm × 1 mm × 1 mm	10^{-9} J/K	3 MHz
1 μm × 0.5 mm × 0.5 mm	0.3 mm × 0.5 mm × 0.5 mm	$0.25 \cdot 10^{-9}$ J/K	3 MHz
1 μm × 1 mm × 1 mm	1 mm × 1 mm × 1 mm	10^{-9} J/K	1 MHz
1 μm × 50 μm × 50 μm	50 μm × 50 μm × 50 μm	$0.25 \cdot 10^{-11}$ J/K	18 MHz
1 μm × 20 μm × 20 μm	20 μm × 20 μm × 20 μm	$4 \cdot 10^{-13}$ J/K	45 MHz

2.4. Generated thermopower

In this section we consider the order of magnitude of the thermopower generated along the detector's sensor upon absorbing a single photon. To be more precise, we estimate the generated maximum thermopower which will occur under the most favorable conditions, namely the value of the thermoelectric figure of merit is maximal, the photon is absorbed near the absorber–thermoelectric sensor border, the thermopower is recorded at the initial time, when it is maximal.

Thus, in accordance with the definition of the Seebeck coefficient, the voltage generated along a thermoelectric material is proportional to the temperature gradient and is defined by the equation:

$$U = S\Delta T. \tag{4}$$

Figure 3 shows the temperature dependence of the Seebeck coefficient of the CeB_6 crystal, according to which $S = 265$ μV/K at 9 K [10]. This is the maximum value of the Seebeck coefficient of the CeB_6 crystal we were able to find in the literature.

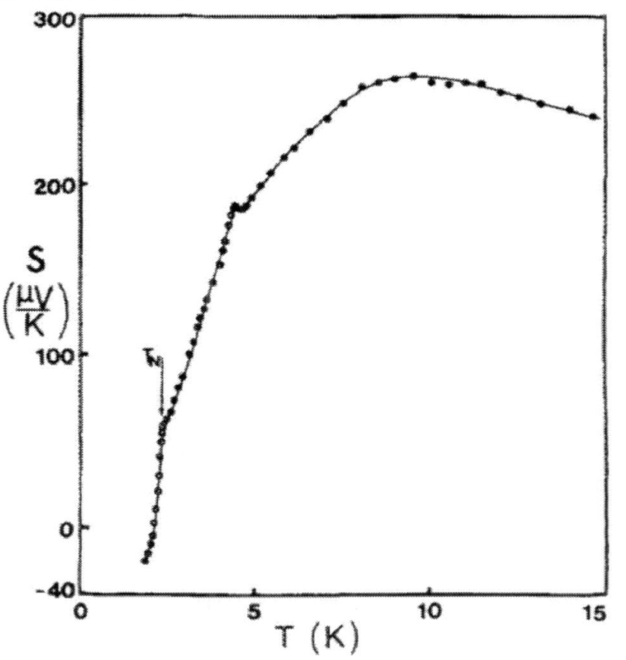

Figure 3. The S(T) dependence of a good CeB_6 crystal from Naushad Ali and S. B. Woods, Transport properties of Kondo lattice CeB_6, J. Appl. Phys. 57, 3182 (1985).

In its turn, the temperature gradient occurring along the thermoelectric detector's sensor upon absorbing a photon with an energy of E_{ph} is

$$\Delta T = E_{ph}/C_{abs}. \qquad (5)$$

For tungsten between 5–15 K we have $C(J/mole \cdot K) = 1.11\times10^{-3}T + 1943(T/378)^3$ [11]. It is not difficult to calculate what temperature gradient occurs along the sensor upon absorbing photons with a certian energy. The values of the temperature gradient occurring along the thermoelectric sensor at 5–15 K temperature upon absorbing a 100 eV photon are shown in Figure 4 (absorber size is 1 μm^3).

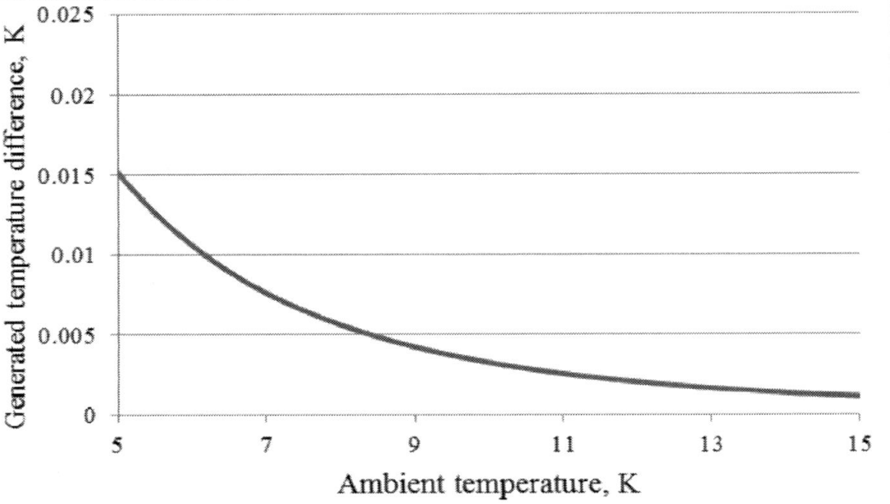

Figure 4. The values of the temperature gradient generated along the thermoelectric sensor at 5–15 K upon absorbing a 100 eV photon.

Assuming the best values of *S*, the thermopower generated upon absorbing a 100 eV ultraviolet photon should be ~1.80 μV at 5 K and ~1.07 μV at 9 K.

3. Conclusion

In this paper we have tried to show both the possibility of creating an actual design of a thermoelectric detector sensor using actual materials and the possibility of registering a UV range single photon absorption. The calculated values of energy resolution and count rate make us hopeful that the thermoelectric detector can be used for solving a number of specific scientific and technical problems. Optimistic calculations show that the voltage arising on the sensor of the thermoelectric detector can reach microvolts. Such voltages are measured with great accuracy by the most common voltmeters. If we assume that the real signal is several orders of magnitude weaker, in that case too its measurement will not be very difficult. The same can be claimed if we consider the time of the photon absorption process. Exact calculations of thermal processes occurring in the sensor of a thermoelectric detector upon the absorption of a single photon will be conducted and published in the near future.

Acknowledgement

The authors thank the CRDF/Government of Armenia/EIF Business Partnership Grant "Sensor for a thermoelectric Single-photon Detector" for support.

References

[1] Irvin K D 2006 "Seeing with superconductors" *Scientific American*, *November* 2006, 86.

[2] Gol'tsman G, Okunev O, Chulkova G, Lipatov A, Semenov A, Smirnov K, Voronov B, Dzardanov A, Williams C and Sobolewski R 2001 "Picosecond superconducting single-photon optical detector," *Appl. Phys. Lett.*, vol. 79, pp.705–707.

[3] Lita A E, Miller A J, Nam S W 2008 "Counting near-infrared single-photons with 95% efficiency" *Opt. Express* 16, 3032-3040.

[4] Fujii G, Fukuda D, Numata T, Yoshizawa A, Tsuchida H, Inoue Sh, Zama T 2010 "Fiber Coupled Single Photon Detector with Niobium Superconducting Nanowire", *Quantum Communication and Quantum Networking, Lecture Notes of the Institute for Computer Sciences, Social Informatics and Telecommunications Engineering,* 36, 220 .

[5] Fritz G G, Wood K S, Van Vechten D, Gyulamiryan A L, Kuzanyan A S, Giordano N J, Jacobs T M, Wu H-D, Horwitz J S, Gulian A M 2000 "Thermoelectric single-photon detector for X-ray/UV radiation", *Proc. SPIE*, 4140, 459.

[6] Gulian A, Wood K, D. van Vechten, Fritzdet G 2004 "Cryogenic thermoelectric (QVD) detectors: Emerging technique for fast single-photon counting and non-dispersive energy characterization," *J. Mod. Opt.*, 51, 1467-1490.

[7] Petrosyan V, "CeB6 as a prospective material for cryogenic thermoelectric single photon detectors," *International Conference on Laser Physics 2010, Ashtarak, October 12-15, 2010, Proceedings of SPIE* Vol.7998(SPIE Bellingham,WA, 2011) 0D.

[8] Kuzanyan A, Petrosyan V, Kuzanyan V, Gulyan A 2009 "Method of obtaining a sensor for a thermoelectric detector," *Patent of Armenia* No. 2304 A .

[9] Harutyunyan S. R., Vardanyan V. H., Kuzanyan A. S., Nikoghosyan V. R., Kunii S., Wood K. S., Gulian A. M., "Thermoelectric cooling at cryogenic temperatures", Applied Physics Letters, 83, Issue 11, id. 2142 (2003).

[10] Naushad Ali and S B Woods 1985 Transport properties of Kondo lattice CeB_6, J. *Appl. Phys*. 57, 3182.

[11] Waite T R, Craig R S, and Wallace W E 1956 "Heat Capacity of Tungsten between 4 and 15°K", *Phys. Rev*. 104, 1240–1241.

Journal of Physics: Conference Series **350** (2012) 012029 doi:10.1088/1742-6596/350/1/012029

Similariton for femtosecond signal analysis

L Kh Mouradian, A S Zeytunyan, G L Yesayan

Ultrafast Optics Laboratory, Faculty of Physics, Yerevan State University
1, Alex Manoogian Street, Yerevan 0025, Armenia

lmouradian@ysu.am

Abstract. The similariton-based spectral interferometry and pulse spectrotemporal imaging in a similariton-induced temporal lens are comparatively experimented, as two applications of similariton to ultrafast optics, particularly, for signal analysis problem on femtosecond timescale. Generation of the 50-THz bandwidth similariton provides a few-femtosecond temporal resolution for accurate and aberration-free measurements of these self-referencing methods.

1. Introduction

The problem of femtosecond signal analysis currently employs the powerful arsenal of contemporary optics, involving the methods of nonlinear and adaptive optics, Fourier optics and holography, spectral interferometry, etc. The nonlinear-optical technique of FROG [1], the most popular, provides complete determination of the temporal amplitude and phase by recording high-resolution spectrograms, which are further decoded by means of iterative phase-retrieval procedures. The approach of spectral interferometry (SI) [2] and its developments to the methods of SPIDER [3] and SPIRIT [4] have the advantage of non-iterative phase retrieval. These methods are based on the spectral phase retrieval, spectrum measurement, and reconstruction of a temporal pulse. For the pulse direct measurement, the method of pulse spectrotemporal imaging (STI) through temporal lensing is more promising, having as a principal limit of resolution the ~1 fs nonlinear response time of silica [5]. Its recent modifications, implemented in the silicon chip [6] and similariton-induced parabolic temporal lens [7], provide accurate, high-resolution direct measurement of a pulse in the spectrometer as in the femtosecond optical oscilloscope (FO). Many modern scientific and technological problems, however, demand both the amplitude and phase information, possible through additional interferometric measurements. This motivates the urgency of SI methods for the complete characterization of femtosecond signal. The classic method of SI is based on the interference of the signal and reference beams spectrally dispersed in a spectrometer, with the spectral fringe pattern caused by the difference of spectral phases [2]. The known spectral phase of the reference permits to retrieve the spectral phase of signal, and, together with the spectrum measurement, to recover the complex temporal amplitude of the signal through Fourier transformation. The SI-measurement is as accurate as any interferometric one, and the setup is rather simple, but its application range is restricted by the bandwidth of the reference. To avoid this restriction, the self-referencing methods of spectral shearing interferometry are developed [3,4], promoting the SI modifications to the class of the most popular methods, compatible with the FROG [1], at the expense of more complicated optical arrangement.

We develop a new method of similariton-based SI, which keeps the simplicity of the principle and configuration of the classic SI [2], along with its self-referencing performance. We experimentally examine the similariton-based SI in comparison with a prototype of the FO based on the pulse STI in the similariton-induced temporal lens. The use of similariton upgrades both the reference-based methods up to self-referencing ones.

2. Similariton-based spectral interferometry versus spectrotemporal imaging

Similaritons, pulses with the distinctive property of self-similar propagation, recently attract the attention of researchers, due to fundamental interest and prospective applications in ultrafast optics and photonics [8,9]. The self-similar propagation of the high-power pulse with parabolic temporal, spectral, and phase profiles was predicted in [10]. In practice, the generation of such parabolic similaritons is possible in active fibers, such as rare-earth-doped fiber amplifiers [11-13] and Raman fiber amplifiers [14], as well as in the laser [15]. The generation of parabolic similariton has been also proposed in a tapered fiber with decreasing normal dispersion, using either passive dispersion-decreasing fiber [16] or a hybrid configuration with Raman amplification [17]. Another type of similariton was generated in a conventional uniform and passive (without gain) fiber under the combined impacts of Kerr nonlinearity and dispersion [18]. In contrast to the parabolic similariton with parabolic amplitude and phase profiles, this nonlinear-dispersive similariton has only parabolic phase but maintains its temporal (and spectral) shape during the propagation, as well. Our SI studies of this type of pulses [18] showed the linearity of their chirp, with a slope given only by the fiber dispersion. This property leads to the spectrotemporal similarity of nonlinear-dispersive similariton, with accuracy given both by spectral broadening and pulse stretching. Both the parabolic similariton of active fiber and nonlinear-dispersive similariton of passive fiber are of interest for applications in ultrafast optics, especially for pulse compression [19,20] and shaping [21], similariton referencing temporal lensing, STI and SI [22]. The SI study provides the complete characterization and mathematical description of the nonlinear-dispersive similariton of up to 5 THz bandwidth [18]:

$$A_f(t) \propto |\tilde{A}_f(\omega)| \exp[-i\,\phi_f(\omega)]\Big|_{\omega=Ct}, \quad \varphi_f(t) = -\phi_f(\omega)\Big|_{\omega=Ct} = \beta_2 f\omega^2/2\Big|_{\omega=Ct} = Ct^2/2, \qquad (1)$$

for the slowly varying amplitude $A_f(t)$ and phase $\varphi_f(t)$, where $\tilde{A}(\omega) \equiv FT[A(t)]$ is the complex spectral amplitude with spectral phase $\phi(\omega)$, $C = (\beta_2 f)^{-1}$ – chirp factor, β_2 – coefficient of second order group velocity dispersion, t – running time, ω – frequency, and f – fiber length. For comparison, the pulse pure dispersive propagation in temporal Fraunhofer zone results in the "spectron" pulse $A_d(t) \propto \tilde{A}_0(\omega)\exp(i\beta_2 f\omega^2/2)|_{\omega=Ct}$, with the same dispersion-induced phase. The applications of similariton demand to check and generalize this key peculiarity for broadband pulses. The studies of broadband similariton serve as a basis for the development of a novel method of similariton-based SI for the femtosecond pulse complete characterization. Below the studies directed to its development in comparison with the method of pulse STI in the similariton-induced temporal lens are presented. Application of similariton to the reference-based methods upgrades them up to the self-referencing ones, substantially improving their performance due to the enlarged application range along with the simplicity of the principle and configuration.

Describing the *principle* of similariton-based SI, the part of signal beam is injected into a fiber to generate the nonlinear-dispersive similariton-reference, with the complex spectral amplitude $\tilde{A}(f,\omega) = |\tilde{A}(f,\omega)|\exp[i\phi(f,\omega)]$. The residual part of the signal, with complex spectral amplitude $\tilde{A}(0,\omega) = |\tilde{A}(0,\omega)|\exp[i\phi(0,\omega)]$, is coupled with the similariton in a spectrometer with an appropriate time delay. The spectral fringe pattern $S_{SI}(\omega) = 2|\tilde{A}(0,\omega)||\tilde{A}(f,\omega)|\cos[\phi(0,\omega) - \phi(f,\omega)]$, on the background of the signal and similariton spectra, completely covers the signal spectrum $S(0,\omega) = |\tilde{A}(0,\omega)|^2$, and the whole phase information becomes available, for any input signal. The known spectral phase of the similariton-reference allows to retrieve the signal spectral phase $\phi(0,\omega)$, and by measuring also the signal spectrum, to reconstruct the complex temporal amplitude $A(0,t)$ of the signal through Fourier transformation. Thus, the method of similariton-based SI joins the advantages of both the classic SI [2] and spectral shearing interferometry [3,4], combining the simplicity of the principle and configuration with the self-referencing performance. Examining the similariton-based SI, we compare its measurements with the ones carried out by a prototype of FO-based on the pulse STI in the similariton-induced temporal lens in the sum-frequency generation process (SFG). Our comparative study, involving also theoretical and autocorrelation check, along with the study of the similariton-based SI, serves also for the inspection of the prototype of similariton-based FO, the measurements of which previously were compared with the autocorrelation only [7].

The method of SFG-STI for direct femtosecond scale measurements is based on the conversion of temporal information to the spectral domain in a similariton-induced parabolic temporal lens [7,22-24]. The setup of the similariton-based SI is modified to FO by replacing the temporal delay with a dispersive delay line in the idle arm, and placing a nonlinear crystal for SFG at the system output. In the spectral domain, the dispersive delay works as a parabolic phase modulator, and the signal $\tilde{A}(0,\omega)$ passed through is described as $\tilde{A}(d,\omega)=\tilde{A}(0,\omega)\exp(i\ddot{\phi}_d\omega^2/2)$, with the given coefficient $\ddot{\phi}_d \approx -C_d^{-1}$. In the fiber arm, we have a nonlinear-dispersive similariton with the known parameters as in the case of similariton-based SI. In both arms of the setup, we have practically linearly chirped pulses, and the temporal and spectral complex amplitudes repeat each other in the temporal Fraunhofer zone, i.e. spectron pulses are formed [7]: $A(d,t)\propto \tilde{A}(d,\omega)$, and $A(f,t)\propto \tilde{A}(f,\omega)$ with $\omega = C_{d,f}t$. Under the conditions of the opposite and same value chirps $C_f = -C_d \equiv C$, and constant similariton spectrum throughout the signal spectrum, the output temporal SFG-signal repeats the input spectral amplitude: $A_{SFG}(t)\propto A(d,t)\times A(f,t) \propto \tilde{A}(0,\omega)$. Accordingly, the output spectral and input temporal amplitudes repeat each other $\tilde{A}_{SFG}(\omega)\propto A(0,t)$, and the output SFG-spectrum displays directly the input temporal pulse: $S_{SFG}(\omega)=|\tilde{A}_{SFG}(\omega)|^2 \propto |A(0,t)|^2 = I(0,t)$, with the scale $\omega = Ct$. The resolution of such a similariton-based FO is given by the transfer function of the similariton's spectrum [7], and FO with a similariton-reference of the bandwidth of a few tens of nanometers provides the direct measurement of temporal pulse in a spectrometer, exceeding the resolution of the achievement of silicon-chip-based ultrafast optical oscilloscope [6] by an order of magnitude.

In the experiment, a standard Coherent femtosecond laser system is used, with the following parameters of radiation: 100 fs pulse duration, 76 MHz repetition rate, 1.6 W average power, 800 nm central wavelength. Different amplitude- and phase-modulated pulses at the setup input are shaped. Afterwards the radiation is splitted into high- and low-power parts. The low-power part is directed to the temporal delay or dispersive delay line (SF11 prism pair with the reverse mirror) for similariton-based SI and STI, respectively. In the second path, the high-power pulse, with average power of up to 500 mW, is injected into a standard single-mode fiber (1.65m, Newport F-SPF PP@820 nm) by a microscope objective ($10\times$) to generate similaritons. For the SI-measurements, these two pulses are coupled directly into the spectrometer and the SI fringe pattern and signal spectrum are registered. To retrieve the spectral phase, the Fourier-transform algorithm of the fringe pattern analysis is used [24]. For the FO-measurements, a BBO nonlinear crystal at the input of spectrometer is placed, and the SFG-spectrotemporal image is registered directly. The similariton-based SI- and FO- measurements are carried out together with the autocorrelation check by a standard APE PulseCheck autocorrelator. First, the similariton-based SI is tested for the laser pulses stretched and chirped in SF11 glasses of different thickness, comparing the results with the autocorrelation measurements. The coefficients of the dispersion-induced parabolic spectral phases are $\alpha \equiv \ddot{\phi}(\omega) = 1.94\times 10^{-3}$, 4.94×10^{-3}, 6.34×10^{-3}, and 10.78×10^{-3} ps^2 for the 0, 2, 3, and 5-cm glasses, respectively. The SI-reconstructed pulses, correspondingly, have durations of 108, 197, 252, and 365 fs, in a good accordance with the measured autocorrelation durations of 156, 298, 369 and 539 fs.

Afterwards, measurements for multi-peak pulses are carried out together with the autocorrelation check. The SI calibrating measurement of the α coefficient for similariton, using the known laser pulse as a reference, gives the value $\alpha = 2.1 \times 10^{-2}\,\mathrm{ps}^2$, in accordance with the expression $\alpha = \beta_2 f$ with the values f = 49 cm and β_2 = 40 fs^2/mm for 850 nm (provided by the fiber manufacturer). Then different multi-peak signal pulses are shaped inserting thin glass plates in parts of the beam. The beam parts passed through the plates obtain time delay with respect to the free-propagated part. The power proportion among the peaks is adjusted by moving the plates. The thicknesses of the plates give the time delay between the peaks; e.g. a 0.12 mm thick glass plate gives a 200 fs delay, if assuming the refractive index of the plate equal to 1.5. Using double- and triple- peak signal pulses, we carry out SI-measurements and compare the results with measured autocorrelation tracks. As Fig. 1 shows, having the spectrum (b, thin solid line) and SI-retrieved spectral phase (b, thick solid) of pulse, its temporal profile (c, solid) is reconstructed through Fourier transformation. To check the precision of our measurements through similariton-based SI, we calculate the autocorrelation of reconstructed pulse (d, thick solid) and compare it with the intensity autocorrelation measured at the input of the system (d, thin solid). The spectral shape of nonlinear-dispersive similariton (a) ensures the fulfillment of necessary conditions for the parabolicity of the spectral phase of similariton. The structure of the similariton spectrum (a), strange at first glance, is typical for short lengths of nonlinear-dispersive interaction, and is observed also for parabolic similaritons generated in fiber amplifiers [25]. The solid and dashed curves correspond to the pulse reconstruction with the spectral phase coefficients $\alpha = 2.1 \times 10^{-2}\,\mathrm{ps}^2$ and $\alpha = 1.995 \times 10^{-2}\,\mathrm{ps}^2$ (5% difference), respectively.

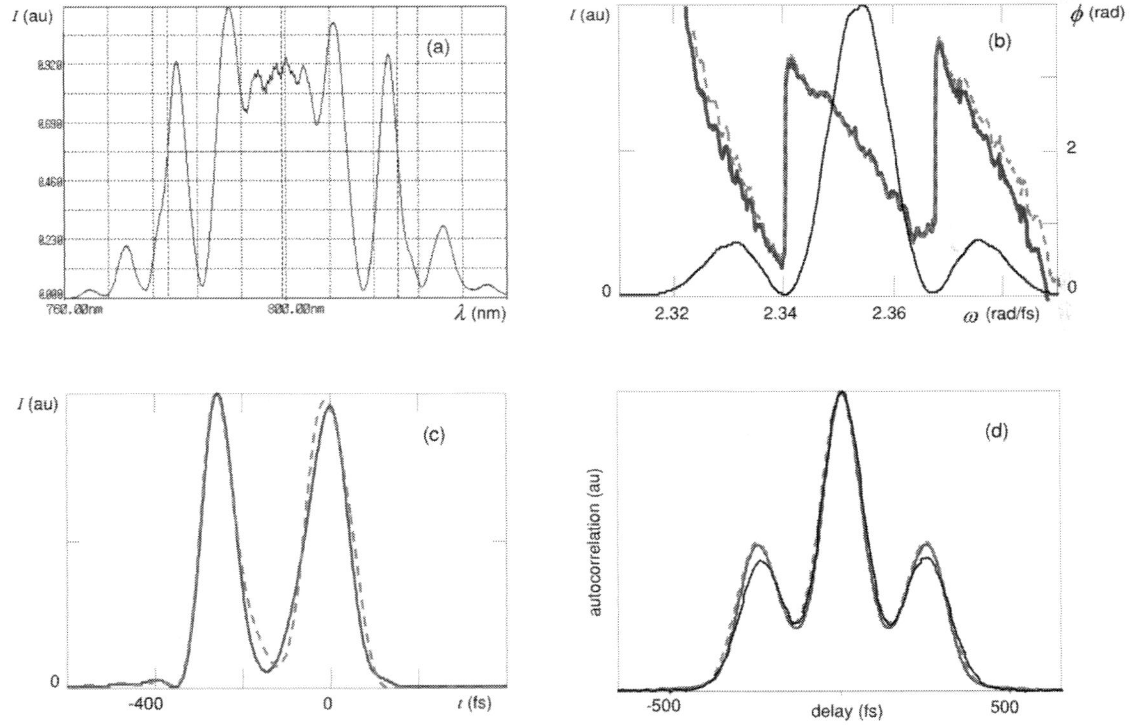

Fig. 1. Reconstruction of double-peak pulse (shaped by a 130-μm thick glass) through similariton-based SI: (a) spectrum of nonlinear-dispersive similariton; (b) retrieved spectral phase and measured spectrum; (c) reconstructed pulse temporal profile; and (d) autocorrelation tracks. Solid and dashed curves are for $\alpha = 2.1 \times 10^{-2}\,\mathrm{ps}^2$ and $\alpha = 1.995 \times 10^{-2}\,\mathrm{ps}^2$ (5% difference), respectively, and the thin solid one in (d) is the measured autocorrelation track.

Fig. 2 shows the analogue procedures for a pulse with more complex sub-structure. The application range and limitations of similariton-based SI are conditioned by the parameters of input radiation and fiber, which are necessary to generate the similariton-reference with parabolic phase. Fig. 1 and 2 illustrate the experiment for the typical regime preceding the similariton shaping: results for double- and triple-peak signal pulses are shown.

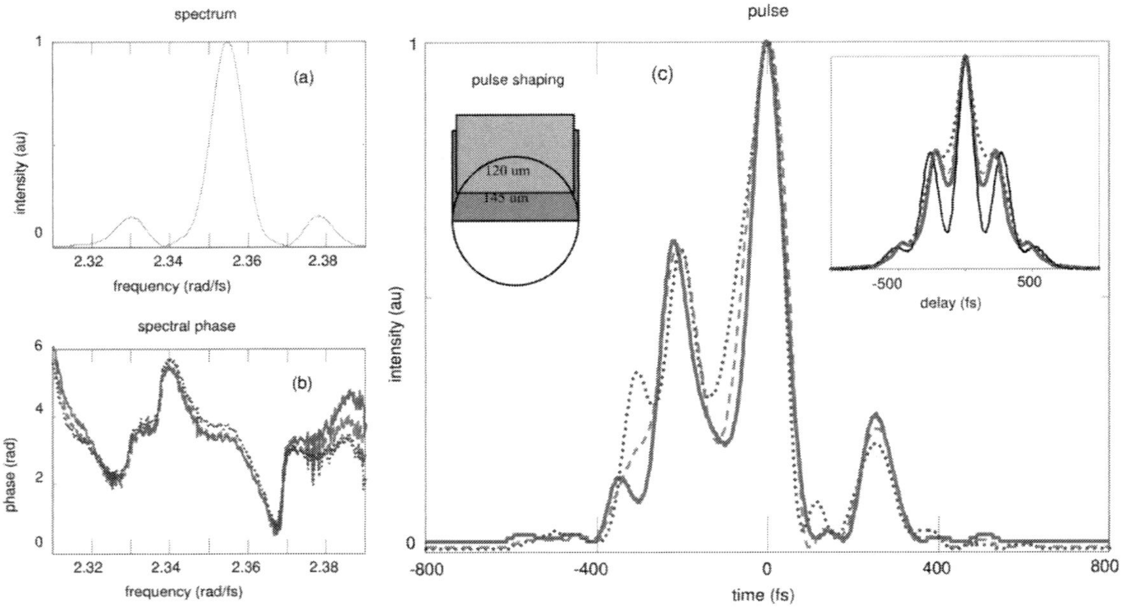

Fig. 2. Reconstruction of three-peak pulse (shaped by means of 145 and 120-um thick glasses) through similariton-based SI: (a) measured spectrum; (b) retrieved spectral phases; (c) reconstructed pulse temporal profiles and autocorrelation tracks (inset). Dotted, dashed and solid curves are for $\alpha =$ $2.1\times 10^{-2}\,ps^2$, $1.995\times 10^{-2}\,ps^2$, and $1.89\times 10^{-2}\,ps^2$, respectively, and the thin solid one of the inset is the measured autocorrelation track.

Finally, we compare the measurements of the similariton-based SI and FO, together with a theoretical check. The double-peak signal pulses are shaped with the spectral domain amplitude- and phase-modulation given by the peaks' temporal distance T and their proportion μ. The temporal amplitude $A(t) = A_0(t) + \mu A_0(t+T)$ corresponds to the complex spectral amplitude $\tilde{A}(\omega) = \tilde{A}_0(\omega)\rho(\omega)\exp[i\phi(\omega)]$, with the $\rho(\omega) = \sqrt{1+\mu^2+2\mu\cos(\omega T)}$ amplitude- and $\phi(\omega) = \arctan[(\sin\omega T)/(\mu^{-1}+\cos\omega T)]$ phase-modulation. To shape such double-peak pulses, the laser beam is expanded and a thin glass plate in its part is placed, as explained above. The similariton-based SI and FO are comparatively experimented using the double-peak signal pulse: the SI-reconstructed pulses are compared with spectrotemporal images of the signal. Fig. 3 illustrates this experiment: the quantitative accordance of the measured spectrum and retrieved spectral phase with the theoretical curves leads to an accurate pulse reconstruction through similariton-based SI (dotted). An accurate STI (solid) is ensured by the similariton of the bandwidths of ≥ 40 nm. The differences between these independent SI- and FO-measurements and theoretical curve (dashed) are hardly seen, evidencing both the accuracy of the mentioned measurements and the potential of similariton-based methods.

Thus, the methods of similariton-based SI and STI are experimentally demonstrated as two applications of similariton. The reference-based methods become self-referencing by the use of similariton. Obviously, the demonstrated methods of femtosecond signal characterization can be implemented also by the use of "standard" parabolic similaritons generated in active or dispersion decreasing fibers. In a recent progress in the generation of parabolic broadband similaritons, bandwidths of up to 11 THz (40 nm at 1050 nm central wavelength) are achieved [25]. However, the use of the nonlinear-dispersive similariton generated in a piece of standard passive fiber currently is more beneficial, providing larger bandwidths and thus larger application ranges with technically simpler experimental arrangement.

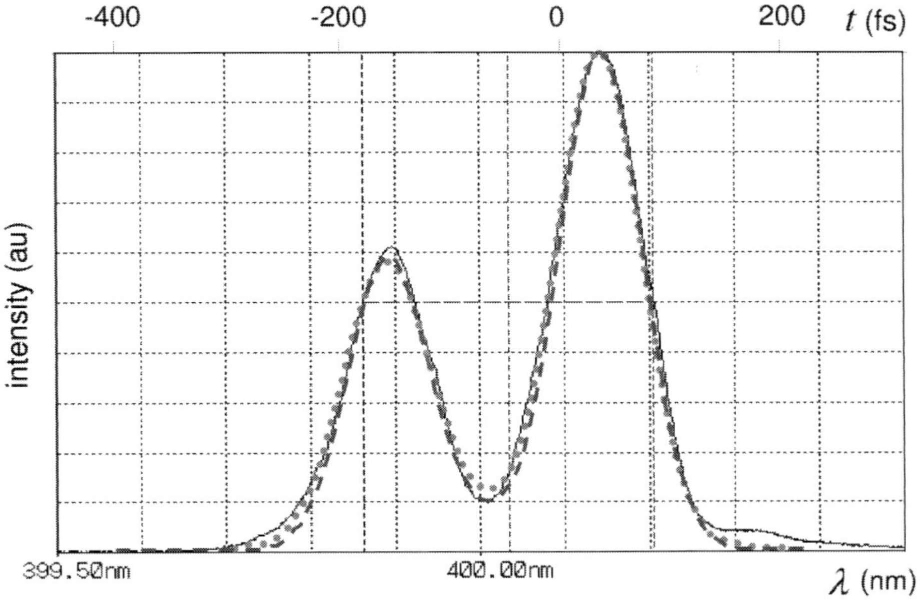

Fig. 3. Comparison of similariton based SI and FO for a double-peak signal pulse. Dotted, solid and dashed curves are for SI, STI and simulation, respectively

3. Characterization of broadband similaritons generated in passive fiber

The applications of similariton to ultrafast optics demand the generation and study of broadband similariton i.e. generalization of Eq. (1). Particularly, the resolution of the femtosecond oscilloscope, based on the similariton-induced parabolic lens, is given by the bandwidth of similariton, and the application range for similariton-based SI is as large as broadband the similariton-reference is [7,22-24]. The pulse compression ratio is also as high as the spectral broadening factor is [19,20].

We applied the chirp measurement technique through spectral compression and frequency tuning in the SFG process [23,24] for broadband similaritons of up to 50-THz bandwidths with the experimental setup of FO. The intensity profile of broadband similariton can be measured by means of the cross-correlation technique, using the SFG-interaction of similariton with the laser pulse as a reference. A spectral detection of the SFG-signal will give also information on the chirp of similariton, modifying the cross-correlation technique to the cross-correlation frequency-resolved optical gating (XFROG) [26]. Our additional modification, the use of a dispersively chirped reference pulse, provides a spectrally compressed SFG-signal in a wider spectral range, and thus, more efficient measurement [23,24]. The SFG-interaction of up- and down-chirped pulses results in the chirp cancellation and spectral compression, and a temporal delay between these two pulses leads to the frequency shift of the SFG-signal, according to the concept of the temporal lens [7]. The temporal delay between the SFG-interacting pulses in the range of $\pm$ 16 ps results in a $\pm$ 20 nm wavelength shift for the 22 times SFG-spectrally compressed signal (down to 0.12 nm at 400 nm central wavelength), corresponding to the chirp measurement of similariton in the span range of 160 nm (75 THz) at 800 nm central wavelength. Resulting experimental 3D pattern completely characterizes the generated broadband similariton, giving by projections the temporal and spectral profiles of the intensity $I(t)$ and $I(\lambda)$, and the curve $\lambda(t)$ connected with the chirp $\omega(t) = \dot{\varphi}_f(t)$ [23,24]. The processing of experimental data gives the polynomial $\dot{\phi}_f(\omega) = -15.06\times10^3\ \mathrm{fs}^3 \times \omega^2 - 81.13\times10^3\ \mathrm{fs}^2 \times \omega$ for derivative of spectral phase. Fig. 4 shows the extracted spectral phase of similariton $\phi_f(\omega)$. The circles in Fig. 4 are the measured experimental points; the dashed and solid curves are for the polynomial and parabolic fits, respectively.

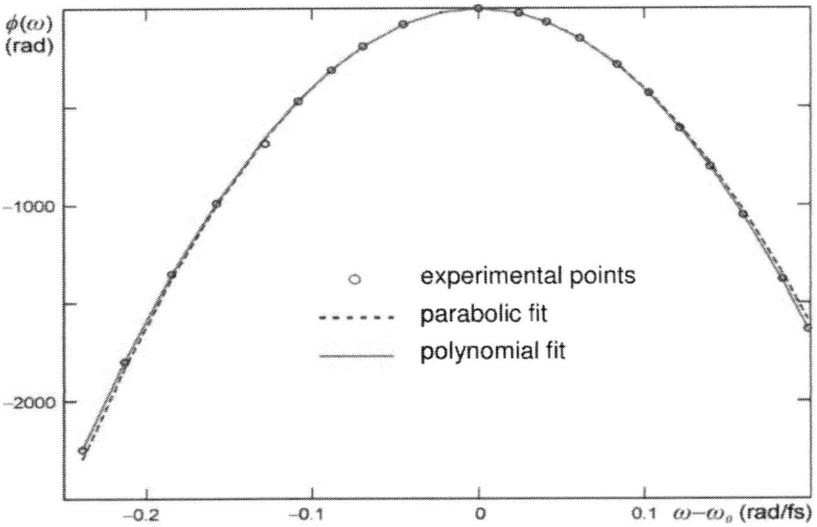

Fig. 4.
The $\phi_f(\omega)$ spectral phase of broadband similariton. The ~1% difference from the parabolic fit in the range of ~50 THz gives the range of applications for the similariton-based methods of femtosecond signal characterization.

Thus, the complete characterization of 50-THz bandwidth nonlinear-dispersive similariton states that only fiber dispersion determines its phase (chirp). The third order dispersion (TOD) of the fiber results in the same additional phase for broadband nonlinear-dispersive similariton and spectron: $\Delta\varphi(f,t) \approx \Delta\phi(f,\omega)|_{\omega=Ct} = -(\beta_3 f/6)\omega^3\big|_{\omega=Ct} \approx -\beta_3(6\beta_2^3 f^2)^{-1}t^3$ (β_3 is the TOD coefficient). The ~1% accuracy of the parabolic fit for the spectral phase of the 50-THz bandwidth similariton gives the range of applications for aberration-free similariton-based STI and SI.

4. Conclusion

We develop and implement a similariton-based self-referencing method of SI for the complete characterization of femtosecond signal. The method is based on the similariton generation from the part of signal and its use as a reference for the interference with the signal in the spectrometer. Therefore, the method of similariton-based SI combines the advantage of the simple principle and configuration with the self-referencing performance. We experiment the similariton-based method of SI in comparison with the measurements carried out with the prototype of femtosecond oscilloscope based on the STI in a similariton-induced temporal lens. Our comparative study, carried out together with theoretical check and autocorrelation measurements, evidences the quantitative accordance and high precision of both the similariton-referencing methods of SI and STI for accurate femtosecond scale temporal measurements. The similariton-based STI has the advantage of direct pulse measurement leading to the development of a femtosecond optical oscilloscope, but it does not give the phase information without additional interferometric measurement. The novel method of similariton-based SI, with a rather simple setup and self-referencing performance, provides the complete (amplitude and phase), high-resolution characterization of femtosecond signal.

Acknowledgments

The work was carried out within the framework of the collaborative programme project IE007 of CNRS, France – SCS, Armenia.

References

1. D.J.Kane, R.Trebino "Single-shot measurement of the intensity and phase of an arbitrary ultrashort pulse by using frequency-resolved optical gating" Opt.Lett.**18**, 823 (1993).
2. J.Piasecki, B.Colombeau, M.Vampouille, C.Froehly, J.A.Arnaud "Nouvelle méthode de mesure de la réponse impulsionnelle des fibres optiques" Appl. Opt. **19,** 3749 (1980).
   F.Reynaud, F.Salin, A.Barthélémy "Measurement of phase shifts introduced by nonlinear optical phenomena on subpicosecond pulses" Opt. Lett. **14**, 275 (1989).
3. C.Iaconis, I.A.Walmsley "Spectral phase interferometry for direct electric-field reconstruction of ultrashort optical pulses" Opt.Lett. **23**, 792 (1998).
4. V.Messager, F.Louradour, C.Froehly, A.Barthélémy "Coherent measurement of short laser pulses based on spectral interferometry resolved in time" Opt. Lett. **28**, 743 (2003).
   M.Lelek, F.Louradour, A.Barthélémy, C.Froehly, T.Mansourian, L.Mouradian, J.-P.Chambaret, G.Chériaux, B.Mercier "Two-dimensional spectral shearing interferometry resolved in time for ultrashort optical pulse characterization" J. Opt. Soc. Am. B **25**, A17 (2008).
5. L.Kh.Mouradian, F.Louradour, V.Messager, A.Barthélémy, C.Froehly "Spectrotemporal imaging of femtosecond events" IEEE J. Quantum Electron. **36**, 795 (2000).
6. M.A.Foster, R.Salem, D.F.Geraghty, A.C.Turner-Foster, M.Lipson, A.L.Gaeta "Silicon-chip based ultrafast optical oscilloscope" Nature **456**, 81 (2008).
7. T.Mansuryan, A.Zeytunyan, M.Kalashyan, G.Yesayan, L.Mouradian, F.Louradour, A.Barthélémy "Parabolic temporal lensing and spectrotemporal imaging: a femtosecond optical oscilloscope" J. Opt. Soc. Am. B **25**, A101 (2008).
8. J.M.Dudley, C.Finot, D.J.Richardson, G.Millot "Self-similarity and scaling phenomena in nonlinear ultrafast optics" Nature Physics **3**, 597 (2007).
9. C.Finot, J.M.Dudley, B.Kibler, D.J.Richardson, G.Millot "Optical parabolic pulse generation and applications" IEEE J. Quantum Electron. **45**, 1482 (2009).
10. D.Anderson, M.Desaix, M.Karlson, M.Lisak, M.L.Quiroga-Teixeiro "Wave-breaking-free pulses in non-linear optical fibers" J. Opt. Soc. Am. B **10**, 1185 (1993).
11. M.E.Fermann, V.I.Kruglov, B.C.Thomsen, J.M.Dudley, J.D.Harvey "Self-similar propagation and amplification of parabolic pulses in optical fibers" Phys. Rev. Lett. **84**, 6010 (2000).
12. V.I. Kruglov, A.C. Peacock, J.M. Dudley, J.D. Harvey "Self-similar propagation of high-power parabolic pulses in optical fiber amplifiers" Opt. Lett. **25**, 1753 (2000).
13. V.I.Kruglov, A.C.Peacock, J.D.Harvey, J.M.Dudley "Self-similar propagation of parabolic pulses in normal-dispersion fiber amplifiers" J. Opt. Soc. Am. B **19**, 461 (2002).
14. C.Finot, G.Millot, C.Billet, J.M.Dudley "Experimental generation of parabolic pulses via Raman amplification in optical fiber" Opt. Express **11**, 1547 (2003).
15. F.Ö.Ilday, J.R.Buckley, W.G.Clark, F.W.Wise "Self-similar evolution of parabolic pulses in a laser" Phys.Rev.Lett. **92**, 213902 (2004).
16. T.Hirooka, M.Nakazawa "Parabolic pulse generation by use of a dispersion-decreasing fiber with normal group-velocity dispersion" Opt. Lett. **29**, 498 (2004).
17. C.Finot, B.Barviau, G.Millot, A.Guryanov, A.Sysoliatin, S.Wabnitz "Parabolic pulse generation with active or passive dispersion decreasing optical fibers" Opt. Express **15**, 15824 (2007).
18. A.Zeytunyan, G.Yesayan, L.Mouradian, P.Kockaert, P.Emplit, F.Louradour, A.Barthélémy "Nonlinear-dispersive similariton of passive fiber" J.Europ. Opt. Soc. Rap.Public. **4**, 09009 (2009).
19. V.I.Kruglov, D.Méchin, J.D.Harvey "High compression of similariton pulses under the influence of higher-order effects" J. Opt. Soc. Am. B **24**, 833 (2007).
20. K.Palanjyan, A.Muradyan, A.Zeytunyan, G.Yesayan, L.Mouradian "Pulse compression down to 17 femtoseconds by generating broadband similariton" Proc. SPIE **7998**, 79980N (2010).
21. C.Finot, G.Millot "Synthesis of optical pulses by use of similaritons" Opt. Express **12**, 5104 (2004).
22. A.Zeytunyan, G.Yesayan, L.Mouradian, F.Louradour, A.Barthélémy "Applications of similariton in ultrafast optics: spectral interferometry and spectrotemporal imaging" *Frontiers in Optics*, OSA Tech. Digest, FWI5 (2009).
23. A.Zeytunyan, A.Muradyan, G.Yesayan, L.Mouradian "Broadband similariton" Laser Phys. **20**, 1729 (2010).
24. A.Zeytunyan, A.Muradyan, G.Yesayan, L.Mouradian, F.Louradour, A.Barthélémy "Generation of broadband similaritons for complete characterization of femtosecond pulses" Opt. Commun. **284**, 3742 (2011).
25. W.H.Renninger, A.Chong, F.W.Wise "Self-similar pulse evolution in an all-normal-dispersion laser" Phys. Rev. A **82**, 021805(R) (2010).
26. D.T.Reid, P.Loza-Alvarez, C.T.A.Brown, T.Beddard, W.Sibbett "Amplitude and phase measurement of mid-infrared femtosecond pulses by using cross-correlation frequency-resolved optical gating" Opt. Lett. **25**, 1478 (2000).

Nonlinear superluminality in cold atoms confined in hollow core photonic crystal fiber

Sh Petrosyan[1] and Yu Malakyan[1,2]
[1]Institute for Physical Research of Armenian National Academy of Sciences, Ashtarak 2, 0203, Armenia
[2]Centre of Strong Field Physics, Yerevan State University, 1 A. Manukian St., Yerevan, 0025, Armenia

Abstract. We propose a scheme for superluminal propagation of a linearly polarized smooth light pulse in cold atoms trapped inside a hollow-core fiber with intensity-dependent negative group velocity. The tight confinement of the atoms inside the fiber core prevents atom-wall collisions, while the Doppler width is negligible as compared to atomic decay rate. We show that with initially prepared Zeeman coherence, the absorption of light is smaller enough to be ignored, while due to nonlinear dependence on laser intensity more intense parts of the pulse travel slower leading to the formation of the extremely sharp trailing edge and to the lengthening of the leading edge of the pulse. Such a behavior is contrast to the case of nonlinear propagation of the pulse in a medium with normal dispersion where a shock wave is generated. It is highly important that with 10^4 atoms inside the fiber the large nonlinearities in the medium refractive index can be achieved at extremely low light levels.

1. Introduction

Propagation of light pulses through the media with a normal dispersion is adequately described by the group velocity, which in this case is the velocity of signal transfer. In the region of anomalous dispersion, for example near an absorption line or nonlinear gain [1], the group velocity of light becomes faster than vacuum speed of light c or even negative, thus losing its role of signal velocity. However, it has been shown in Ref. [2] that the group velocity is a useful mathematical concept to describe the propagation of sufficiently smooths pulses, such as Gaussian wave packets, in media with anomalous dispersion. Since then a number of theoretical [3-8] and experimental [9-11] works have been published reporting superluminal propagation of a probe field in different schemes of anomalous dispersion in gaseous and solid media.

The results of these papers are based on linear response of the medium. Similar results have been obtained, when a weak probe beam propagates with an effective index of refraction induced by a nonlinear interaction of the medium with a strong pump field. Under these conditions both the probe light ultraslow propagation [12-14] at the electromagnetically induced transparency (EIT) [15] and the negative "group-delay" propagation associated with electromagnetically induced opacity [16, 17] have been recently observed experimentally. However, from the point of view of the dependence of medium optical properties on probe field intensity, these systems may be clearly considered as effectively linear optical media with an extraordinary large dispersion.

Intriguing questions arise whether the nonlinear superluminality occurs in a medium, the refractive index of which is modified by the propagating light pulse itself. Here the main requirement is to observe the pulse superluminal propagation in a medium with an intensity-dependent refractive index, but without accompanying other nonlinear processes, which lead to pulse reshaping as well. The simplest system displaying these properties is a three-level Λ atom with ground-state Zeeman sublevels, which interacts with the right (RCP) and left (LCP) circularly polarized components of the optical light pulse in the presence of weak magnetic field applied along the direction of light-pulse propagation (Fig1). This system has been first analyzed for room-temperature Rb vapor [18] and it has been shown that the relaxations due to Doppler broadening and the collisions between warm alkali-metal atoms impose severe limitations on the field intensity, which represents a fundamental obstacle for observation of the proposed effect. In this paper, we propose to use hollow core photonic-crystal fibers [19], where these obstacles are overcome. The advantage of hollow core fibers is that the strong confinement of cold atoms prevents atom-wall collisions inside the fiber core, while the Doppler broadening is negligibly small as compared to the linewidth of atomic transitions, so that the effect of the nonlinear superluminality can be observed for small probe fields thus making the hollow core scheme feasible for experimental realization.

2. Basic equations

We consider a near-resonant interaction of a linearly polarized probe field E of frequency ω and wave vector k parallel to the static longitudinal magnetic field B. Its two circularly left- and right-polarized components $E_{1,2}$ act on the atomic transitions 1→3 and 2→3, with the detuning $\Delta_{1,2} = \omega - \omega_0 \mp \Delta$, where ω_0 is the frequency of the unshifted atomic resonance and $\Delta = \mu_B B/\hbar$ is the Zeeman shift, which is induced by the magnetic field for the ground-state sublevels 2 and 1 with magnetic quantum numbers ±1. Here g is the ground-state gyromagnetic factor and μ_B is Bohr's magneton.

Due to the nearly maximal Zeeman coherence induced by the field, large nonlinear susceptibilities are created at the resonance frequency, analogous to those observed in EIT experiments (for relationship between EIT and nonlinear magneto-optics see Ref. [14]). The initial preparation of Zeeman coherence is necessary to avoid the pulse distortion due to the energy loss needed to initiate the EIT. The Zeeman coherence suppresses the absorption of the medium, while it creates a highly anomalous and intensity-dependent dispersion in the vicinity of the resonance line center. This results in a negative nonlinear group velocity at the values of light intensity, which are many times smaller than the saturation intensity of the corresponding atomic transitions.

The Hamiltonian of the system is written as

$$H_{\text{int}} = \hbar\Delta_1\sigma_{11} + \hbar\Delta_2\sigma_{22} - \hbar\left(\Omega_1\sigma_{31} + \Omega_2\sigma_{32} + H.c.\right) \tag{1}$$

Here $\sigma_{ij}=|i\rangle\langle j|$ are atomic operators. The Rabi frequencies of field polarization components are defined as $\Omega_{1,2}=\mu E_{1,2}/\hbar$, where the dipole electrical moments of the transitions $|1\rangle \to |3\rangle$ and $|2\rangle \to |3\rangle$ are taken equal: $\mu_{13}=\mu_{23}=\mu$.

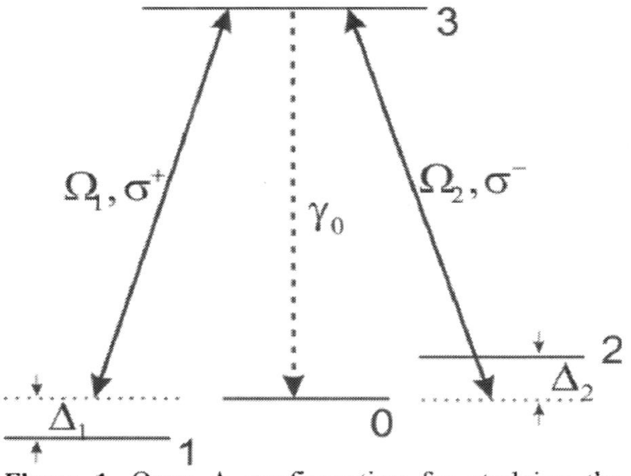

Figure 1. Open Λ configuration for studying the nonlinear superluminal propagation.

$$\frac{\partial}{\partial t}\rho = -\frac{i}{\hbar}\left[H_{\text{int}}, \rho\right] + \Lambda\rho \tag{2}$$

The matrix $\Lambda\rho$ accounts for all atomic relaxations

$$\Lambda\rho = \begin{pmatrix} \gamma\rho_{33} & -\gamma_c\rho_{12} & -\Gamma\rho_{13} \\ -\gamma_c\rho_{21} & \gamma\rho_{33} & -\Gamma\rho_{23} \\ -\Gamma\rho_{31} & -\Gamma\rho_{32} & -(\gamma_0 + 2\gamma)\rho_{33} \end{pmatrix} \tag{3}$$

where γ and γ_0 are spontaneous decay rates of the upper state 3 on the transitions $|1\rangle \rightarrow |3\rangle$, $|2\rangle \rightarrow |3\rangle$ and $|0\rangle \rightarrow |3\rangle$, respectively, γ_c is a ground-state decoherence rate and $\Gamma=(2\gamma+\gamma_0)/2$.
The evolution of the slowly varying field amplitudes $E_{1,2}$ along the z axis is determined by Maxwell's equations

$$\left(\frac{\partial}{\partial z} + \frac{1}{c}\frac{\partial}{\partial t}\right)E_{1,2}(z,t) = 2\pi i\frac{\omega}{c}P_{1,2}(z,t) \tag{4}$$

where $P_{1,2}(z,t)$ are the field induced polarizations

$$P_i(z,t) = -N\left\langle\frac{\partial H_{\text{int}}}{\partial E_i^*}\right\rangle = N\mu\rho_{3i} \tag{5}$$

with N the atomic number density. As has been shown in [18], $P_i(z,t)$ is expressed through field amplitudes $E_i(z,t)$ via Tailor series

$$P_i(z,t) = \chi_i(\omega,z)E_i(z,t) + i\frac{\partial\chi_i}{\partial\omega}\frac{\partial E_i}{\partial t} - \frac{1}{2}\frac{\partial^2\chi_i}{\partial\omega^2}\frac{\partial^2 E_i}{\partial t^2} + \ldots \tag{6}$$

where

$$\chi_i(\omega,z) = N\mu^2\rho_{3i}/\hbar\Omega_i \tag{7}$$

The third term in Eq.(6) describes the group velocity dispersion, which is responsible for the pulse spreading during its propagation through the medium. In what follows we neglect the pulse distortion due to this and higher dispersion terms in order to observe the pure pulse self-steepening without pulse spreading. This approximation constrains the sample length to a value for which the pulse broadening caused by the self-steepening does not exceed the medium absorption line width Γ, more detailed discussion of this point can be found in [18], where it is shown that even in the presence of Doppler broadening the upper bound of sample length, for reasonable values of the parameters, is few hundreds of cm, while in our case L does not exceed several centimeters.
Since in this work we restrict our attention to the superluminal propagation of the pulse, so we will consider equation only for the dimensionless intensities defined as: $I_i = \Omega_i^2/\Gamma\Delta$. By substituting Eqs.(5-7) into Eq.(4), we obtain the following equation for the intensities.

$$\left(\frac{\partial}{\partial z} + \frac{1}{v_i}\frac{\partial}{\partial t}\right)I_i(z,t) = -\alpha_i I_i(z,t) \tag{8}$$

where the nonlinear absorption coefficients α_i and group velocities v_i are given by

$$\alpha_i = \frac{4\pi\omega}{c}\operatorname{Im}\chi_i \tag{9}$$

$$v_i = \frac{c}{n_{gi}} = c\left[1 + 2\pi\omega\operatorname{Re}\frac{\partial\chi}{\partial\omega}\right]^{-1} \tag{10}$$

Below we omit the subscripts 1 and 2, as the absorption and group velocities of the two components are the same.

We solve Eq. (8) assuming that the initial laser beam is a Gaussian pulse with duration T

$$I(0,t) = I_0 \exp\left[(-t/T)^2\right] \tag{11}$$

We suppose that the incident pulse does not lose energy for preparation of EIT and its reshaping in the medium is caused solely by the nonlinear dispersion. Besides, as has been mentioned above, the tight confinement of cold atoms in hollow core fiber prevents atom-wall collisions and, hence the ground states decoherence rate γ_c is taken zero. Then the steady-state solution of Eq.(2) is given as

$$\begin{aligned}
&\operatorname{Im}\frac{\rho_{3i}}{\Omega_i} = \Gamma\Delta^2\delta\rho / D \\
&\operatorname{Re}\frac{\rho_{31}}{\Omega_1} = -\Delta\left(\varepsilon\Delta + \Omega^2\right)\delta\rho / D \\
&\operatorname{Re}\frac{\rho_{32}}{\Omega_2} = -\Delta\left(\varepsilon\Delta - \Omega^2\right)\delta\rho / D \\
&D = \Delta^2\left(\varepsilon^2 + \Gamma^2\right) + \Omega^4
\end{aligned} \tag{12}$$

Where $\Omega^2=|\Omega_1|^2=|\Omega_2|^2$. This solution is obtained under the assumptions $\Delta<<\Gamma$ and $\Gamma^2>> \Omega_0^2$. We suppose that the ground atomic states are equally populated and are not altered during the interaction with the field, so that the population difference $\delta\rho$ between the ground Zeeman sublevels and the upper state |3> is equal 1/3.
From the Eqs(9), (10) and (12) for the group index and the absorption coefficient we obtain

$$n_g = 1 - \frac{2\pi\omega N\mu^2\delta\rho}{\hbar\Gamma^2}\frac{1}{1+I^2} \tag{13}$$

$$\alpha = \frac{4\pi\omega N\mu^2\delta\rho}{\hbar c\Gamma}\frac{1}{1+I^2} \tag{14}$$

In our simulations for the model of the atomic system we have chosen parameters corresponding to the D_1 absorption line of ^{87}Rb. The total decay rate of the upper level is $\Gamma=\pi\times5\times10^6$ s^{-1}, the wavelength of optical transitions λ=800 nm, for the hollow core fiber length L~3 cm, diameter ~10μm and number of atoms ~10^4 the atomic density is ~10^9 cm^{-3}. For these parameters, $\alpha L<0.3$ at $I=0$, so that the absorption is negligible, while the group index n_g~$-40/(1+I^2)$ ensuring sensitive dependence of group velocity on light intensity. The external magnetic field is taken B=30mG, which corresponds to Δ=0.002Γ for the ground state of rubidium.

A negative velocity can be understood by comparing the times it would take for identical pulses of light to cover the distance L in a vacuum (traveling at velocity c) and in a superluminal medium (traveling at velocity v). The difference in transit times $\Delta T = L/v_g - L/c$ is a negative quantity, if the velocity is superluminal. If v_g has a negative value, then ΔT can become sufficiently negative that the peak of the pulse emerges from the medium at an instant earlier than when the peak of the pulse enters. Moreover, with nonlinear dependence of v_g on laser intensity, as is the case here, the more intense parts of the pulse travel slower leading to controllable shaping of output pulse.

Fig.2 shows the calculated output pulse (red) traversed through a distance $z\Gamma/c$~0.2 in the rubidium vapor. For comparison we depict also the input Gaussian pulse (in blue), which propagates the same distance in vacuum with light speed c. It is seen that nonlinear propagation in an anomalous dispersive medium leads to the strong self-steepening of trailing edge of the pulse.

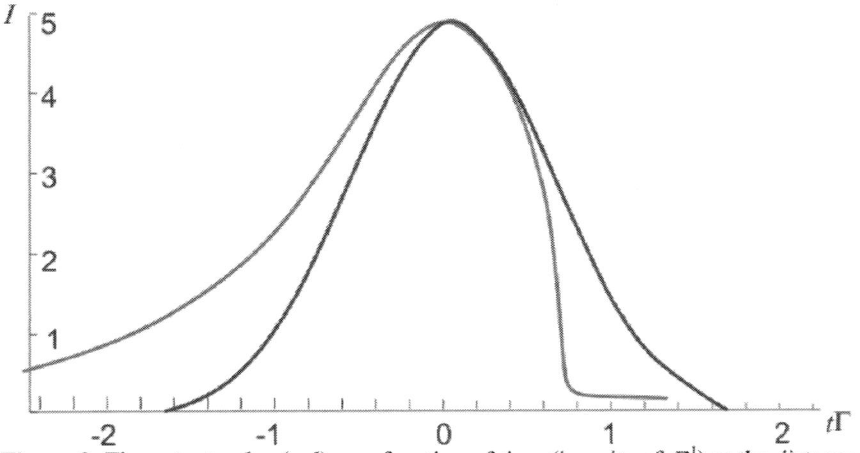

Figure 2. The output pulse (red) as a function of time (in units of Γ^{-1}) at the distance $z/(c\Gamma^{-1})$=0.7 for I_0=5 corresponding to the intensity ~ 0.2mW/cm^2. The input pulse propagating at speed c through the same distance in vacuum is shown in blue

Meanwhile, the parameters for numerical simulation are chosen such that the group velocity at maximal intensity is c, so that the peak points of real and unperturbed pulses coincide.

3. Summary

In this paper we have analyzed the superluminal propagation of a light pulse in a nonlinear medium and showed that necessary conditions for these effects to be observable are realized in a three-level λ-system interacting with a linearly polarized laser beam in the presence of a static magnetic field. We have demonstrated that similar to the linear case the group velocity for smooth pulses is physically meaningful even when it is negative. But, it cannot be interpreted as a velocity of real signal, which could propagate no faster than light without violating the Einstein causality [21]. It is highly important that in our case the nonlinearity of the refractive index of the medium arises in a very low power regime, when all other nonlinear processes are negligible. We have shown that this regime is realized in cold atoms confined in hollow-core photonic crystal fiber, where a light pulse experiences a strong pulse shaping with formation of an extremely sharp trailing edge and lengthening of the leading edge of the pulse. Such a behavior is opposite to the case of nonlinear propagation of the pulse in a medium with normal dispersion [21], where a shock wave is generated.

Acknowledgments

This research has been conducted in the scope of the International Associated Laboratory (CNRS-France SCS-Armenia) IRMAS. We also acknowledge the support from the Scientific Research Foundation of the Government of the Republic of Armenia.

Referances

[1] Marangos J 2000 *Nature* **406** 243-244.
[2] Garrett C G B and Mc Cumber D E 1970 *Phys. Rev. A* **1** 305.
[3] Chu S and Wong S 1982 *Phys. Rev. Lett.* **48** 738
[4] Segard B and Macke B 1985 *Phys. Rev. Lett.* A **109** 213
[5] Steinberg A M, Kwiat P G and Chiao R Y 1993 *Phys. Rev. Lett.* **71** 708
[6] Steinberg A M and Chiao R Y 1994 *Phys. Rev. A* **49** 2071
[7] Bolda E L, Garrison J C and Chiao R Y 1994 *Phys. Rev. A* **49** 2938
[8] Dogariu A, Kuzmich A and Wang L J 2001 *Phys. Rev. A* **63** 053806
[9] Wang L J, Kuzmich A and Dogariu A 2000 Nature (London) **406** 277
[10]Bigelow M S, Lepeshkin N N, Boyd R W 2003 Science **301** 200
[11]Gehring G M et al, 2006 Science **312** 895
[12]Hau L V, Harris S E, Dutton Z and Behroozi C H 1999 *Nature London* **397** 594
[13]Kash M M et al. 1999 *Phys. Rev. Lett.* **82** 5229
[14]Budker D, Kimball D F, Pochester S M and Yashchuk V V 1999 *Phys. Rev. Lett.* **83** 1767
[15]Harris S E 1997 *Phys. Today* **50** (7) 36
[16]Akulshin A M, Barreiro S and Lezama A 1998 *Phys. Rev. A* **57** 2996
[17]Lezama A, Barreiro S and Akulshin A M 1999 *Phys. Rev. A* **59** 4732
[18]Ghulghazaryan R and Malakyan Yu 2003 *Phys. Rev. A* **67** 063806
[19]Russel P St J 2006 J.Lightwave Technol. **24**, 479
[20]Budker D, Kimball D F, Pochester S M, and Yashchuk V V, 1999 Phys. Rev. Lett. **83**, 1767
[21]Brillouin L, *Wave Propagation and Group Velocity, Academic* 1960, New York
[22]Grischkowsky D, Courtens E, and Armstrong J A 1973 Phys. Rev.Lett. **31** 422

Dispersion surfaces and light propagation in homogeneous dielectric-magnetic uniaxial medium

M. S. Rafayelyan[1] and A. H. Gevorgyan
Yerevan State University, 1 Alex Manoogyan St., Yerevan, 0025, Armenia

E-mail: mrafayelyan@gmail.com

Abstract. We investigated light propagation in a homogeneous medium having double anisotropy (i.e. with anisotropy of both dielectric and magnetic permittivities) and an arbitrary orientation of its optical axis in the plane of incidence. We investigated wave surfaces for the considered medium. We showed that only some groups of surfaces can arise. We investigated the conditions for total independent of polarization reflection and the conditions for total transmission, as well as possibilities of the subject system of serving as: an omnidirectional reflector, a beam splitter and a phase retarder.

1. Introduction

Metamaterials are artificial composites containing sub-long-wave structures and manifesting such linear and non-linear optical properties as: negative refraction, reverse Doppler effect, electromagnetic wave energy propagation in the direction opposite to the wave vector, etc. [1, 2]. They also have other extraordinary applications, such as: perfect lenses [3], invisible cloaks [4, 5], perfect absorbers [6], etc. Though the easiest way of having negative refraction is the application of the isotropic metamaterials, this refraction can also be observed in anisotropic metamaterials. Moreover, in the last general case it is not necessary to require that all the elements of the dielectric and magnetic permittivities be negative [7]. Recently the investigation of such anisotropic metamaterials has been of great interest [8-24]. However, the literature mainly considers the cases when the dielectric and magnetic tensor principal elements are either parallel or perpendicular to the boundaries of the system. In [10], the peculiarities of super-light propagation in an anisotropic metamaterials were investigated, for an arbitrary orientation of the optical axis in the incidence plane. In [13], the case of the omnidirectional total transmission and possibility of existence of a negative Brewster's angle – at the boundary, *isotropic medium-anisotropic medium* – for an arbitrary orientation of the optical axis when $\hat{\mu}=\hat{I}$ ($\hat{\mu}$ is the magnetic permittivity tensor, $\hat{I}$ is the unit matrix) are investigated. In [19], the possibilities of total reflection at the boundary, *isotropic medium-anisotropic medium,* are investigated and a condition for total reflection is obtained. In [20], the possibilities of total negative reflection at the boundary, *isotropic medium-anisotropic medium,* are investigated for an arbitrary orientation of the optical axis, again, for $\hat{\mu}=\hat{I}$. In [16], the dispersion equations for anisotropic metamaterials are classified.

In the present paper, the dispersion surface peculiarities and dependences of the dispersion curves on the optical axis orientation are investigated. We also investigated the conditions for independent of polarization total reflection and the conditions of the total transmission, as well as the possibilities of use of the subject system as: an omnidirectional reflector, a beam splitter and a phase retarder.

Published under licence by IOP Publishing Ltd

2. Dispersion Surfaces

Let us consider peculiarities of the dispersion surfaces of an anisotropic medium having arbitrary orientated optical axis in the incidence plane. We assume that the dielectric and magnetic permittivity local tensors of the medium can both be diagonalized, and the tensors $\hat{\varepsilon}_0$ and $\hat{\mu}_0$ have the following form in the corresponding frame:

$$\hat{\varepsilon}_0 = \begin{pmatrix} \varepsilon_1 & 0 & 0 \\ 0 & \varepsilon_2 & 0 \\ 0 & 0 & \varepsilon_3 \end{pmatrix}, \ \hat{\mu}_0 = \begin{pmatrix} \mu_1 & 0 & 0 \\ 0 & \mu_2 & 0 \\ 0 & 0 & \mu_3 \end{pmatrix}, \tag{1}$$

i.e. it is assumed that the principal axes of the dielectric and magnetic permittivity tensors coincide. Below we also assume that the medium is uniaxis, i.e. $\varepsilon_2=\varepsilon_3\neq\varepsilon_1$ and $\mu_2=\mu_3\neq\mu_1$. In the laboratory frame $\hat{\varepsilon}$ and $\hat{\mu}$ have the following form:

$$\hat{\varepsilon} = \hat{T}[y,\phi]\hat{\varepsilon}_0\hat{T}[y,\phi]^{-1}, \ \hat{\mu} = \hat{T}[y,\phi]\hat{\mu}_0\hat{T}[y,\phi]^{-1}, \tag{2}$$

where $\hat{T}[y,\phi]$ is the rotation matrix for the y axis, at ϕ angle.

A plane electromagnetic wave of ω frequency and $\vec{k}$ wave vector propagates in the mentioned medium. From Maxwell's equations, we obtain the following dispersion equations for refractive index:

$$\left(n^2(1-\delta_\varepsilon)+\mu_m\,\varepsilon_m\left(\delta_\varepsilon^2-1\right)\left(1-\delta_\mu\right)+\delta_\varepsilon\xi\right)\left(n^2(1-\delta_\mu)+\mu_m\,\varepsilon_m\left(\delta_\mu^2-1\right)\left(1-\delta_\varepsilon\right)+\delta_\mu\xi\right)=0, \tag{3}$$

where $n^2=n_x^2+n_y^2+n_z^2$, $\xi=2\left(n_x\cos\phi+n_z\sin\phi\right)^2$, $\varepsilon_m=\left(\varepsilon_1+\varepsilon_2\right)/2$, $\mu_m=\left(\mu_1+\mu_2\right)/2$,

$\delta_\varepsilon=\left(\varepsilon_1-\varepsilon_2\right)/\left(\varepsilon_1+\varepsilon_2\right)$, $\delta_\mu=\left(\mu_1-\mu_2\right)/\left(\mu_1+\mu_2\right)$, $n_x=\frac{\lambda}{2\pi}k_x$, $n_y=\frac{\lambda}{2\pi}k_y$, $n_z=\frac{\lambda}{2\pi}k_z$, k_x, k_y, k_z - are the wave vector components, and λ is the wavelength in the medium. The dispersion equation for plane waves in such a material can be factorized into two terms: One – dispersion equation for electric modes and the other dispersion equation for the magnetic modes.

Let us reduce the dispersion equation of electric modes to the canonic form. To do it we represent n_x, n_y and n_z in the following forms:

$$n_x=a\,n_1+n_2+n_3,\ n_y=n_1+b\,n_2,\ n_z=n_1+n_2+c\,n_3,$$

where

$$a=1,\ b=2\frac{1+\delta_\varepsilon\sin 2\phi}{\delta_\varepsilon-1},\ c=\frac{1+\delta_\varepsilon\cos 2\phi+\delta_\varepsilon\sin 2\phi}{-1+\delta_\varepsilon\cos 2\phi-\delta_\varepsilon\sin 2\phi} \tag{4}$$

Then the dispersion equation for electric modes takes the form:

$$\frac{n_1^2}{\lambda_1}+\frac{n_2^2}{\lambda_2}+\frac{n_3^2}{\lambda_3}=1, \tag{5}$$

where

$$\begin{gathered}\lambda_1=\varepsilon_m\mu_m\left(3-\delta_\varepsilon+2\delta_\varepsilon\sin 2\phi\right)\left(\delta_\varepsilon^2-1\right)\left(1-\delta_\mu\right),\ \lambda_2=2\varepsilon_m\mu_m\left(1+\delta_\varepsilon\sin 2\phi\right)\left(1-\delta_\mu\right),\\ \lambda_3=2\varepsilon_m\mu_m\left(1+\delta_\varepsilon\sin 2\phi\right)\left(3-\delta_\varepsilon+2\delta_\varepsilon\sin 2\phi\right)\left(1+\delta_\varepsilon\right)\left(1-\delta_\mu\right).\end{gathered} \tag{6}$$

Dispersion equation for magnetic modes also has the same form, but in this case λ_1, λ_2, λ_3, n_1, n_2, n_3 are obtained by the interchanges: $\delta_\varepsilon\to\delta_\mu$ and $\mu_m\to\varepsilon_m$, in (4) and (6).

Dispersion surfaces characterize the dependence of the electromagnetic wave refraction in the medium on the light propagation direction. Electromagnetic plane waves propagating inside the material, depending on the values of λ_1, λ_2 and λ_3, can exhibit dispersion surfaces in the form of ellipsoids of revolution, hyperboloids of one sheet, or hyperboloids of two sheets. Furthermore, depending on the optical axis orientation, the intersections of these surfaces with the propagation plane can be circles, ellipses, hyperbolas, or straight lines. Now let us go into the details of the problem.

I. if in (5) λ_1, λ_2, λ_3 are positive, that is, when:

$f=\varepsilon_m\mu_m\left(\delta_\varepsilon-1\right)\left(\delta_\mu-1\right)>0$, $g=\left(\delta_\mu+1\right)\left(3-\delta_\mu+2\delta_\mu\sin 2\phi\right)>0$ and $h=\left(\delta_\mu-1\right)\left(1+\delta_\mu\sin 2\phi\right)<0$, the dispersion surface of electric modes is an ellipsoids of revolution, with semiaxes along the directions: $\vec{n}_1$, $\vec{n}_2$ and $\vec{n}_3$, i.e. along the directions:

$$\vec{n}_1 = (n_x\hat{x} + n_z\hat{z})(1+\delta_\varepsilon \sin 2\phi) + (n_x\hat{x} - n_z\hat{z})\delta_\varepsilon \cos 2\phi + n_y\hat{y}(1-\delta_\varepsilon)$$
$$\vec{n}_2 = (n_x\hat{x} + n_z\hat{z} - 2n_y\hat{y})(1+\delta_\varepsilon \sin 2\phi) + (n_x\hat{x} - n_z\hat{z})\delta_\varepsilon \cos 2\phi,\ \vec{n}_3 = n_x\hat{x} - n_z\hat{z} \quad (7)$$

where $\hat{x}$, $\hat{y}$ and $\hat{z}$ are the unit vectors of the x, y and z axes.

II. If $\lambda_1 < 0$, $\lambda_2 < 0$ и $\lambda_3 < 0$, i.e. for $f<0$, $g>0$ and $h<0$, the mode is evanescent.

III. If one of λ_1, λ_2, λ_3 is negative, and the others are positive, i.e. for $f<0$ and $g<0$, or $f<0$ и $h>0$, then the dispersion surface is a hyperboloids of one sheet, with semiaxes along the directions $\vec{n}_1$, $\vec{n}_2$ and $\vec{n}_3$.

IV. If one of λ_1, λ_2, λ_3 is positive, and the others are negative, i.e. for $f > 0$ and $g<0$, or for $f>0$ and $h>0$, then the dispersion surface is a hyperboloids of two sheet, with semiaxes along the directions $\vec{n}_1$, $\vec{n}_2$ and $\vec{n}_3$.

V. If $\delta_\varepsilon = 1$, the dispersion surface is a plane, and for $\delta_\varepsilon = 1$, we have for electric modes: $n_x \cos\phi + n_z \sin\phi = 0$, i.e. the dispersion surface becomes a plane. It should be noted, that for $\delta_\varepsilon = -1$, the dispersion surface has the form: $n_y^2 + (n_z \cos\phi - n_x \sin\phi)^2 = 0$. From this follows, that for $n_y = 0$, the dispersion surface is the straight line, $n_z = n_x tg\phi$. In the opposite case the mode is evanescent. Let us note, that the plane arises only for $\delta_\varepsilon = 1$, and the straight line for $\delta_\varepsilon = -1$.

On figure 1, we present (for various parameters of the medium) the possible (in the general case) pairs of dispersion surfaces (one – for the electric modes, the other for the magnetic modes). They can be defined from dispersion equation (3).

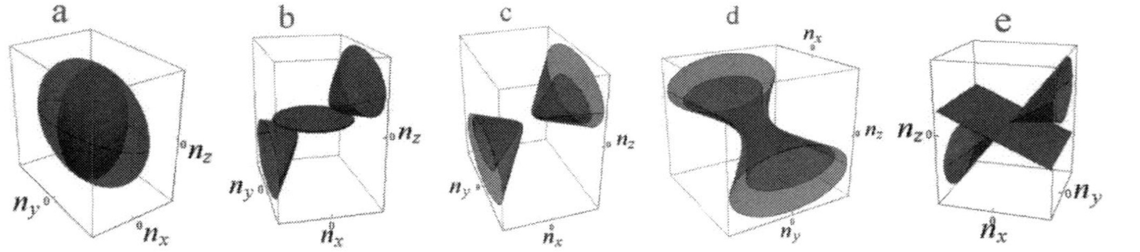

Figure 1. The dispersion surfaces for various parameters of the system.
a: $\varepsilon_1 = 2.5, \varepsilon_2 = 1.5,\ \phi = \pi/3,\ \mu_1 = 1.7, \mu_2 = 2.9$. b: $\varepsilon_1 = 3.1, \varepsilon_2 = 2.5,\ \phi = \pi/4,\ \mu_1 = -1.3, \mu_2 = 2.2$.
c: $\varepsilon_1 = 1.2, \varepsilon_2 = -1.5,\ \phi = \pi/5,\ \mu_1 = 1.3, \mu_2 = -1.1$. d: $\varepsilon_1 = -2.2, \varepsilon_2 = 3,\ \phi = \pi/4,\ \mu_1 = 1.3, \mu_2 = -2.2$.
e: $\varepsilon_1 = 2.5, \varepsilon_2 = 0,\ \phi = \pi/4,\ \mu_1 = -0.9, \mu_2 = 3.7$.

Let us note that in the general case the following pairs are impossible: an ellipsoid of revolution with a hyperboloid of one sheet, a hyperboloid of one sheet with a hyperboloid of two sheets, one evanescent mode with a hyperboloid of two sheets. It is natural, for one can show that $\lambda_1\lambda_2\lambda_3\lambda_4\lambda_5\lambda_6 > 0$, that is, only even numbers of negative λ_i can exist. Here λ_4, λ_5 and λ_6 are the corresponding coefficients for the canonical dispersion equation of magnetic modes. They are obtained by the interchanges $\delta_\varepsilon \to \delta_\mu$ and $\mu_m \to \varepsilon_m$ in the λ_1, λ_2, λ_3. Let us also note that if the dispersion surface of one of the modes is a plane, then the other is either a conical surface (turning either into a plane or a straight line in particular cases) presented in figure 1e, or an evanescent. And if the dispersion surface of one of the modes is a straight line, then the other can be: a revolution ellipsoid; a hyperboloid of one sheet; a hyperboloid of two sheets; a straight line (figure 2).

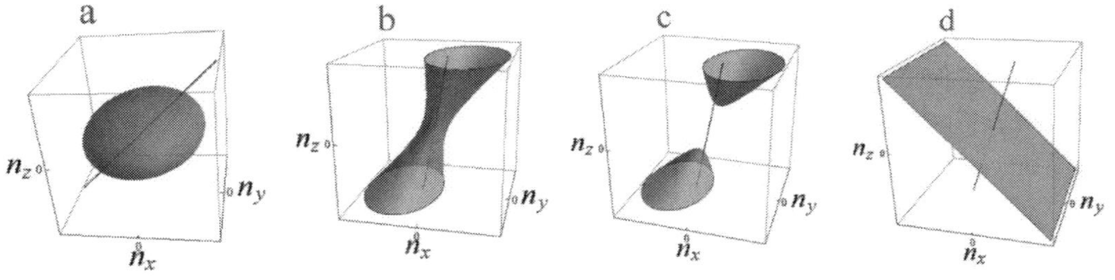

Figure 2. The dispersion surfaces for the case when one of them is a straight line.

a: $\varepsilon_1 = 0$, $\varepsilon_2 = 0.7$, $\phi = \pi/3$, $\mu_1 = 1.2$, $\mu_2 = 1.1$. b: $\varepsilon_1 = 0$, $\varepsilon_2 = 0.7$, $\phi = \pi/3$, $\mu_1 = 1.2$, $\mu_2 = -1.5$.
c: $\varepsilon_1 = 0$, $\varepsilon_2 = 0.7$, $\phi = \pi/3$, $\mu_1 = -1.2$, $\mu_2 = 1.5$. d: $\varepsilon_1 = 0$, $\varepsilon_2 = 0.7$, $\phi = \pi/3$, $\mu_1 = 1.5$, $\mu_2 = 0$.

Now we pass on to detailed analysis of the dispersion equation (3) for $n_y = 0$, that is for fixed incident plane. Let's investigate the dependences of dispersion curves on the optical axis orientation.

Figure 3a presents the dependences of dispersion curves of electric modes on the parameter, ϕ, for the same parameters of the problem for which the dispersion curve is an ellipse. If the rotation angle, ϕ, of the optical axis is equal to $\pi k/2$, then the ellipse semiaxes are directed along the directions: $\hat{n}_x$ and $\hat{n}_z$. For the other values of this angle, the ellipse semiaxes are shifted from the directions, $\hat{n}_x$ and $\hat{n}_z$.

Figure 3b presents the dependences of the dispersion curves of electric modes on the parameter, ϕ, for the same parameters of the problem for which the dispersion curve is a hyperbola.

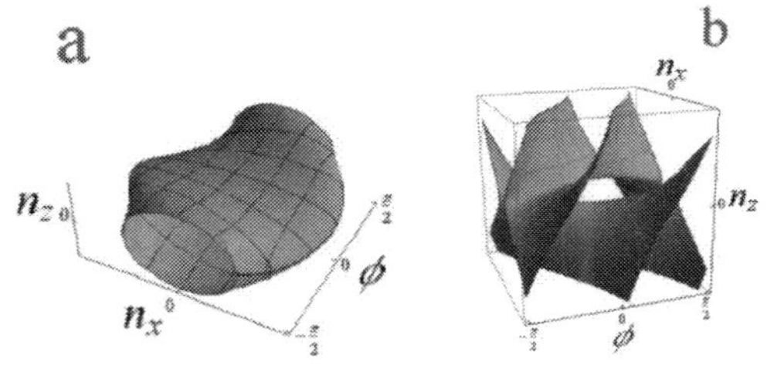

Figure 3. The dependences of dispersion curves on the optical axis orientation.
a: $\varepsilon_1 = 1.2$, $\varepsilon_2 = 3.8$, $\mu_1 = 1.5$, $\mu_2 = 1.1$.
b: $\varepsilon_1 = 1.2$, $\varepsilon_2 = -0.7$, $\mu_1 = 1.5$, $\mu_2 = 1.2$.

As it is seen from that picture, if the angle, ϕ, changes, the dispersion curves, which are hyperbolas, rotate in the plane, $n_x n_z$, and (for some specific value of that angle) these directions become asymptotes.

At the end of this section, let us note that the above considerations remain true also for the magnetic modes.

3. Omnidirectional reflection and total transmission

Let us consider light reflection and refraction on the border of an *anisotropic medium – uniaxial anisotropic metamaterial.* The medium is uniaxial and it occupies the half-space, $z \geq 0$, i.e. the medium border is parallel to the plane, *xy*, and the incident plane coincides with the plane *xz* (*xyz* is the laboratory system). An electromagnetic wave of ω frequency is incident, at the incidence angle α,

Journal of Physics: Conference Series **350** (2012) 012031 doi:10.1088/1742-6596/350/1/012031

from isotropic and homogenous medium, having the parameters, ε_l and μ_l (the dielectric and magnetic permittivities of the medium), on the subject half-space.

Now we consider the possibility of obtaining total reflection on the base of metamaterials such that it does not depend on the incidence angle and polarization and total transmission independent of polarization – for certain incidence angles.

Total reflection condition for p- polarization $\Re_p = 1$ (reflection coefficient for p- polarization) has the following form:

$$\left(1-\delta_\varepsilon^2\right)\left(n_x^2-\varepsilon_m\mu_m\left(1-\delta_\mu\right)\left(1-\delta_\varepsilon\cos 2\phi\right)\right)>0 . \tag{8}$$

The condition of total reflection, for p- polarization, for an arbitrary incident angle has the following form:

$$\begin{cases} 1-\delta_\varepsilon^2>0, \\ 0>\varepsilon_m\mu_m\left(1-\delta_\mu\right), \end{cases} \text{or} \begin{cases} 1-\delta_\varepsilon^2>0, \\ \varepsilon_m\mu_m\left(1-\delta_\mu\right)\left(1-\delta_\varepsilon\cos 2\phi\right)>\varepsilon_l\mu_l, \end{cases} \tag{9}$$

Doing the interchanges $\delta_\varepsilon \to \delta_\mu$ and $\mu_m \to \varepsilon_m$ in the (8) and (9) we will obtain analogous conditions for s- polarization, consequently, requiring the conditions for s- and p- polarizations simultaneously, we can obtain omnidirectional reflection. Our calculations show that, in particular, for $\varepsilon_1 = 1.7$, $\varepsilon_2 = 2.8$, $\mu_1 = -1.0$, $\mu_2 = -1.4$, $\phi = \pi/4$ omnidirectional reflection takes place.

For the condition:

$$n_x^2-\varepsilon_m\mu_m\left(1-\delta_\mu\right)\left(1-\delta_\varepsilon\cos 2\phi\right)=\frac{\mu_l}{\varepsilon_l}\varepsilon_m^2\left(\delta_\varepsilon^2-1\right)\cos^2\alpha , \tag{10}$$

we have $\Im_p = 1$ (transmission coefficient for p- polarization).

From (10) follows that total transmission for p- polarization is possible for certain incidence angles, which are defined from (10) (the Brewster angle for p- polarization):

$$\alpha_B^p = \sin^{-1}\left(\sqrt{\frac{\varepsilon_x\varepsilon_y - \mu_l/\varepsilon_l\,(\varepsilon_y\mu_y\cos^2\phi+\varepsilon_x\mu_y\sin^2\phi)}{(\varepsilon_x\varepsilon_y-\varepsilon_l^2)}}\right). \tag{11}$$

It is to be noted that, in contrast to the reflection on the border of two isotropic media when there is only one Brewster angle, here we have two of them (for p- and s- polarizations). Having same conditions for s- polarization, we can calculate the condition of $\alpha_B^s = \alpha_B^p = \alpha_B$:

$$\sin^2\phi = \frac{(\varepsilon_y\mu_y-\varepsilon_l\mu_l)(\varepsilon_x\varepsilon_y\mu_l^2-\mu_x\mu_y\varepsilon_l^2)}{\varepsilon_l^2\mu_x\mu_y{}^2(\varepsilon_x-\varepsilon_y)-\mu_l^2\varepsilon_x\varepsilon_y{}^2(\mu_x-\mu_y)+\varepsilon_l^2\mu_l^2(\varepsilon_y\mu_x-\varepsilon_x\mu_y)}, \tag{12}$$

Consequently, there is total transmission at the incident angle α_B, regardless of the polarization.

Now let us consider possibilities of anisotropic metamaterials as beam splitters. The refraction angles of the two forward eigen waves ($P_{z1} > 0$ and $P_{z2} > 0$, $\vec{P}_i$ is the Poynting vector of the i-th eigen wave) and split angle are defined as follows:

$$\theta_1 = \tan^{-1}\left(\frac{P_{x1}}{P_{z1}}\right),\ \theta_2 = \tan^{-1}\left(\frac{P_{x2}}{P_{z2}}\right) \text{ and } \Delta\theta = \left|\theta_2-\theta_1\right| . \tag{13}$$

Figure 4 presents the dependence of $\Delta\theta$ on the ε_1. It is to be noted that for each ε_1 $\alpha = \alpha_B$ and the ϕ angle is chosen in such a way that satisfies (12), i.e. here $\Delta\theta$ is the beam splitting angle for total transmission. As it is seen from the figure 4, $\Delta\theta \to \pi$, for the certain parameters. Consequently, it is possible to design miniature beam splitters without any intensity loss on the base of metamaterials. At $\Delta\theta = 0$, both refraction angles are the same. At this condition, the system can work as a phase retarder. From figure 4 we can see, that for certain parameters $\Delta\theta = 0$, therefore, on this parameters of the problem, the system can work as an ideal phase retarder, again, without any intensity loss.

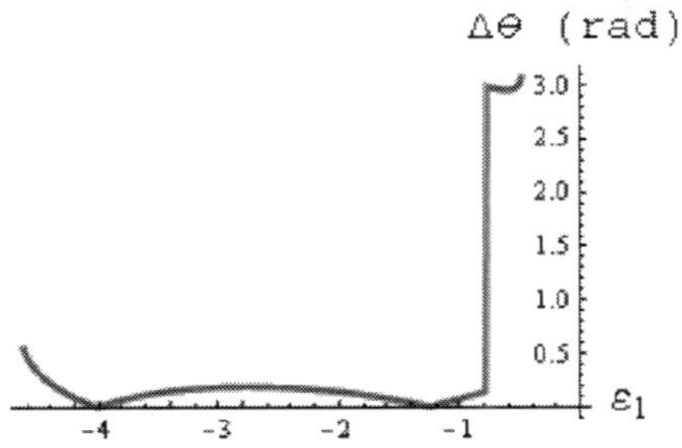

Figure 4. The dependence of split angle, $\Delta\theta$, on the dielectric permittivity, ε_1. The problem parameters are: $\varepsilon_2 = 1.1$, $\mu_1 = 1.5$, $\mu_2 = -3.4$, $\varepsilon_l = 1$, $\mu_l = 1$, $\phi = \phi_B$, $\alpha = \alpha_B$

4. Conclusion

Concluding, let us to note that we investigated dispersion surfaces for the anisotropic metamaterials with dielectric and magnetic anisotropies. We showed that for an arbitrary orientation of the optical axis in the incidence plane, some groups of surfaces can arise. The conditions for independent of polarization total reflection and the conditions for total transmission are obtained, as well as obtained the conditions of the subject system of serving as: an omnidirectional reflector, a beam splitter and a phase retarder.

5. References

[1] Veselago V G 2002 *Sov.Phys.Usp.* **45** 509
[2] Lee S H, Park C M, Seo Y M and Kim C K 2010 *Phys. Rev. B* **81** 241102
[3] Pendry J B 2000 *Phys. Rev. Let.* **85** 3966
[4] Alu A and Engheta N 2005 *Phys. Rev. E* **72** 016623
[5] Leonhardt U 2006 *Science* **312** 1777
[6] Landy N I, Sajuyigbe S, Mock J J, Smith D R.and Padillia W J 2008 *Phys. Rev. Lett.* **100** 207402
[7] Lindell I V, Tretyakov S A, Nikoskinen K I and Ilvonen S 2001 *Microw. Opt. Technol. Lett.* **31** 129
[8] Xiang Y, Dai X, Wen S 2007 *Opt. Commun.* **274** 248
[9] Shen N H, Wang H T, Tian Y 2008 *Europhys. Lett.* **83** 67007
[10] Luo H, Hu W, Shu W, Li F, Ren Z 2006 *Europhys. Lett.* **74** 1081
[11] Gevorgyan A H 2002 *Mol. Cryst. Liq. Cryst.* **382** 1-19
[12] Gevorgyan A H 2002 *Opt. Spectrosc.* **92,** 207
[13] Chen H, Xu Sh, Li J 2009 *Opt. Express.* **17** 19791
[14] Rafayelyan M S and Gevorgyan A H 2010 *Proc. SPIE*, 7998, 79980K; doi:10.1117/12.891315
[15] Rafayelyan M S and Gevorgyan A H 2010 *Modern Problems in Optics & Photonics. Proc. of Int. Adv. Res. Workshop* 44-47
[16] Depine R A, Inchaussandague M E and Lakhtakia 2006 A *J. Opt. Soc. Amer. A* **23** 949
[17] Smith D R and Schurig D 2003 *Phys. Rev. Lett.* **90** 077405
[18] Pishnyak O P and Lavrentovich O D 2006 *Appl. Phys. Lett.* **89** 25110
[19] Xiang Y, Dai X and Wen S 2007 *Opt. Commun.* **274** 248
[20] Yonghua L, Pei W, Peijun Y, Jianping X and Hai M 2005 *Opt. Commun.* **246** 429
[21] Luo H, Ren Z, Shu W and Li F 2007 *Appl. Phys. A* **87** 245
[22] Lu Y, Pei W, Yao P, Xie J and Ming H 2005 *Opt. Commun.* **246** 429
[23] Markel V A, Schotland J C 2010 *J. Opt.* **12** 01510
[24] Hu L, Chui S T 2002 *Phys. Rev. B* **66** 085108

Analysis of the moiré pattern of moving periodic structures using reciprocal vector approach

S. Rasouli[1, 2,4,*] and M. T. Tavassoly[2,3]

1- Department of Physics, Institute for Advanced Studies in Basic Sciences, Zanjan 45195-1159, Iran

2-Optics Research Center, Institute for Advanced Studies in Basic Sciences, Zanjan 45137-66731, Iran

3-Physics Department, University of Tehran, Kargar shomally Ave., Tehran 14399-66951, Iran

4-The Abdus Salam ICTP, Strada Costiera 11, 34151 Trieste, Italy

* The corresponding author's e-mail address: rasouli@iasbs.ac.ir

Abstract. Several mathematical approaches have been used to explore the moiré pattern. All of them have been considered for superimposing the static structures. In this paper, we have presented a theoretical approach to take into account the relative motion in superimposed gratings and its effect on the moiré fringe patterns. We have used a reciprocal vector approach and consequently obtained a comprehensive description of the dynamic behavior of the moiré patterns in a cinematic viewpoint. Formulations of the rotation and parallel moiré patterns of superposition of static and dynamic periodic structures have been derived in an unified form. Besides, some applications of moiré technique that have been already carried out in dynamical phenomena will be briefly reviewed.

1. Introduction

The moiré phenomenon has been known for a long time; it was already used by the Chinese in ancient times for creating an effect of dynamic patterns in silk cloth. However, modern scientific research into the moiré technique and its application started only in the second half of the 19th century. The word moiré seems to be used for the first time in scientific literature by Mulot [1].

The moiré technique has been applied widely in different fields of science and engineering, such as metrology and optical testing. It is used to study numerous static physical phenomena such as refractive index gradient [2, 3]. In addition, it has a severe potential to study dynamical phenomena such as atmospheric turbulence [4-7], wave-front sensing [8, 9], nonlinear refractive index measurements [10, 11], vibrations [12], displacements and stress [13, 14], velocity measurement [15], acceleration sensing [16], etc. Application of the moiré techniques to the displacement measurements and light deflection improves precision remarkably.

The moiré pattern can be created, for example, when two similar grids (or gratings) are overlaid at a small angle, or when they have slightly different mesh sizes. In many applications one of the superposed gratings is the image of a physical grating [3-11, 17]. When the image forming lights propagate in a perturbed medium, the image grating is distorted and the distortion is magnified by the moiré pattern.

Several mathematical approaches can be used to explore the moiré phenomenon. The classical geometric approach [18, 19] is based on a geometric study of the properties of the superposed grids, their periods and angles. By considering relations between triangles, parallelograms, or other geometric entities generated between the superposed layers, this method leads to formulae that can predict, under certain limitations, the geometric properties of the moiré patterns. Another widely used classical approach is the parametric equations method [20]; this is a purely algebraic approach, based on the equations of each family of lines in the superposition, which also yields the same basic formulae. The best adapted approach for investigating this phenomenon in superposing of periodic structures is the spectral approach, which is based on the Fourier theory. This approach has been largely developed by Isaac Amidror [21]. The same enables one to analyze properties not only in the original grids and in their superposition but also in their spectral representations. First, one considers the use of Fourier series decompositions, purely in the image domain, for representing the original repetitive structures, their superposition's and their moirés. Second, the use of the Fourier theory for the interpretation of the moiré pattern in spectral terms as an aliasing phenomenon is introduced.

Calculating the moiré pattern using Ewald's sphere of reflection, has also been suggested [22-25]. The information about the fringe structure is obtained by adding spatial frequencies vectorially. The period and orientation of the structures can be represented by a vector in the spatial frequency plane. The length of the vector is the frequency and its angle is the orientation of the grating lines. In this approach the influence of the moiré fringes profile of the grid period is not considered [23].

It must be noted that a rigorous approach to the moiré pattern is presented by R. Gevers [26]. He succeeded in performing a treatment of the moiré effect taking into account the dynamical effects wholly by using the dynamical theory of electron diffraction.

Experimental moiré techniques have been utilized to investigate a variety of dynamic problems, including production of sinusoidal phase grating in photo-refractive crystals [27, 28], study of rapidly changing stress fields and stress pulse propagation [29, 30], propagating cracks [31], studies of atmospheric turbulence[6-8], and so on. Theoretical consideration of the behavior of the moiré fringes in the dynamical phenomena is not comprehensively demonstrated. In this paper we attempt to extend the analysis of the moiré patterns of moving gratings. Formulation of the moiré pattern of superposition of moving periodic structures is derived from the reciprocal vector approach. This approach proves to be very useful in the investigation of superposed periodic layers and their moiré effects. The individual structures are characterized by spatial frequency, orientation, profile (variation of transmittance over one period), and velocity parameters. As we will see, all the specific moiré formulae are derived using the presented approach.

2. Formulation

In the moiré phenomenon, when one of the superposed linear patterns is translated on top of the other in the direction perpendicular to its lines' direction, the moiré fringes move in a direction perpendicular to the fringes much faster than the moving grating. Thus, the moiré pattern has a velocity much larger than the velocity of the individual patterns. In the following we consider this kind of superposition of the periodic structures.

For simplicity of the calculations, transmittances of the superposed structures are assumed sinusoidal functions. Gratings periods, d_1 and d_2, and their reciprocal vectors $\mathbf{k}_1$ and $\mathbf{k}_2$ are related together as (see Fig. 1)

$$\mathbf{k}_1 = \frac{2\pi}{d_1}\hat{\mathbf{k}}_1, \qquad \mathbf{k}_2 = \frac{2\pi}{d_2}\hat{\mathbf{k}}_2 \tag{1}$$

where, $\hat{\mathbf{k}}_1$ and $\hat{\mathbf{k}}_2$ are the unit vectors directed along the perpendicular to the gratings' lines, respectively. Transmittance functions of the gratings can be written as

$$T_1(\mathbf{r}) = a_1[1 + V_1 \cos(\mathbf{k}_1.\mathbf{r} + \varphi_1)], \qquad (2)$$

$$T_2(\mathbf{r}) = a_2[1 + V_2 \cos(\mathbf{k}_2.\mathbf{r} + \varphi_2)], \qquad (3)$$

where, $\mathbf{r}$ is the position vector in the plane of the gratings, V_i, a_i, and φ_i are the visibility, transmission coefficient, and initial phase of the transmittance function of the individual gratings (i = 1; 2). V_i and a_i satisfy the following constraints:

$$0 < a_i(1 + V_i) \le 1, \qquad 0 \le V_i \le 1, \qquad i = 1,2.$$

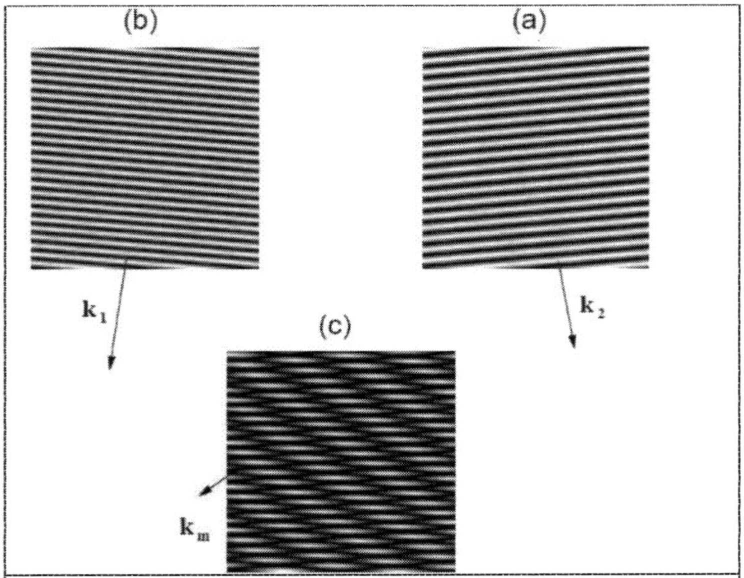

Figure 1. (a) and (b), typical sinusoidal gratings with reciprocal vectors $\mathbf{k}_1$ and $\mathbf{k}_2$ (c) corresponding multiplication moiré pattern with reciprocal vector $\mathbf{k}_m$

Now, consider the gratings to move in-plane by the velocities $\mathbf{v}_1$ and $\mathbf{v}_2$, respectively, in directions perpendicular to their lines' directions, i.e. $\mathbf{v}_1 = v_1\hat{\mathbf{k}}_1$ and $\mathbf{v}_2 = v_2\hat{\mathbf{k}}_2$. In this case the transmittance functions of the gratings can be written as:

$$T_1(\mathbf{r},t) = a_1[1 + V_1 \cos(\mathbf{k}_1.\mathbf{r} - \omega_1 t + \varphi_1)], \qquad (4)$$

$$T_2(\mathbf{r},t) = a_2[1 + V_2 \cos(\mathbf{k}_2.\mathbf{r} - \omega_2 t + \varphi_2)], \qquad (5)$$

so

$$\omega_1 = \mathbf{v_1}.\mathbf{k_1} = 2\pi \frac{v_1}{d_1}(\hat{\mathbf{k}}_1.\hat{\mathbf{k}}_1) = \frac{2\pi}{\tau_1},$$

$$\omega_2 = \mathbf{v_2}.\mathbf{k_2} = 2\pi \frac{v_2}{d_2}(\hat{\mathbf{k}}_2.\hat{\mathbf{k}}_2) = \frac{2\pi}{\tau_2},$$

where, τ_1 and τ_2, respectively, are the time required to transfer the gratings equal to their spatial periods. Now, to get the velocities of the phase translations we equate the differentials of Eqs. (4) and (5) equal to zero:

$$d(\mathbf{k_1}.\mathbf{r} - \omega_1 t + \varphi_1) = 0.$$

For constant $\mathbf{k_1}$, ω_1, and φ_1 it is equivalent to

$$\mathbf{k_1}.d\mathbf{r} - \omega_1 dt_1 = 0,$$

or

$$k_1 dr_1 - \omega_1 dt_1 = 0, \tag{6}$$

where r_1 is the component of $\mathbf{r}$ in the direction of $\mathbf{k_1}$, and $k_1 = |\mathbf{k_1}|$. Thus, the magnitude of the first grating velocity v_1, is

$$\frac{dr_1}{dt} = \frac{\omega_1}{k_1} = v_1. \tag{7}$$

By the same token, for the second grating we obtain:

$$v_2 = \frac{\omega_2}{k_2}. \tag{8}$$

The moving moiré pattern is produced by the multiplication of the transmittance functions of the moving gratings:

$$\begin{aligned} T = T_1 T_2 &= a_1 a_2 \left[1 + V_1 \cos(\mathbf{k_1}.\mathbf{r} - \omega_1 t + \varphi_1) + V_2 \cos(\mathbf{k_2}.\mathbf{r} - \omega_2 t + \varphi_2)\right] \\ &+ \frac{a_1 a_2 V_1 V_2}{2} \cos\left[(\mathbf{k_1} + \mathbf{k_2}).\mathbf{r} - (\omega_1 + \omega_2)t + (\varphi_1 + \varphi_2)\right] \\ &+ \frac{a_1 a_2 V_1 V_2}{2} \cos\left[(\mathbf{k_1} - \mathbf{k_2}).\mathbf{r} - (\omega_1 - \omega_2)t + (\varphi_1 - \varphi_2)\right] \end{aligned} \tag{9}$$

Because, the moiré pattern is the pattern with the lowest spatial frequency, the transmittance function of the moiré pattern can be deduced as

$$T_m = a_1 a_2 \left\{ 1 + \frac{V_1 V_2}{2} \cos\left[(\mathbf{k_1} - \mathbf{k_2}).\mathbf{r} - (\omega_1 - \omega_2)t + (\varphi_1 - \varphi_2) \right] \right\}. \tag{10}$$

Thus, visibility V_m, transmission coefficient a_m, and initial phase φ_m of the moiré fringes are

$$V_m = \frac{V_1 V_2}{2}, \quad a_m = a_1 a_2, \quad \varphi_m = \varphi_1 - \varphi_2. \tag{11}$$

It should be mentioned that, here the formulation for the visibility of the fringes is introduced considering full coherence of the illumination or immediate superimposing of the gratings. However, in many applications such as imaging through the atmosphere a certain degree of coherence should be considered in the formulation [32].
The transmittance function of the moiré pattern can be expressed as

$$T_m = a_m \left[1 + V_m \cos(\mathbf{k_m}.\mathbf{r} - \omega_m t + \varphi_m) \right], \tag{12}$$

where

$$\mathbf{k_m} = \mathbf{k_1} - \mathbf{k_2}, \tag{13}$$

and

$$\omega_m = \omega_1 - \omega_2. \tag{14}$$

By squaring both sides of the Eq. (13) we obtain the following expression

$$k_m^2 = k_1^2 + k_2^2 - 2k_1 k_2 \cos\theta, \tag{15}$$

where θ is the angle between $\mathbf{k_1}$ and $\mathbf{k_2}$ (see Fig. 2). Using $k_1 = \frac{2\pi}{d_1}$ and $k_2 = \frac{2\pi}{d_2}$, the period of the moiré pattern is obtained as:

$$d_m = \frac{d_1 d_2}{\left(d_1^2 + d_2^2 - 2d_1 d_2 \cos\theta\right)^{1/2}}, \tag{16}$$

In addition, according to Fig. 2 we have

$$\mathbf{k_2}.\hat{\mathbf{k}}_0 = -\mathbf{k_m}.\hat{\mathbf{k}}_0,$$

or

$$k_2 \sin\theta = k_m \sin\gamma, \tag{17}$$

where, γ is the angle between $\mathbf{k_1}$ and $\mathbf{k_m}$, and $\hat{\mathbf{k}}_0$ is a unit vector directed along the perpendicular to $\mathbf{k_1}$ in the $\mathbf{k_1}$ and $\mathbf{k_2}$ plane, as illustrated in Fig. 2. Substitution of Eq. (15) into (17) leads to the moiré fringes' orientation formula

$$\sin\gamma = \frac{d_1 \sin\theta}{\left(d_1^2 + d_2^2 - 2d_1 d_2 \cos\theta\right)^{1/2}}. \qquad (18)$$

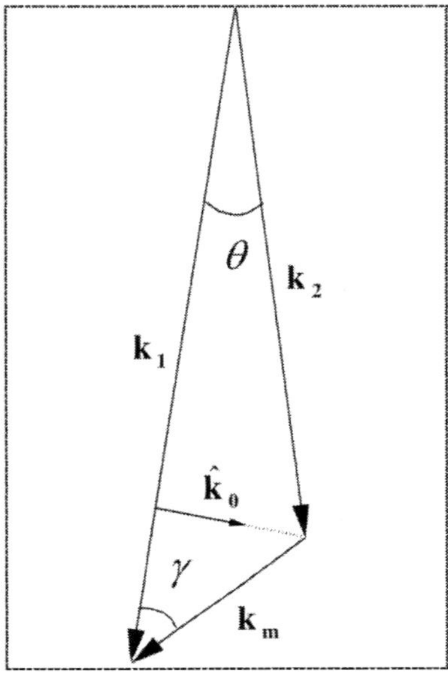

Figure 2. Representation of the gratings reciprocal vectors $\mathbf{k}_1$ and $\mathbf{k}_2$, and the corresponding reciprocal vector of the moiré pattern $\mathbf{k}_m$. $\hat{\mathbf{k}}_0$ is a unit vector directed along the perpendicular to $\mathbf{k}_1$ in the $\mathbf{k}_1$ and $\mathbf{k}_2$ plane.

Now ,we set the differential of the argument in Eq. (10) equal to zero:

$$d\left[(\mathbf{k}_1 - \mathbf{k}_2).\mathbf{r} - (\omega_1 - \omega_2)t + (\varphi_1 - \varphi_2) = 0\right],$$

For constant $\mathbf{k}_i$, ω_i, and φ_i $(i = 1,2)$ it leads to

$$(\mathbf{k}_1 - \mathbf{k}_2).d\mathbf{r} - (\omega_1 - \omega_2)dt = 0,$$

or

$$k_m dr_m - \omega_m dt = 0, \qquad (19)$$

where r_m is the scalar component of $\mathbf{r}$ in the direction of $\mathbf{k}_m$. Thus, the magnitude of the velocity of the moiré fringes v_m, is

$$v_m = \frac{\omega_m}{k_m}. \qquad (20)$$

Now, if we express the velocity of the moiré fringes in terms of the gratings `velocities and constants we get

$$\begin{aligned} v_{\mathrm{m}} &= \frac{k_1 v_1 - k_2 v_2}{\left(k_1^2 + k_2^2 - 2k_1 k_2 \cos\theta\right)^{1/2}} \\ &= \frac{d_2 v_1 - d_1 v_2}{\left(d_1^2 + d_2^2 - 2d_1 d_2 \cos\theta\right)^{1/2}} \\ &= \frac{d_2 v_1 - d_1 v_2}{d_1 d_2} d_m . \end{aligned} \tag{21}$$

Special cases:

a) The period and orientation of the moiré fringes in the parallel moiré pattern are derived by considering $d_1 \neq d_2$, $\theta = 0$ in Eqs. (16) and (18):

$$d_m = \frac{d_1 d_2}{|d_1 - d_2|}, \tag{22}$$

and

$$\gamma = 0. \tag{23}$$

Thus, in this case the moiré fringes are parallel to the lines of the individual gratings.

b) The period and orientation of the moiré fringes in the rotation moiré pattern are derived by considering $d_1 = d_2 = d$ and $\theta \neq 0$ in Eqs. (16) and (18):

$$d_m = \frac{d}{2\sin(\theta/2)}, \tag{24}$$

and

$$\sin\gamma = \cos(\theta/2),$$

or

$$\gamma = (\frac{\pi}{2} - \frac{\theta}{2}). \tag{25}$$

That means, the moiré fringes are perpendicular to the bisector of the gratings' lines.

c) For the case $d_1 \neq d_2$, $\theta = 0$, $v_1 \neq 0$, $v_2 = 0$ the velocity of the moiré fringes becomes:

$$v_m = \frac{v_1}{d_1} d_m = v_1 \frac{d_2}{|d_1 - d_2|}. \tag{26}$$

This means moiré fringes move in the same direction as the grating, but much faster.

d) For the case $d_1 = d_2 = d$, $\theta \neq 0$, $v_1 \neq 0$, $v_2 = 0$, the velocity of the moiré fringes becomes:

$$v_m = \frac{v_1}{d_1} d_m = \frac{v_1}{2\sin(\theta/2)}. \tag{27}$$

In this case moiré fringes move in the direction almost perpendicular to the grating motion.

It should be noted that the presented approach is valid for non-relativistic velocities and for relativistic velocities some modification must be considered.

3. Conclusion

In many experiments, the moiré technique has been utilized to investigate a variety of dynamical problems. At the formation of the moiré patterns sometimes one of the superposed gratings is the image of a physical grating. When the image forming lights propagate in a dynamic medium, the image grating appears as a dynamic grating and dynamical behavior of the medium is magnified by the moiré pattern.

Theoretical consideration of the behavior of the moiré fringes in the dynamical phenomena is not completely demonstrated. In this paper a general theory for the moiré patterns of the moving periodic structures is derived from the reciprocal vectors approach. Formulation of the rotation and parallel moiré patterns in the superposition of static and dynamic periodic structures are unified. The individual structures are characterized by spatial frequency, orientation, profile (variation of transmittance over one period), and velocity parameters. All the specific moiré formulae are derived using the presented approach.

Acknowledgements

The authors would like to thank B. Farnudi from IASBS for the linguistic editing of the paper.

References

[1] Patorski K and Kujawinska M 1993 *Handbook of the moiré fringe technique*, Elsevier, Amsterdam.
[2] Karny Z and Kafri O 1982 Refractive-index measurements by moiré deflectometry *Appl. Opt.* **21** 3326-3328.
[3] Ranjbar S, Khalesifard H R, and Rasouli S 2006 Nondestructive measurement of refractive index profile of optical fiber preforms using moiré technique and phase shift method *Proc. SPIE* **6025** 602520.
[4] Rasouli S and Tavassoly M T 2006 Application of moiré technique to the measurement of the atmospheric turbulence parameters related to the angle of arrival fluctuations *Opt. Lett.* **31** 3276-3278.
[5] Rasouli S and Tavassoly M T 2006 Measurement of the refractive-index structure constant, C_n^2, and its profile in the ground level atmosphere by moiré technique *Proc. SPIE* **6364** 63640G.
[6] Rasouli S and Tavassoly M T 2008 Application of the moiré deflectometry on divergent laser beam to the measurement of the angle of arrival fluctuations and the refractive index structure constant in the turbulent atmosphere *Opt. Lett.* **33** 980-982.
[7] Rasouli S 2010 Use of a moiré deflectometer on a telescope for atmospheric turbulence measurements *Opt. Lett.* **35** 1470 – 1472.
[8] Rasouli S, Ramaprakash A N, Das H K, Rajarshi C V, Rajabi Y, and Dashti M 2009, Twochannel wavefront sensor arrangement employing moiré deflectometry *Proceedings of SPIE* **7476** Optics in Atmospheric Propagation and Adaptive Systems XII, pp. 74760K-1, doi: 10.1117/12.829962.
[9] Rasouli S, Dashti M, and Ramaprakash A N 2010 An adjustable, high-sensitivity, wide dynamic range two channel wavefront sensor based on the moiré deflectometry *Opt. Exp.* **18** 23906 – 23915.
[10] Rasouli S, Ghasemi H, Tavassoly M T, and Khalesifard H R 2011 Application of "parallel"

moiré deflectometry and the single beam Z-scan technique in the measurement of the nonlinear refractive index *Appl. Opt.* **50** 2356–2360.

[11] Jamshidi-Ghaleh K and Mansour N 2004 Nonlinear refraction measurements of materials using the moiré deflectometry *Optics Communications* **234** 419–425.

[12] Harding K G and Harris J S 1983 Projection moiré interferometer for vibration analysis *Appl. Opt.* **22** 856-861.

[13] Post D, Han B, and Ifju P 1993 *High Sensitivity moiré: experimental analysis for mechanics and materials*, Springer, Berlin, Germany.

[14] Walker C A 2004 *Handbook of moiré measurement*, Institute of Physics, London.

[15] Tay C J, Quan C, Fu Y, and Huang Y 2004 Instantaneous velocity, displacement, and contour measurement by use of shadow moiré and temporal wavelet analysis *Appl. Opt.* **43** 4164-4171.

[16] Oberthaler M K, Bernet S, Rasel E M, Schmiedmayer J, and Zeilinger A 1996 Inertial sensing with classical atomic beams *Phys. Rev. A* **54** 3165-3176.

[17] Rasouli S and Tavassoly M T 2006 Moiré Deflectometer for measuring distortion in sheet glasses *Proc. SPIE* **6024** 60240E.

[18] Nishijima Y and Oster G 1964 Moiré patterns: Their application to refractive index gradient measurements *J. Opt. Soc. Am.* **54** 1-5.

[19] Tollenaar D 1992 Moiré in halftone printing interference phenomena, English translation published in G. Indebetouw and R. Czarnek (Eds.), Selected papers on optical moiré and applications. SPIE Milestone Series, Vol. MS64, pp. 618-633, SPIE Optical Engineering Press., Washington.

[20] Oster G, Wasserman M, and Zwerling C 1964 Theoretical interpretation of moiré patterns *J. Opt. Soc. Am.* **54** 169-175.

[21] Amidror I 2000 *The Theory of the Moiré Phenomenon* Kluwer Academic Publishers, Netherlands.

[22] Rogers G L 1959 A simple method of calculating moiré patterns *Proc. Phys. Soc.* **73** 142-144.

[23] Bryngdahl O 1974 Moiré: formation and interpretation *J. Opt. Soc. Am.* **64** 1287-1294.

[24] Bryngdahl O 1975 Moiré and higher grating harmonics *J. Opt. Soc. Am.* **65** 685-694.

[25] Rogers G L 1959 A geometrical approach to moiré pattern calculation *Optica Acta* **24** 1-13.

[26] Gevers R 1962 Dynamical theory of moiré fringe patterns *Philosophical Magazine* **82** 1681-1720.

[27] Santos P A M and Oliveira G N 2005 Holographic Fourier synthesis of dynamic moiré like patterns in the photorefractive Bi12TiO20 crystal *Optical Engineering* **44** 125803.

[28] Santos P A M and Tagliaferri A A 1997 Dynamic moiré patterns produced by sinusoidal gratings superposition in the Bi12TiO20 crystal *Optics & Laser Technology* **29** 345-347.

[29] Deason V A, Epstein J S, and Abdallah M 1990 Dynamic diffraction moiré: theory and applications *Optics and Lasers in Engineering* **12** 173-187.

[30] Bruck H A 1997 The effects of motion on dynamic moiré interferometry *Optics and Lasers in Engineering* **27** 343-354.

[31] Kokaly M T, Lee J, and Kobayashi A S 2003 Moiré interferometry for dynamic fracture study, Optics and Lasers in Engineering 40 231-247.

[32] Rasouli S , Madanipou K, and Tavassoly M T 2006 Measurement of modulation transfer function of the atmosphere in the surface layer by moiré technique *Proc. SPIE* **6364** 63640K.

Luminescence spectrum thermal properties of Rhodamine 6G doped polymethyl metacrylate film sandwiched between cholesteric liquid crystal layers

R B Alaverdyan[1], T K Dadalyan[1], A S Karapetyan[1] and N S Torosyan[1]
Yerevan State University, Alex Manoogyan 1, 0025 Yerevan, Armenia

E-mail: tigranxf@gmail.com

Abstract. We experimentally investigate fluorescence spectrum properties of multilayer system consisting of Rhodamine 6G doped polymethyl metacrylate thin film sandwiched between two right handed cholesteric liquid crystal layers at various temperatures and pumping powers. We observe fluorescence intensity enhancement while increasing sample temperature from 16 to 22°C when pumping sample with constant power laser irradiation. At constant temperature fluorescence intensity also increases with increasing pumping power.

1. Introduction

Development of compact and tunable optical elements that provide control of light radiation and parameters is a very important problem of contemporary science. Thanks to their optical properties Photonic Crystals (PC)-s have attracted a great deal of interest. In the recent years a considerable number of studies are devoted to investigation of Photonic Band Gap materials (PBG). These are structures with periodic modulation of refractive index and prohibit the propagation of light within certain frequency band [1].

Studies of PBG materials have started with the works of Bykov [2], Yablonovich [3] and John [4] where existence of PBG-s in media with periodic modulation of refractive index were predicted and since that PBG materials have found many practical applications. The ability of PC-s to inhibit spontaneous emission within band gap may be used to create low threshold lasers [5].

Much attention is paid to studies of PC-s that assume control of their properties by external influence. This enables one to control characteristics of propagating light. This can be easily met with both liquid crystals and their dye-doped polymeric film containing structures.

Cholesteric Liquid Crystals have self organized helical structure, with periodical modulation of refractive index and are reflecting circularly polarized light with the same handedness as CLC helix (selective reflection of the light); this allows us to consider CLC-s as 1D photonic crystals[6,7]. For normal incident light with wavelength within $n_o p \leq \lambda \leq n_e p$ region, and circularly polarization with the same handedness of helix reflection coefficient is 1, whereas reflection coefficient for light with opposite handedness propagates without reflection. Here n_o and n_e are ordinary and extraordinary refractive indices of CLC and p is the pitch of the CLC helix [22]. In addition to selective reflection property the sensitivity of CLC pitch to external factors such as temperature, electric field is making them very interesting object of intensive studies and predicts many interesting practical applications, such as optical diodes, filters and optical windows and other compact optical components[8-10,15] .

In combination with laser dyes CLC-s have found one of their very interesting applications. Unique properties of CLC-s enabled one to create low threshold and mirror less tunable dye lasers [11-14]. CLC PBG sensitivity to the external factors provides tunability of the wavelength of lasing from these systems. The CLC-s in these systems could be used as dielectric mirrors or distributed feedback gratings [25]. In the case, when CLC is used as dielectric mirror, the active medium is separated from CLC; this could be implemented with various types of polymers [16-19], which are good alternative for gain medium of liquid state dye lasers. These types of solid state dye lasers have several advantages of small size of active medium and use of inflammable and safe for environment materials.

There are several studies devoted to improvement of lasing characteristics of these systems [20, 21]. The spectral peculiarities of systems containing isotropic defect layer sandwiched between two CLC layers are widely investigated theoretically [26-30].

Because of easy processability and other physical properties polymethyl metacrylate provides great opportunity to conduct experiments on multilayer systems consisting of Liquid Crystalline materials and polymer films.

During experiments we have used Rhodamine 6G laser dye, which is efficient laser dye and covers the wavelength region from 500 to 700 nm. It is very popular dye for experiments and applications. It has remarkable photostability and solubility in the most known solvents [24]. Usually Rhodamine 6G gain medium is pumped by the 2-nd harmonic of Nd: YAG laser (532 nm).

In the present work we study spectral properties of the system consisting of Rhodamine 6G doped polymethyl metacrylate thin films sandwiched between two right handed CLC layers. We experimentally investigate thermal peculiarities of the system fluorescence, excited with the laser irradiation with 532 nm wavelength at various pumping powers.

2. Experimental setup and materials

For obtaining solid state matrix for Rhodamine 6G we have used polymethyl Metacrylate (PMMA), which could be solved in chloroform or dichlorethane.

The CLC used in experiment was a mixture of 18 wt. % cholesteryl oleate (5-Cholesten-3b-ol 3-Oleate), 35 wt. % of cholesteryl pelargonate and 47 wt. % of commercial Nematic Liquid Crystal E7.

The sample consists of glass substrates, R6G doped PMMA thin film sandwiched between CLC layers (Figure 1). The thickness of R6G doped PMMA layer is 25 μm, concentration of R6G is $4*10^{-2}$ mol/liter. The distance of dye doped PMMA layer from glass substrates was given by means of spacers with thickness of 10 μm. The dimensions of PMMA layer were $10*10\ mm^2$. For avoiding PMMA layer deformations and non-uniformities, the film was partially covered by spacers, so 5*5 mm^2 sized central area of R6G doped PMMA was used for measurements.

For obtaining CLC uniform aligning the glass substrates were coated with polymer, and rubbed in anti-parallel directions. The CLC mixture was heated into the isotropic phase and droplet filled into the cell, containing R6G doped PMMA film on spacers. Then it was left for two days in the room temperature. The right handed helical CLC structure formed. Then four sides of the cell were glued.

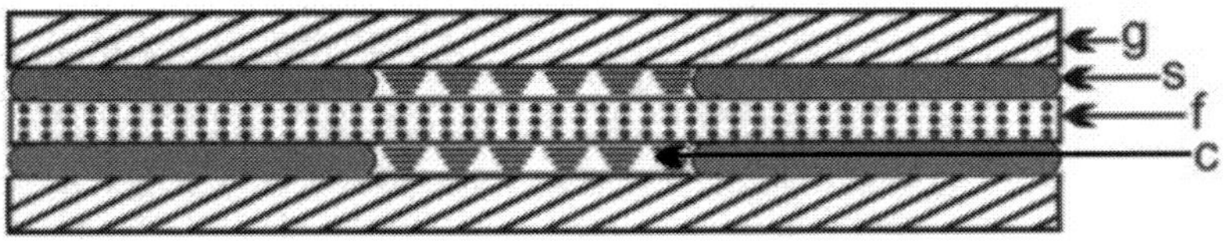

Figure 1 The sketch of the CLC-R6G-CLC multilayer system, g-glass substrate, s-spacers, which kept the thickness of CLC layers (c) 10 μm, f-R6G doped PMMA film with the 25 μm thickness

The R6G doped PMMA film was obtained by adding R6G into the PMMA dichlorethane solution and spin coated on polyethylene film. After volatilization of dichlorethane the PMMA film formed and then was separated from polyethylene film.

The temperature of the cell was adjusted by a temperature controller which is tuning the temperature of Peltier thermoelectric module. The temperature of controller was set with accuracy of 0.5^{o} C. Temperature of the sample was measured by means of thermocouple, which was placed adjacent to the sample.

For fluorescence spectrum study, the Rhodamine 6G doped film was pumped with various powers of cw laser, with the wavelength of the second harmonic of Nd: YAG laser emission (532nm). For sample pumping the oblique angle geometry of pumping beam incidence was used. The angle between pumping beam and the cell normal was 54°, the pumping laser spot had 2 mm diameter. We have measured fluorescence of the system at 15, 12 and 8 mW pumping powers. The experimental setup is shown on Figure 2.

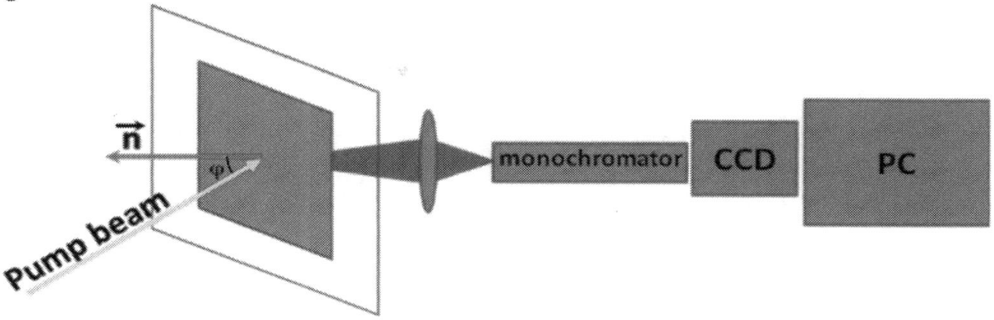

Figure 2. Experiment geometry. Pump beam is incident to the cell normal at φ=54°, the fluorescence is collected by means of lens with 6cm focal length.

Fluorescence spectrum was registered by homemade apparatus, consisting of monochromator ISP 51; the spectrum was registered using CCD camera. The measurements were carried out with image processing special program working in Lab View environment, which measures the values of the grayscale image pixels along given line. The optical resolution of the spectrophotometer was less than 1 nm.

3. Results and discussions

During experiments we have measured fluorescence form R6G doped PMMA 25μm thin film which was sandwiched between two right handed CLC layers at various temperatures and pumping powers. Particularly the measurements were carried out within 16-22°C temperature region. Pumping powers were 15, 12 and 8 mW of cw laser irradiation with 532 nm wavelength.

The luminescence spectrum of the multilayer system under various pumping powers is depicted on Figure 3. Three curves correspond to the 15, 12 and 8mW pumping powers. The temperature of the system is 22°C. From the chart it could be seen that fluorescence of Rhodamine 6G covers wavelength region 535-640 nm. With increase of pumping power luminescence increases.

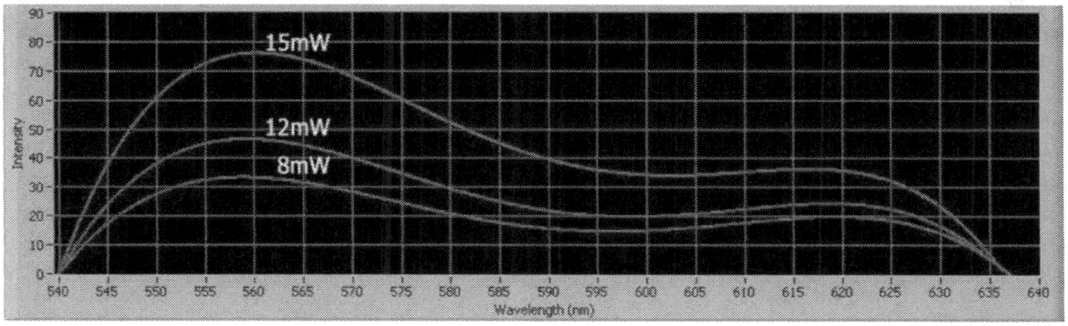

Figure 3. Fluorescence from CLC-R6G-CLC system increases with the pumping power. Temperature is 22°C; the pumping powers are given on the curves.

In figure 4 the CLC-R6G doped PMMA film-CLC system fluorescence spectrum thermal dependences are represented. In this graph we have combined fluorescence from multilayer system at the same pumping power, but different temperatures of the sample.

From figure 4 it could be noticed that fluorescence of the system is increasing with increase of temperature.

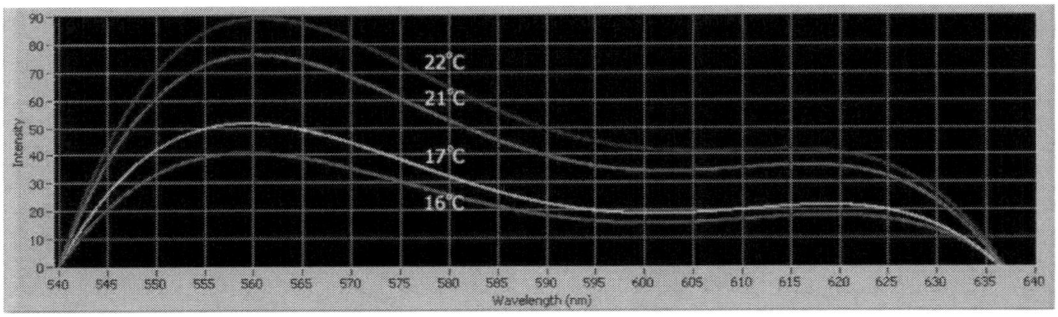

Figure 4. The temperature dependence of fluorescence spectrum. The pumping power is 15mW

This result most likely could be explained by temperature shift of selective reflection band which causes reflection of pumping laser irradiation. As it was mentioned CLC could be interpreted as distributed feedback grating, with period sensitive to external temperature. In our mixture the raise of temperature leads to decrease in the helical pitch of CLC. The central wavelength of selective reflection band is proportional to helical pitch and is given with $\lambda_c \sim p(T)$ where p is helical pitch of CLC and T is temperature of the system. So the central wavelength of selective reflection band shifts with change of temperature. In our experiment we used oblique incident pumping beam so at 16°C the selective reflection takes place and a part of pumping laser energy is reflected and R6G is excited with less pumping power. The further increase of temperature shifts selective reflection band toward short wavelength region and at 22°C pumping light is out of selective reflection band and pumping beam passes through the system without any reflection and this causes the increase of fluorescence intensity. Because of non-uniformities of the CLC pitch the selective reflection band edges could be not sharp, so this might be the reason that reflection changes not dramatically as expected.

The Transmission spectrum measurement of the system is complicated because of the strong absorption of R6G in visible wavelength region [23].

4. Conclusions

In this study we've investigated fluorescence spectrum of CLC – R6G doped PMMA thin film – CLC multilayer system, which was excited by 532 nm wavelength cw laser irradiation with various powers at different temperatures of the sample. The pumping powers were 15, 12 and 8mW. We observe increase of fluorescence intensity corresponding to pumping power increase when temperature is constant. The fluorescence spectrum intensity at constant pumping power and increasing temperature also increases. Increasing fluorescence intensity at constant pumping power and various system temperatures could be caused by temperature shift of selective region band. However this interpretation of obtained results must be confirmed by additional experiments which are in progress. Obtained results are demonstrating easy tunability of the system which makes the system promising for further investigation.

Acknowledgments

This work was supported by State Committee of Science Republic of Armenia, grant 11-1c194.

REFERENCES

[1] Sibilia C, Benson T M, Marciniak M, Szoplik T 2008 Photonic Crystals: Physics and Technology *Springer-Verlag Italia*

[2] Bykov V P 1972 Spontaneous emission in a periodic structure *Sov. Phys. JETP* **35** 269

[3] Yablonovitch E 1987 Inhibited spontaneous emission in solid state physics and electronics *Phys. Rev. Lett* **58** 2059-2062

[4] John S 1987 Strong localization of photons in certain disordered dielectric superlatices *Phys.Rev. Lett* **58** 2486-2489
[5] Ozaki M Kasano M Kitasho T Ganzke D Haase 2003 W Electro-Tunable Liquid Crystal Laser *Adv. Matter.* **15** 974-977
[6] de Gennes P E 1974 *The physics of Liquid Crystals* Oxford
[7] Khoo Iam-Choon 2007 *Liquid Crystals 2-nd edition* John Wiley & Sons New Jersey, ()
[8] Song M H Park B Takanishi Y 2006 Simple electro-tunable optical diode using photonic and anisotropic liquid crystal films *Thin Solid Films* **509** 49 – 52
[9] Song M H Park B 2006 Electrotunable Non-reciprocal laser emission from a liquid crystal photonic device *Adv. Funct. Mater.* **16** 1793–1798
[10] Hwang J., Song M. H., et.al., "Electro-tunable optical diode based on photonic band gap liquid-crystal heterojunctions nature materials", nature materials 4, (2005)
[11] Coles H Morris H 2010 Liquid Crystal lasers *Nature Photonics* **4** 676-685
[12] Takanishi Y Youko O 2010 Low threshold lasing from dye-doped cholesteric liquid crystal multi-layered structures *Optics Express* **18** 12909-12914
[13] Wang Chun-Ta Lin Tsung-Hsien 2008 Multi wavelength laser emission in dye-doped photonic liquid crystals *Optics Express* **16** 18334-18339
[14] Palto S P 2006 Lasing in LC thin films *JETP* **103** 472
[15] Lukishova S G Belyaev S V 1996 Behavior of nonlinear liquid crystal mirrors, made of a nonabsorbing cholesteric, in the cavity of an Nd:YAG laser operating in the cw regime and at a high pulse repetition frequency *Quantum Electron* **26** 796-798
[16] Shaposhnikov A A Kuznetsova R T 2004 Absorption, luminescent and lasing properties of laser dyes in silica gel matrices and thin gel films *Quantum Electron* **34** 715-721
[17] Bezrodnii V I Derevyanko N A 2001 Dye laser on base of polyurethane matrix *JTP* **71**
[18] Korobkin Yu V Sidorov O 1996 Gain media of lasers on dye doped epoxides and other thirosin containing materials *JTP* **67**
[19] Takahashi Y Maeda A Kojima K Uchida K 2000 Luminescence of dyes doped in a sol-gel coating film", *Journal of Luminescence* **87 - 89** 767-769
[20] Shirvani-Mahdavi H Fardad Sh Mohajerani E Wu Shin-Tson 2010 High efficiency cholesteric liquid crystal lasers with an external stable resonator *Optics Express* **18** 13593-13599
[21] Chilaya G Chanishvili A Petriashvili G 2006 Enhancing cholesteric liquid crystal laser stability by cell rotation *Optics Express* **14** 9939- 9943
[22] Belyakov V A Sonin A S 1982 *Optics of Cholesteric Liquid Crystals* Nauka Publ Moscow
[23] Alaverdyan R Dadalyan T Hayrapetyan N 2011*Proc of SPIE* **8114** Liquid Crystals XV(San Diego)
[24] Shankarling G S Jarag K J 2010 Laser dyes *Resonance* 804-818
[25] Chilaya G Chanishvili A et. al. 2011 *Materials Sciences and Applications* **2** 116-129
[26] Belyakov V A 2006 *Mol. Cryst. Liq. Cryst.***43** 453
[27] Gevorgyan A H Harutyunyan M Z 2007 Chiral photonic crystals with anisotropic defect layer *Phys. Rev. E* **76** 031701
[28] Gevorgyan A H Harutyunyan M Z 2009 Tuning of Lasing wavelength in Chiral Photonic Crystals with an anisotropic defect layer *J. Mod. Opt.* **56** 1163-1173
[29] Gevorgyan A H Oganesyan K B 2011 Defect modes of Chiral Photonic Crystals with an anisotropic defect *Opt. Spectrosc.* **110** 952-960
[30] Gevorgyan A H Oganesyan K B Harutyunyan E M Arutyunyan S O 2010 *Opt. Communn.* **283** 3707-3713

Journal of Physics: Conference Series **350** (2012) 012034 doi:10.1088/1742-6596/350/1/012034

Some Light-Induced Orientational Effects in Nematic Liquid Crystals

R.S. Hakobyan, M.R.Hakobyan, A.K. Aleksanyan, G.S. Gevorgyan, R.B. Alaverdyan,
Yerevan State University, Faculty of Physics, Department of Optics,
1 Alex Manoogian str., Yerevan, 0025, Republic of Armenia

Email: rhakob@web.am

Abstract. Some new light-induced orientational-hydrodynamical nonlinear optical effects in nematic liquid crystals (NLC) are presented. Among them jump-like light-induced hydrodynamic reorientation caused by direct volume expansion, thermomechanical oscillations in hybrid NLC and thermomechanical mechanism of orientational optical nonlinearity in twist aligned NLC as well as in azobenzene liquid crystals. It is also shown that the orientational optical nonlinearity of revealed thermomechanical effect may be as strong as well-known giant optical nonlinearity. Besides, at normal incidence of laser the giant optical nonlinearity does not exist, but thermomechanical effect exists at any angle of incidence of laser radiation and the nonlinearity of thermomechanical effect remains in the same order.

1. Introduction

Nematic liquid crystals consist of elongated molecules that tend to aligned parallel to each other. The unit vector along preferred orientation of molecules is called as director.

In 1980 it was predicted and observed a very high optical nonlinearity of liquid crystals in [1]. This orientational optical nonlinearity was billion times larger than that of classical nonlinear media such as carbon-bisulfur (CS_2). That is why it was called as giant optical nonlinearity (GON). The mechanism of this nonlinearity is that the electrical field of light wave can as well reorient liquid crystal (LC) molecules as the direct current electrical field. Of course, the establishment time of the nonlinearity proved to be the same times larger, too. Therefore, the nonlinear optical phenomena are displayed, at first, at very low powers (~ mW) of laser radiation and, in the second place, they proceed on a real time scale (0.001-10 s).

In the further works of many groups in the world it was proved that important peculiarity of NLC is the great number of accessible mechanisms of nonlinearities such as pure orientational, thermal, conformational, orientational-hydrodynamical, etc [1-7]. Moreover, it is always possible to invent such kind of experimental schemes, when only one mechanism predominates. It is important to note that LC illustrates its specific effects at all mechanisms of optical nonlinearities.

As indicated in 1984 [8], an extremely small absorption may provide energy sufficient for remarkable NLC director reorientation. The problem was to find a suitable mechanism transforming the absorbed energy into LC director deformation energy. Then the nonlinearities, which are conditioned by the absorption, may prove to be much stronger than all other "dynamical" nonlinearities which were known before. The simplest group of phenomena is conditioned by heating of the medium in case of light absorption. One of such kind of mechanisms is the "photo-conformational" mechanism of optical nonlinearity [10]. At present many new mechanisms of optical nonlinearities of NLC are predicted and studied [11-16].

More specific to LC's is the hydrodynamically induced reorientation where the hydrodynamic motions are generated under the influence of light waves [8]. These so called opto-hydrodynamic orientational effects were investigated in a number of articles and proved to provide perhaps the strongest mechanism for optical nonlinearity of LC's [17-20] due to the stronger coupling of hydrodynamic motion with the director orientation [21-23]. It was studied theoretically and experimentally three mechanisms of light induced hydrodynamic motions: gravity or Rayleigh-Benard mechanism, thermocapillary or Marangoni mechanism, and volume expansion mechanism.

Published under licence by IOP Publishing Ltd

Rayleigh-Benard mechanism of laser induced hydrodynamic convection is conditioned by temperature dependence of NLC density, Marangoni mechanism – by temperature dependence of NLC surface tension coefficient and finally third mechanism is owned to usual volume expansion due to the laser radiation absorption.

In 1990 Hungarian scientist Jannosy observed a new mechanism of transformation of laser absorption energy to NLC orientational energy [23]. Jannosy effect exists in NLC doped with dye, molecules of which have orientational interaction with NLC molecules. Light absorption makes transformation of dye molecules, which leads to additional reorientation of NLC.

Recently in [24] it was shown that the constants of nonlinear refraction of photoinduced isotropic (PI) phase of azobenzene LCs are comparable to that of LC mesophases. The basic process of obtaining the PI state of azo LC consists in subjecting it to radiation in the blue-green region of wavelengths; *trans-cis* photoisomerization of azobenzene molecules reduces the order parameter of azo LCs thus driving them into their isotropic phase upon accumulation of a critical concentration of the nonmesogenic *cis* form of azo molecules. These cis isomers transform to trans due to absorption of red radiation of He-Ne laser beam, thus inducing the order parameter and refractive index of NLC.

Very strong interaction of laser radiation with azobenzene LC caused to the cell mechanical deformation. This effect is a good example of light energy transformation to the mechanical energy. The mechanism for the photoresponse of azo-LC is determined by the wavelength of the radiation source. UV radiation (365 nm) efficiently isomerizes the azobenzene linkage from trans to cis. Generation of cis-isomers in LC induces the mechanical response through a relative volume change from a reduction of (a) the order parameter of the network and (b) the molecular length of the azobenzene unit from 9A° to 5.5A°.

Another possibility of transformation of light energy to mechanical one gives us thermomechanical effect in deformed nematic LC [6]. Let us remember what they are? The first-direct effect is the onset of hydrodynamic flow due to the temperature gradient, the second-inverse effect is the onset of a temperature drop in an inhomogeneous flow, and the third is the additional reorientation of director by the heat flow. For example, first thermomechanical effect can be realized in the NLC cell with hybrid or cylindrical-hybrid alignment of molecules. In the hybrid cell we have hydrodynamical flow in the presence of temperature gradient and in the cylindrical-hybrid cell we get hydrodynamical rotation [25, 26].

In the case when the hydrodynamic flow tends to reduce the curvature of the "flexible ribbon", an oscillatory motion was observed [25, 27]. In the hybrid aligned cell an oscillatory motion was observed. And in the cylindrical-hybrid aligned cell a clockwise and counterclockwise oscillatory rotations were observed and studied.

2. Experimental part

Third thermomechanical effect was studied in twist-aligned cell [28]. In the presence of transversal temperature gradient molecules go out from their initial orientation plane. In the case when this transversal temperature gradient is conditioned by laser radiation with, for instance, Gaussian transversal distribution of intensity, we can get reorientation of NLC director, and thus change of refractive index of the medium. This refractive index perturbation is proportional to light intensity. Coefficient of proportionality is the constant of thermomechanical mechanism of orientational optical nonlinearity. Theoretical prediction and experimental measurements and studies have shown that this constant is as high as the well known nonlinear coefficient of GON.

Experimental setup for investigation of the abovementioned thermomechanical effect is given in fig.1. A Nd:YAG laser with 1.06 μm wavelength, power of 3 W and linear polarization was used to generate a two-dimensional temperature gradient in LC-cell. The laser beam was focused on a twist-aligned nematic liquid crystal cell by a lens with 4cm focal length. The image of laser was registered on the screen by digital camera after passing through the cell. The thickness of the used cell was 100 μm. 5CB liquid crystal was used for the investigation. Experiment was done at a room temperature.

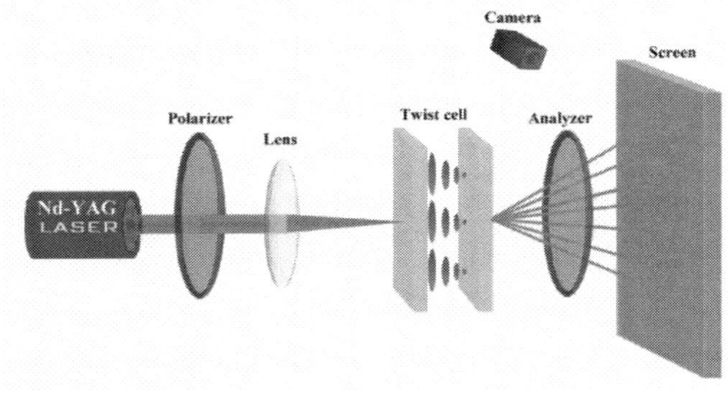

Figure 1. Nd-YAG laser beam of 3W power passing through this system is registered on the screen by digital camera

Before discussion of the measured data, it is necessary to notice that due to absorption of light the temperature distribution in the sample becomes close to radial distribution of laser intensity, which has a Gaussian profile (fig.2a). Accordingly, for the absolute value of derivative of this distribution we will have a profile shown in fig.2b, which must coincide with distribution of reorientation angle. However, because of interaction between LC molecules the reorientation is transferred also to the central part of the eradiated area of the cell and real distribution of reorientation of molecules gets a form shown in fig.2c. As is seen, the distribution of reorientation angle of NLC director in thermomechanical effect differs from distributions for the cases of GON and nonlinearities, studied in [1-5]. In these nonlinearities the distribution has a form close to the form of radial distribution of laser intensity, instead of its derivative.

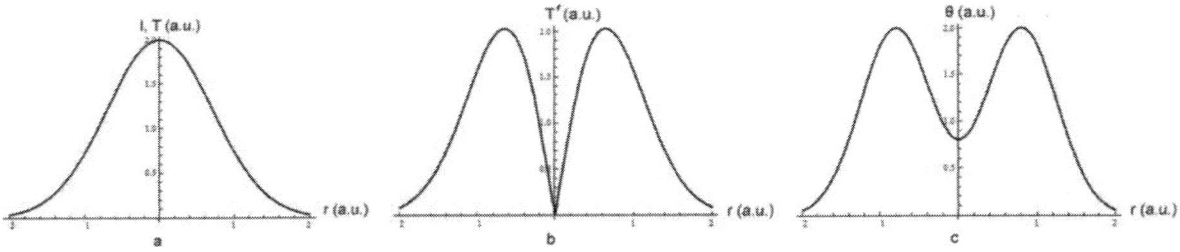

Figure 2. a) temperature distribution, b) temperature gradient, c) director reorientation in a sample

According to distribution of fig.2c and depending on mutual position of polarizers it is possible to predict the image on the screen (fig.3).

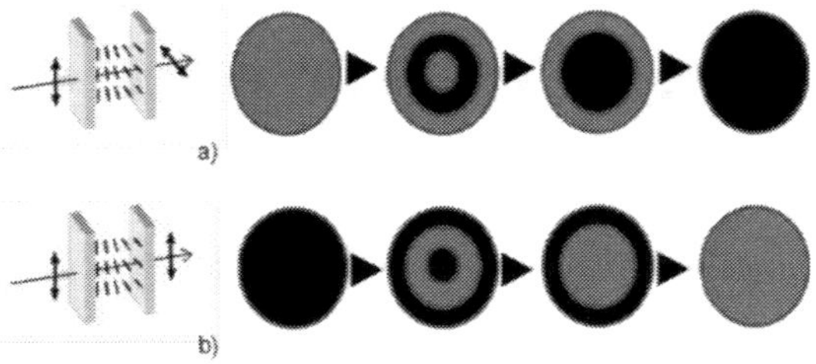

Figure 3. Predicted ring patters on the screen

With increase of intensity of exciting IR laser at first a reorientation should occur in those parts of the beam, where the temperature gradient is maximal. Further increase of intensity shall lead to reorientation of molecules in the central part of the beam, too. By continuing the increase of laser intensity an isotropic phase transition should occur. Correspondingly, in the case of twist-cell placed between crossed polarizers, the images shown in fig.3a shall be observed and in the case of twist-cell placed between parallel polarizers the images shown in fig.3b should be observed as increasing the laser intensity.

Images shown in fig.4 are observed on the screen in our experiment with increase of laser intensity at various temperatures. At first, the effect is observed at T=26°C room temperature. In this case, at low intensities, full transmission of laser beam is observed (fig.4a). Then, the image shown in fig.4b is observed on the screen after increasing the laser intensity. Consequently, reorientation of molecules has occurred only in those parts where the temperature gradient is big. Further increase of intensity led to reorientation of molecules of central part, too (fig.4c). By continuing the increase of laser intensity an isotropic phase transition and no transmission of light should have occurred (fig.4d).

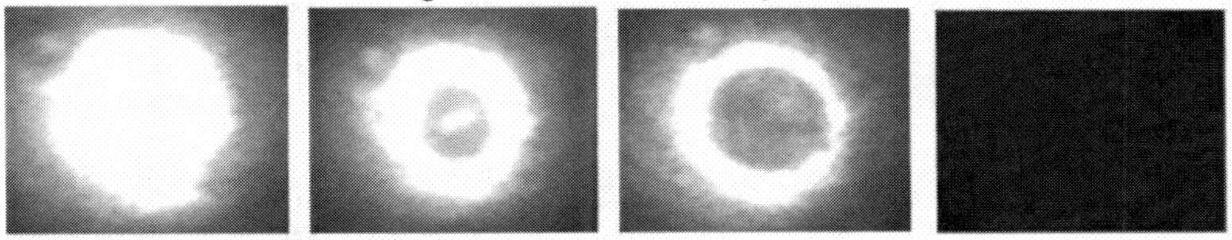

a b c d

Figure 4. Images observed on the screen with the increase of laser intensity at various temperatures

In the case of azobenzene NLC we have observed two mechanisms of nonlinearity: first is thermomechanical and second due to the trans-cis transition. In the fig.5 there are shown images of He-Ne laser on the screen at 0° (a), 45° (b) and 90° (c) angles between laser polarization and direction of director at a forward wall of the cell.

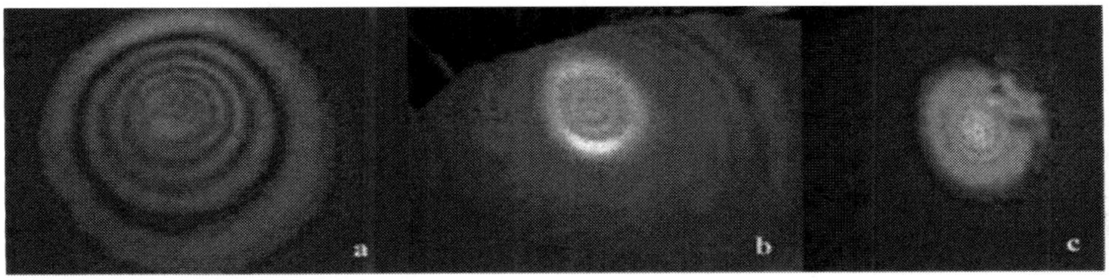

Figure 5. He-Ne laser on the screen at 0° (a), 45° (b) and 90° (c) angles between laser polarization and direction of director at a forward wall of the cell

The dependences of number and stabilization time of the rings on laser power for twist aligned cell are given in fig.6.

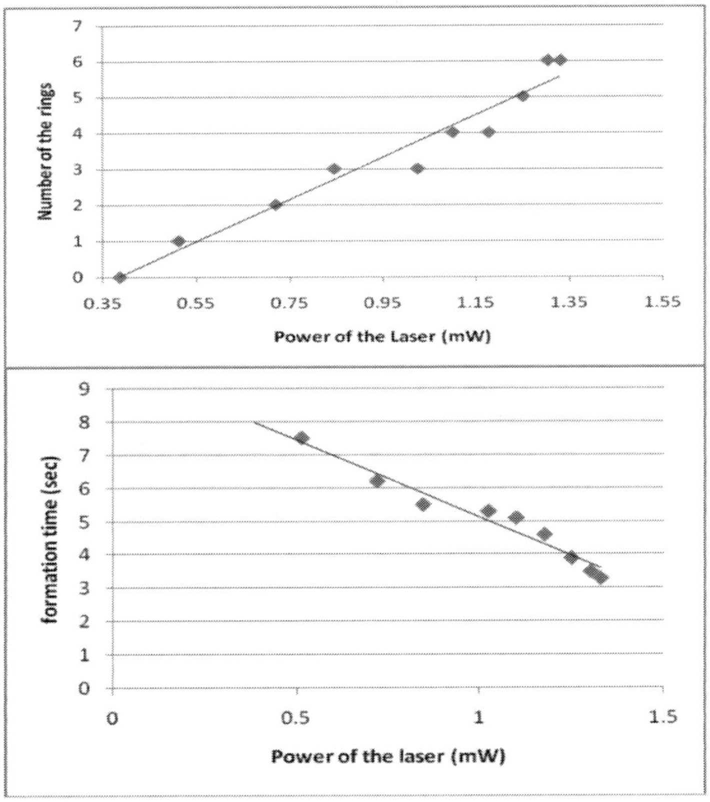

Figure 6. Number and stabilization time of rings for various laser powers

3. Theoretical part

In order to write equations for above mentioned effects, we have to create "dissipation" function and free energy density for thermomechanical effects. They require thermal flux (temperature gradient) and inhomogeneous director distribution in a finite medium. Thus, dissipation function must be proportional to the gradients of temperature, director and velocity.

In our calculations density of free energy is taken in usual Frank's form.

$$F = \frac{1}{2}K_1(div\mathbf{n})^2 + \frac{1}{2}K_2(\vec{n}\cdot curl\mathbf{n})^2 + \frac{1}{2}K_3[\vec{n}\times curl\mathbf{n}], \quad (1)$$

Where K_1, K_2, K_3 are Frank's constants, n – director.

After some scientific notation we can write "dissipative" function R in the most general form.

$$\begin{aligned} R^{TM} &= \xi_1(\mathbf{NA})M_0 + \xi_2\mathbf{n}[\mathbf{NA}]P + \xi_3 N_i A_k M_{ik} + \xi_4(\mathbf{NM})A_0 + \\ &\xi_5(\mathbf{DA})M_0 + \xi_6\mathbf{n}[\mathbf{D}\times\mathbf{A}]P + \xi_7 D_i A_j M_{ij} + \xi_8(\mathbf{DM})A_0 + \xi_9(\mathbf{AM})D_0 \quad . \quad (2) \\ &+ \xi_{10}D_0 A_0 M_0 + \xi_{11}A_i M_j D_{ij} + \xi_{12}A_0 D_{ij} M_{ji} \end{aligned}$$

where ξ_i are thermomechanical constants. Other variables in equation (2) have a complicated form and can be found in [8].

Variation of the "dissipative" function by velocity gradient gives us thermomechanical part of hydrodynamic stress tensor and variation of "dissipative" function by temperature gradient gives us thermomechanical part of the heat flow.

$$
\begin{aligned}
\sigma_{ik}^{TM} &= 1/4(2\xi_1 - \xi_3 + 2\xi_5 - \xi_7)n_i\nabla_k T\mathrm{div}\,\mathbf{n} - 1/4(2\xi_1 - \xi_3 - 2\xi_5 + \xi_7)n_k\nabla_i T\mathrm{div}\,\mathbf{n} \\
&- 1/2(\xi_2 + \xi_6)(\mathbf{n}\,\mathrm{rot}\,\mathbf{n}) + n[\mathbf{n}\times\nabla]_k T + 1/2(\xi_2 - \xi_6)(\mathbf{n}\,\mathrm{rot}\,\mathbf{n})n_k[\mathbf{n}\times\nabla]_i T \\
&- 1/2(\xi_3 - \xi_4 + \xi_7 - \xi_8 + \xi_{11} + 2\xi_{12})n_i m_{kj} n_j(\mathbf{n}\nabla)T \\
&+ 1/2(\xi_3 - \xi_4 - \xi_7 + \xi_8 - \xi_{11} - 2\xi_{12})n_k m_{ij} n_j(\mathbf{n}\nabla)T \\
&+ 1/2(-2\xi_5 + \xi_7 + 2\xi_{10} + \xi_{12})n_i n_k \mathrm{div}\,\mathbf{n}(\mathbf{n}\nabla)T + 1/2(-2\xi_7 + \xi_{11})n_i n_k n_j m_{jp}\nabla_p T \\
&+ 1/2\xi_{11} n_j(m_{jk}\nabla_i T + m_{ji}\nabla_k T) + \xi_{12} m_{kj}(n\nabla)T
\end{aligned}
\tag{3}
$$

$$
\begin{aligned}
-T^{-1}q_i^{TM} &= \frac{\delta R_{\mathrm{TM}}}{\delta(\partial T/\partial x_i)} = 1/2(2\xi_1 - \xi_3)\mathrm{N}_1\mathrm{div}\,\mathrm{n} + \xi_2(\mathbf{n}\,\mathrm{rot}\,\mathbf{n})\times[\mathbf{nN}]_i \\
&+ \xi_3 m_{ik}N_k - (\xi_3 - \xi_4)n_i N_j m_{kj} + 1/2(2\xi_5 - \xi_7)d_{ij}n_j \mathrm{div}\,\mathbf{n} \\
&- 1/2(2\xi_5 - \xi_7 - 2\xi_{10} - \xi_{12})n_i n_j n_k d_{jk}\mathrm{div}\,\mathbf{n} - \xi_6(\mathbf{n}\,\mathrm{rot}\,\mathbf{n})e_{ijk}n_k n_p d_{pj} + \xi_7 m_{ik}d_{kj}n_j \\
&+ \xi_{12}n_i d_{kj}m_{jk} + 1/2(-2\xi_7 + 2\xi_9 + \xi_{11})m_{ij}n_j n_k n_p d_{kp} - (\xi_7 - \xi_8 + 2\xi_{11} + 2\xi_{12})n_i n_p d_{pj}m_{jk}n_k
\end{aligned}
\tag{4}
$$

Variation of "dissipative" function by the director change velocity gives us thermomechanical "force" acting on the director.

$$
\begin{aligned}
\mathrm{f}_i^{\mathrm{TM}} &= \frac{\delta \mathrm{R}^{\mathrm{TM}}}{\delta(\partial \mathrm{n}_i/\partial \mathrm{t})} = \frac{1}{2}(2\xi_1 - \xi_3)\nabla_i \mathrm{T}\mathrm{div}\vec{\mathrm{n}} - \xi_2(\vec{\mathrm{n}}\cdot\mathrm{curl}\,\vec{\mathrm{n}})[\vec{\mathrm{n}}\times\vec{\nabla}]_i\mathrm{T} + \xi_3 \mathrm{m}_{ik}\nabla_k\mathrm{T} \\
&- (\xi_3 - \xi_4)\mathrm{m}_{ik}\mathrm{n}_k(\vec{\mathrm{n}}\vec{\nabla})\mathrm{T}
\end{aligned}
\tag{5}
$$

Euler-Lagrange-Raleigh variation equations are used to get equations describing director, velocity and temperature dynamic.

$$
\frac{\delta F}{\delta(dq_i/dt)} - \frac{\partial}{\partial x_k}\frac{\delta F}{\delta(\partial q_i/\partial x_k)} = -\frac{\delta R}{\delta(dq_i/dt)} + \frac{\partial}{\partial x_k}\frac{\delta R}{\delta(d^2 q_i/dt\partial x_k)}
\tag{6}
$$

Director equation in conjunction with Navier-Stokes and heat transfer equations give the complete picture of processes in liquid crystal. All these equations have the following form

$$
\Pi_{ij}\left[\frac{\delta \mathrm{R}}{\delta(\partial \mathrm{n}_i/\partial \mathrm{t})} - \frac{\partial}{\partial \mathrm{x}_k}\frac{\delta \mathrm{R}}{\delta(\partial^2 \mathrm{n}_i/\partial \mathrm{x}_k\partial \mathrm{t})} + \frac{\partial \mathrm{F}}{\partial \mathrm{n}_i} - \frac{\partial}{\partial \mathrm{x}_k}\frac{\delta \mathrm{F}}{\delta(\partial \mathrm{n}_i/\partial \mathrm{x}_k)}\right] = 0\ ,\ \Pi_{ij} = \delta_{ji} - n_j n_i
\tag{7}
$$

$$
\begin{aligned}
&\rho\left(\frac{\partial \mathrm{v}_i}{\partial t} + (\mathbf{v}\nabla)\mathrm{v}_i\right) = \frac{\partial \sigma_{ki}}{\partial x_k}, \\
&\sigma_{ki} = -p\delta_{ki} + \sigma'_{ki} + \sigma_{ki}^{TM}, \quad \mathrm{div}\,\mathbf{v} = 0,
\end{aligned}
\tag{8}
$$

$$
\rho c_p\left(\frac{\partial T}{\partial t} + \mathbf{v}\nabla T\right) - \mathrm{div}\,\mathbf{q} = U\,,
\tag{9}
$$

where v is the velocity, p – pressure, T – temperature, c_p – specific heat, ρ – density of the medium.

For the problem of thermomechanical hydrodynamic flow in hybrid oriented NLC the velocity in the cell with L=100μm, ΔT=10K and for NLC 5CB is $v = \xi\Delta T/L\eta \approx 0.2$μm/sec (η is viscosity, L – cell thickness, ξ – thermomechanical coefficient).

For the problem of the cell with cylindrical-hybrid orientation the maximum flow velocity in the cell with thickness L=100μm and temperature difference between top and bottom surfaces ΔT=5°C was $v_{\varphi\,\max} = \xi\Delta T/8\eta L = 0.5$μm/s.

The third type of thermomechanical effect which is the emergence of director reorientation due to the temperature gradient has also been studied [29]. For Gaussian laser beam, director equation in twist-aligned nematic liquid crystal cell has the following form

$$\gamma\frac{\partial\delta n}{\partial t}=K\Delta\delta n+K\frac{\pi^2}{4L^2}\delta n+\frac{\pi}{2L}(\xi_2+\frac{1}{2}\xi_3)(\frac{\partial T}{\partial x}\sin\frac{\pi z}{2L}-\frac{\partial T}{\partial y}\cos\frac{\pi z}{2L}),\qquad(10)$$

Temperature distribution in the sample due to absorption of laser radiation by the medium has the following form.

$$\Delta T(r,t)=\frac{2P\alpha}{\pi c\rho_0 a^2}\int_0^t\frac{1}{1+\frac{2t'}{t_c}}\exp\Big(-\frac{\frac{2r^2}{a^2}}{1+\frac{2t'}{t_c}}\Big)dt'\qquad(11)$$

This result and the range of its applicability are presented in works [30, 31].

Numerical solution of above-mentioned director equation for 5CB nematic liquid crystal and for 40 mW laser beam is presented in Fig. 7. Absorption coefficient of the sample is $3cm^{-1}$.

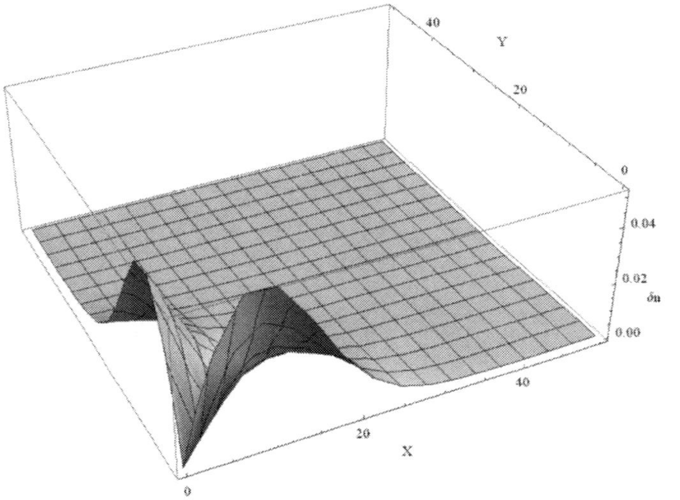

Figure 7. Director distribution for 40mW excitation laser beam in 5CB nematic liquid crystal twist aligned cell

Using a simple equation for phase shift we get

$\delta\Phi=2\pi L\delta n(ref)/\lambda\approx6\pi$, where L is the cell thickness, λ – wavelength (632nm).

Nonlinear coefficient in this case is about $0.14cm^3/erg$, which is at the same order as that of the giant optical nonlinearity.

4. Discussion and Conclusions

Thus, in the present paper we studied some mechanisms of orientational optical nonlinearity and among them nonlinearity connected with third thermomechanical effect. We would like to point out here that the thermomechanical stress tensor in Ref. [6] is invariant under time reversal. Thermomechanical effects are not dissipative. That is, the work of thermomechanical forces doesn't turn into heat (into the system's inner energy), but turns, for example, into the additional kinetic energy of hydrodynamic flow (in the case of direct thermomechanical effects). Although, the change of energy for a unit of NLC volume on account of the thermomechanical forces is made as a result of external sources, it is reversible. Due to the above-mentioned the change of entropy is made. Of course, if we introduce here a "dissipative" function corresponding to thermomechanical effects, it will not be invariant under time reversal. But, this introduction by the authors of Ref. [6] is an artificial method for obtaining correct equations describing the thermomechanical effects. And if we consider here the entropy equation: $\partial s/\partial t + div(s\mathbf{v}+\mathbf{q}/T) = 2R/T$, the appearance of new, antisymmetric under time reversal function ("dissipative" function) will not make any contradiction. To see this, all one has to do is to remember that our system is open and the heat source is outer. And the entropy of such kind of adiabatically nonisolated system can be changed in any manner.

To conclude, the reversible thermomechanical effects do exist, and their investigation produces and will produce new important information about the molecular dynamics, elastic properties and structure peculiarities of NLC mesophase.

This research was made possible by Armenian Award No 11-1c252.

References

[1] Zel'dovich B Ya, Pilipetskii N F, Sukhov A V, Tabiryan N V 1980 *JETP Letters* **31** 263
[2] Arakelyan S M, Chilingaryan Yu S 1984 *Nonlinear optics of liquid crystals (in Russian)*; Arakelyan S M, Chilingaryan Yu S 1984 *IEEE J. of Quantum Electronics* **QE-22** 1276
[3] Zel'dovich B Ya, Tabiryan N V 1985 *Sov. Phys. Usp.* **28** 1059
[4] Khoo I C, Wu S T 1993 *Optics and nonlinear optics of liquid crystals* World Scientific
[5] Simoni F 1994 *Nonlinear optical properties of LC and PDLC* World Scientific
[6] Akopyan R S, Zel'dovich B Ya 1984 *Sov. Phys. JETP* **60** 953; Akopyan R S, Alaverdyan R B, Nersisyan S Ts, Santrosyan E A, Chilingaryan Yu S 1999 *ZhTF* **69** 122; 1999 *Pisma v ZhTF* **25** 71; Akopyan R S, Alaverdyan R B, Chilingaryan Yu S, Santrosyan E A 1997 *Pisma v ZhTF* **23** 77; 2000 *Quantum Electronics* **30** 653
[7] Akopyan R S, Nersisyan S R 1993 *JETP* **78** 68
[8] Akopyan R S, Zel'dovich B Ya 1984 *Sov. Phys. JETP* **59** 311
[9] Tabiryan N V, Zel'dovich B Ya, and A.V.Sukhov 1986 *Mol. Cryst. Liq. Cryst.* ***136*** 1
[10] Odulov S G, Reznikov Yu A, Soskin M S and Khizhniak A I 1982 *Sov. Phys. JETP* **55** 854
[11] Rudenko E V, Sukhov A V 1994 *JETP* **78** 875
[12] Hakobyan R S, Tabiryan N V, Zel'dovich B Ya 1983 *Opt. Commun.* **46** 249
[13] Chilingaryan Yu S, Nersisyan S R, Tabiryan N V 1985 *J. Physique* **46** 955
[14] Tabiryan N V, Nersisyan S R, Mkrtchyan A R 1997 *Proceedings of SPIE* **2949** 76
[15] Janossy I and Szabados L 1998 *Phys.Rev.E* **58** 4598
[16] Rudenko E V, Sukhov A V 1994 *JETP Lett.* **59** 142; 1994 *JETP* **78** 875; Khoo I C, Li H, and Liang Y 1994 *Opt. Lett.* **19** 1723; Khoo I C 1995 *Opt. Lett.* **20** 2137; Tabiryan N V, Umaton C 1998 *Phys.Rev.E* **58** 4619
[17] Akopyan R S, Zel'dovich B Ya, and Tabiryan N V 1988 *Opt. Spectrosc.* **65** 637
[18] Akopyan R S and Khosrovyan G R 1991 *Sov. Phys. Tech. Phys.* **36** 1203
[19] Akopyan R S, Zel'dovich B Ya, and Tabiryan N V 1988 *Opt. Spectrosc.* **65** 344
[20] Akopyan R S, Tabiryan N V, and Tschudi T 1994 *Phys. Rev. E* **49** 3143
[21] Pieranski P, Guyon E 1974 *Phys. Rev.* **A9** 404; Manneville P and Dubois-Violette E 1976 *J.Physique* **37** 1115-1124; Jannossy P, Pieranski P and Guyon E 1976 *J.Physique* **37** 1105; Dubois-Violette E, Durand G, Guyon E, Manneville P and Pieranski P 1978 *Liq. Cryst., Solid State Phys. Suppl.* **14** 147
[22] Chilingaryan Yu S, Hakobyan R S, Tabiryan N V, Zel'dovich B Ya 1984 *J.Physique* **45** 413; Akopyan R S, Alaverdyan R B, Chilingaryan Yu S 1998 *Opt. i Spectr.* **64** 762; Akopyan R S, Alaverdyan R B, Chilingaryan Yu S 2000 *Quantum Electronics* **30** 745
[23] Janossy I 1990 *MCLC* **179** 1-12
[24] Tabiryan N, Hrozhyk U, and Serak S 2004 *Phys. Rev. Lett.* **93** 113901-1-4
[25] Akopyan R S, Alaverdian R B, Santrosian E A, and Chilingarian Yu S 2001 *J. App. Phys.* **90(7)** 3371-3376
[26] Hakobyan R S, Zeldovich B Ya, and Seferyan H 2004 *JETP* **99** 1039-1043
[27] Hakobyan R S, Yesayan G L, Zeldovich B Ya 1998 *Phys. Rev. E* **73** 061707-1-4
[28] Poursamad J B, Aleksanyan A K, Hakobyan R S 2010 *J. App. Phys.* **108** 123113-1-5
[29] Aleksanyan A K, and Hakobyan R S 2010 *Theoretical study of laser-induced thermomechanical effect and new type of orientational optical nonlinearity in twist aligned nematic liquid crystal : Proc. International Conference on Laser physics 2010 (Ashtarak, Armenia, 12 October 2010)* doi:10.1117/12.890864
[30] Shen J., Lowe R D and Snook R D 1992 *Chem. Phys.* **165** 385-396
[31] Shen J, Snook R D 1993 *J. Appl. Phys.* **73** 5286–5288

AUTHOR INDEX